From Ron Christmas 1995
JOLIE

Woodhouse's Textbook of Naval Aeronautics

Part of the United States Atlantic Fleet "somewhere in the United States," photographed from the seaplane of which part is shown in the photo.

Allied ships photographed from a French coast patrol dirigible, the pilot of which is shown in the photograph.

Woodhouse's Textbook of Naval Aeronautics

HENRY WOODHOUSE

With an Introduction by Clark G. Reynolds

Naval Institute Press
Annapolis, Maryland

This book was originally published in New York by The Century Co. in 1917.

Library of Congress Cataloging-in-Publication Data

Woodhouse, Henry, 1884–?
[Textbook of naval aeronautics]
Woodhouse's textbook of naval aeronautics
Henry Woodhouse; with an introduction by Clark G. Reynolds
p. cm.
Reprint. Originally published: New York: Century, 1917.
Includes index.
ISBN 1-55750-931-X (alk. paper)
1. Naval aviation. 2. World War, 1914–1918—Aerial operations.
I. Title. II. Title: Textbook of naval aeronautics.
VG90.W6 1991
359.9′4—dc20 90-32023

Printed in the United States of America on acid-free paper ∞
9 8 7 6 5 4 3 2
First printing

CONTENTS

INTRODUCTION

HENRY WOODHOUSE—WRITER, SELF-PROMOTER, CON MAN

CLARK G. REYNOLDS

When the *Textbook of Naval Aeronautics* was published in June 1917, the United States had been at war for two months, making the book an invaluable summation of the state of the art just at the moment naval aviation was being transformed from its adolescence into a major instrument of national power. At the time, the handsome volume provided an excellent reference tool for the lay public as well as for politicians and even naval officers who had virtually ignored their tiny air arm. The many photographs were so revealing of the nearly three years of fighting in Europe that none of them had been allowed to be printed "over there," leading the Office of Naval Intelligence to question the author closely on his sources and even to demand that the book be withdrawn from circulation. This did not happen, and a second printing occurred in 1918.

The author, Henry Woodhouse, had unimpeachable credentials. Not only was he a governor of the Aero Club of America—the official organization of the sport of flying in the United States—but also a well-known aviation author and editor who boasted sizeable biographical entries in *Who's Who in America* and the *National Cyclopedia of American Biography*. And he talked a good line—energetic, optimistic, pleasant, always smiling, suave, well-read, and clever, *very* clever. He was not an aviator.

To the historian, the *Textbook* is indeed an excellent historical document of early naval aviation, which makes entertaining reading today. Another useful piece of Woodhouse writing is an article in the U. S. Naval Institute *Proceedings* of February 1942 entitled "U. S. Naval Aeronautic Policies, 1904–42." I have used both in my own research and thus welcomed the opportunity to write this introduction. Examination of a file on Woodhouse in the biographical papers of the American Institute of Aeronautics and Astronautics at the Library of Congress and the pages of the *New York Times* revealed quite another image of this man, interesting in its own way and yet disturbing.

For Henry Woodhouse emerges from the documents, newspapers, and journal accounts as what we now call an operator. A physically small man, he was an immigrant with a quick mind, facile pen, unbridled ambition, and unabashed nerve; he exploited opportunities and people to advance his position in the aviation community. He twisted facts to suit his own purposes and overwhelmed unsuspecting readers with long lists of accomplishments and associations as seen on the title page of this book. Apparently he never married, nor did he list a religious affiliation. He was, however, a convicted killer. What follows will hopefully set the record straight, for Woodhouse deserves no closer study than this essay.

Born Mario Terenzio Enrico Casalegno in Turin, Italy, on 23 June 1881, he was one of seven children. According to the two abovementioned encyclopedia entries, both of which he virtually wrote, his father was Ludwig Casalegno, "a prominent manufacturer." Post–World War I entries, however, give the father's first name as Lodovico, obviously the son's precaution to eliminate any possible German connection. After the father's death during the boy's youth, the young man apparently drifted about Europe, reading and becoming an accomplished linguist—certainly in English, which he developed with only a slight Italian accent after emigrating to the United States from Scotland aboard the SS *Majestic* in 1904. He never claimed any formal education beyond high school in Turin.

Employed as an assistant cook in Troy, New York, Casalegno took his third name Enrico as the English Henry and sharpened his writing skills by day while cooking at night. He got into a scrap over an open window with the head cook and stabbed the man in the ribs, killing him but claiming self-defense. The judge was not persuaded and sentenced him to four years and two months at hard labor for first-degree manslaughter, citing Casalegno's youth in not giving him the maximum twenty years. Casalegno apparently cooled off while in Clinton Prison, Dannemora, New York, and was released one year early, in 1908.

Taking a job as chef at the Hotel Wendell in Pittsfield, Massachusetts, not far from Troy, Casalegno began to doctor history in order to serve his own purposes. He translated his Italian last name to the English Woodhouse, a legitimate enough change, but he erased his three years as a convict by listing his date of birth as 24 June 1884 when he applied for citizenship in 1910. With his boyish face, the "new" Henry Woodhouse easily passed for twenty-six instead of his true twenty-nine years.

Dazzled by the balloonists of the Pittsfield Aero Club, Woodhouse wrote articles on ballooning for the club president, Luke J. Minahan, who happened to be proprietor of the Hotel Wendell. This was his first real contact with aeronautics, but he would lie—at age sixty—in the February 1942 *Proceedings* article that he "had been with" the celebrated Brazilian balloonist Alberto Santos-Dumont "since 1900 and acted as his aide while he was in the United States" in 1904. The first person Woodhouse did influence was aviation publisher Alfred W. Lawson, who coined the word "aircraft" and used it for his new journal of that name. Woodhouse moved to New York to write articles for *Aircraft* in 1911, becoming its associate editor in September, managing editor in October, and being dropped in December.

Undaunted, Woodhouse published an essay on progress in aviation in the 27 January 1912 *Bulletin* of the Aero Club of America, which he had joined as a member the previous August. He sufficiently impressed club president Robert J. Collier, the celebrated patron of aviation and owner of *Collier's* magazine, to appoint him associate editor of the *Bulletin* in April 1912 and engage him to manage Aero Club parties. One can imagine how the ambitious Italian exploited the fashionable social environment to advance himself. Indeed, he would soon boast in the *National Cyclopedia* that he accumulated $25,000 in capital assets by the time he was twenty-eight—by his new reckoning, 1912. In October 1912 the *Bulletin* was changed to *Flying* and turned over to professional journalists to manage, with Woodhouse receiving half the profits. But when the new owners quickly lost money, they sold out to Woodhouse early in 1913. And the Aero Club gave him free room and board at its clubhouse on Madison Avenue to run the journal.

No doubt the energetic entrepreneur used his new forum and first of many titles—publisher of *Flying*—to become a spokesman for aviation and to obtain underwriters for more projects. In 1914 he opened a book shop called Aeronautical Library and in March 1915 started a second journal, *Aerial Age Weekly,* to replace the respected *Aero and Hydro.* His avowed editorial aim in the new magazine was "to protect the interests of the industry and to fight the constructors' fights"—an allusion to the patent struggles between the Wright brothers and all other American airplane builders. To generate financial support for both journals and those "fights," Woodhouse founded the National Airplane Fund in May 1915 and eventually raised over $180,000.

On the face of it, these accomplishments appear innocuous enough, though in fact Woodhouse was entrenching himself for greater glory. Simply as an editor of two journals that published dramatic announcements, he began to claim a leading role as a major participant and catalyst of progress in aviation, even a prophet, when he was nothing more than a paid publicist. The *National Cyclopedia* and *Who's Who* entries are little more than self-serving fabrications of the truth and which Woodhouse even dramatized in the February 1942 *Proceedings* essay. As editor of the two journals, he was nominally a member of the Aero Club Board of Governors but by 1916 had become the ruling power. This meant he also controlled the club's funds, as, for example, the receipt that year of some $60,000 in paid advertisements, $13,000 from the Curtiss company alone.

So powerful did Woodhouse become that when, in the spring of 1916, a group of seaplane enthusiasts organized the New York Flying Yacht Club, Woodhouse sought to wreck it as an unwelcome competitor in assisting the U.S. government with the development of flying boats. A more promising vehicle to promote himself was the preparedness movement as the United States drifted toward participation in World War I, already two years old in Europe. As explained in this book (pp. 172–75) Woodhouse was present at the founding dinner of the National Aerial Coast Patrol late in 1915, to which famed arctic explorer Admiral Robert E. Peary affixed his name. The Coast Patrol became instrumental in training young men like those of the First Yale Unit for eventual wartime navy service; Woodhouse says he personally gave the Yale men a Lewis aerial machine gun. And he became a delegate to the Conference Committee on National Preparedness. Under the name of the Aero Club, Woodhouse seems to have raised $380,000 in donations for aerial preparedness.

Against this image of patriotism, Woodhouse engaged in other practices that would later lead to accusations that he was a virtual spy for the Central Powers—Germany, Italy, and Austria-Hungary. Although he had applied for American citizenship, he did not receive it until 28 May 1917—three months after his name was officially changed from the Italian. And his father's name had then been Ludwig, suggesting German sympathies. During 1916 Woodhouse introduced Lyman J. Seeley of the Curtiss company to German agents eager to buy seaplanes, negotiations which the company broke off when the objective of the Germans became apparent. He also became involved in a pro-German lumber company that supplied bad lumber for British cargo ships, three later splitting open at sea with heavy loss of life. He installed listening devices in his offices to record discussions of aeronautical advances and collected large numbers of valuable combat photographs. But none of the above seem to reflect a spy so much as an opportunist, seeking ways to improve his own situation.

Judging by the contents of this book, it was probably conceived by Woodhouse early in 1917 when America's entry into the war became imminent. With his immense energy, he drew upon his considerable file of publications and personal contacts to produce the manuscript. To lend respectability to such a presumptuous "textbook," he solicited the introduction and chapters 3 and 4 from retired Admiral Bradley A. Fiske, a frustrated reformer and inventor always eager to advance naval aviation. For a technical treatise, chapter 33, he reprinted an essay by Commander Holden C. Richardson, with Admiral Peary discussing the Aerial Coast Patrol in chapter 27, and Ralph H. Upson explaining kite balloons in chapter 31. Because he received a commission for assisting the Sperry Gyroscope Company (among other companies) and knew Elmer Sperry, he called upon the latter's son "Gyro" to describe naval aviation training and aerial navigation over water in chapters 15 and 16. Ensign Lawrence B. Sperry was already an accomplished pilot and inventor whose flying skills helped train many wartime naval airmen. And Secretary of the Navy Josephus Daniels proved willing to provide an innocuous letter for the front matter on the eve of publication in June 1917.

The immediate reaction to the book came in the form of two navy and two army intelligence officers who separately interrogated Woodhouse on the source of his remarkable photographs. The several Allied countries had censored the release of such classified materials and now tried to prevent the export of the book abroad. But it was too late. Furthermore, Woodhouse circulated many free copies, apparently to the commanding officer of each major American warship and to each ship's library. This service did indeed publicize the role of aviation in the navies of the world and was probably Woodhouse's greatest contribution to the rise of naval aviation. But the release of such sensitive information did not ingratiate him to official Washington, a resentment deepened in September 1917 when Woodhouse published the secret British gunfire code. He was summoned to Washington to explain, while the British frantically changed their code.

Given the intense national fervor over aviation, Woodhouse rose to new heights, which generally increased resentment against him within the Aero Club, especially as he manipulated proxy votes and culled mail critical of his policies. He reacted with two counterattacks during the summer of 1917. He attacked in print the creation of the Manufacturers Aircraft Association (M.A.A.), organized by the government to end the patent disputes by using cross-licensed patents to stimulate wartime aircraft development and production. Instead of becoming champion of a popular cause, he found little support, for aviation mushroomed to the delight of most builders. His second move was to found a patriotic organization of his own, the Aerial League of America, to promote aviation. He developed elaborate bylaws and designed an impressive stationery letterhead that included several "honorary members" like President Woodrow Wilson, Alexander Graham Bell, and General John J. Pershing. Even though some, like Pershing, ordered their names removed, the League generated dues-paying memberships from naive aviation enthusiasts. The real airmen knew better, among them pioneer flyer J. C. "Bud" Mars, who published a tract exposing Woodhouse's spurious activities to date and calling the League "the most impudent subscription scheme ever put across by anybody anywhere."

While Woodhouse enlarged his scope of conning the American public, he concurrently tried to cover his own past, now brought to light by his critics. Late in 1917 he claimed exemption from the draft on the grounds of "essential labor," when in fact he was ineligible as an ex-convict. Since he had attained his citizenship by hiding his prison record, he stood to lose it. But he engaged noted attorney Bernard S. Sandler to convince New York's reform governor Charles S. Whitman to protect the citizenship of certain ex-cons, including himself. Woodhouse joined several patriotic and professional organizations and kept a coterie of sycophants at hand to ensure his control of the Aero Club. To guarantee re-election to his several positions in the Aero Club in the autumn of 1918, he solicited the necessary proxies from distant members ignorant of the criticism about him. But Bud Mars sent out a warning to many members, comparing him to czarist Russia's conniving monk Rasputin; in November, however, Woodhouse refused to provide the nominating committee with a list of the membership. He was re-elected.

All the while, Woodhouse continued to grind out so many books and essays on aviation that his public image soared. In 1917 he published a well-illustrated *Aircraft of All Nations,* with an introduction by Admiral Peary. Early in 1918, at the suggestion of Admiral Fiske, he submitted an article on Fiske's beloved torpedo plane for publication in the *Proceedings,* but when navy censors balked, Fiske interceded with Assistant Secretary of the Navy Franklin D. Roosevelt to release it—although it did not appear until May 1919, after the war had ended. Meanwhile, encouraged by the success of his *Textbook of Naval Aeronautics,* he had a second printing done in 1918 and ground out a companion volume, *Textbook of Military Aeronautics,* the same year with a second printing in 1919. And the December 1918 issue of the *Proceedings* carried an essay on wartime antisubmarine operations by aircraft. In 1919 Woodhouse produced the *Aero Blue Book and Directory of Aeronautic Organizations* and in 1920 the *Textbook of Applied Aeronautic Engineering* and the *Textbook of Aerial Laws and Regulations for Aerial Navigation,* his "advisory editors" including his own lawyer, Sandler. His editing and writing in *Flying, Aerial Age Weekly,* and other journals continued apace.

The end of World War I in November 1918 heralded the new Air Age, and Henry Woodhouse cashed in on it with his many publications. The problem was that the giants of the wartime aircraft industry—represented by the M.A.A.—and the returning flood of army and navy pilots desiring to join the Aero Club and to fly in air meets threatened Woodhouse's power and financial base. He put up his customary front, but on a grander scale, by organizing three so-called Pan-American aeronautical congresses over 1919–20. When he was not invited to attend an Aircraft Dinner in January 1919, he simply showed up and sat down at a guest table. Finally, while the Aero Club only talked of holding air meets, a group of veteran military flyers founded the American Flying Club and held two sporting races over 1919–20: New York to San Francisco and back and New York to Toronto. Woodhouse countered by hawking a vague Round-the-World race and sending representatives abroad, where, according to C. G. Grey, editor of the respected British journal *The Aeroplane*, they "talked the most arrant nonsense that one has ever heard from human lips on the subject of flying. Not even an English Member of Parliament could have shot off as much pure guff. . . ."

Battle lines were thus developing between the real airmen and this one imposter with his small cohort of followers known as the "Woodhouse gang" who ensured his control of the Aero Club. The challenge to Woodhouse, in April 1920, came internally when club officers appointed a committee to study the possibility of a merger with the American Flying Club. Criticism of Woodhouse mounted, and when club members found monies missing from the National Aeroplane Fund that Woodhouse administered, they prevailed upon the New York district attorney's office to investigate. Now the movement for amalgamation took off. The struggle reached the public eye on 6 June 1920 when the *New York Evening Post* published a biographical article exposing Woodhouse's shady past. Woodhouse immediately sued the paper for libel, a case that would linger until the end of 1923 when the courts rejected the suit.

Woodhouse launched his counterattack against the real airmen by linking the merger attempt to a conspiracy by the M.A.A. to cover up wartime frauds and postwar dumping of excess planes and to monopolize control over all aviation, including the Aero Club. Wartime graft had the attention of Congress, and conspiracies were much in the wind. In April Woodhouse made his claim about the M.A.A. to Attorney General A. Mitchell Palmer, whose anti-Red "witch hunt" had recently ended with the deportation of communists and radicals. Woodhouse took his case to the readers of his *Aerial Age Weekly* (issue of June 21) against the half-dozen individuals "indulged in a most malicious conspiracy to wreck the aeronautics movement so as to get control of it." One month later, July 21, he filed for an injunction against several opposing members of the Aero Club's board of governors in order to prevent the merger with the American Flying Club and thereby to protect the files he said would expose the aircraft scandals. This was a pure smokescreen, for no link existed between the M.A.A. and the Flying Club, and Woodhouse had no meaningful documents. What he was trying to do was prevent club membership files, finances, and trophies from passing out of his hands. The request for the injunction was denied.

In retaliation for trying to sue several of the twenty-four governors, all but two—Woodhouse and a confederate—voted on 4 August 1920 to suspend him from the club, an action that four days later made the front page of the *New York Times*. Now former U.S. Army Air Service Captain Albert P. Loening of the Flying Club organized the final coup d'état by garnering 440 proxy votes and 150 persons of the total Aero Club membership to vote for amalgamation at the Aero Club meeting on the 16th. Woodhouse, electing not to face the music, assigned Lieutenant Rafe Emerson, a navy balloon and airship pilot, and a non-flying army major to argue against the merger. They got five votes against Loening's 590! The two organizations then merged under the name of the Aero Club and moved to the more spacious Flying Club building on East 38th Street.

In reporting the takeover, the venerable British editor Grey in the 29 September 1920 *Aeroplane* applauded the defeat of Woodhouse, a "hot-air merchant" who had "helped a crowd of similar self-advertising gas-bags to foist themselves on the American Public and on the Aero Club as authorities on aviation when they know nothing about the subject." Grey further decried Woodhouse's editing of "that amazingly humorous publication *Aerial Age*, a species of aeronautical *Comic Cuts*, and of . . . a paper called *Flying*, which is as dull as any official organ can aspire to be." Needless to say, Woodhouse was now dropped by the amalgamated Aero Club as editor of both journals. He responded by announcing formation of a new Aerial League of the World, which Grey passed off as nothing but "a very feeble joke" and which got nowhere. But Grey regretted that Woodhouse "intends to carry on a *guerrilla* [war] against his conquerors by heckling and litigating ... in order to bother everybody as much as possible."

That he did. Not only did Woodhouse delay surrendering Aero Club files to the organization throughout 1921 and 1922, but he continued his attacks on the aircraft industry with articles in the Hearst press and then enlarged his scope to include the "oil interests" of America. He founded yet another journal, *Scientific Age,* as the successor to *Air Power,* and made it into another pseudo-expert repository of general progress in all aspects of science and technology. Although stripped of access to his former journals, Woodhouse maintained his role as president of the Aerial League of America, which he had founded in 1917. But when he began soliciting funds for it early in 1922, his treasurer was arrested and convicted of swindling, whereupon Woodhouse became treasurer as well. Worst of all, however, his tactics of delay—two years of legal wrangling—seriously undermined the credibility of the restructured Aero Club, which not only lost many discouraged members but reached insolvency by the end of 1921, with liabilities of over $25,000. The name of the Aero Club had become so tarnished that its members formed an entirely new organization, the National Aeronautic Association.

Desperate, the former Aero Club officers in January 1922 called upon a group of sympathetic legal, business, and naval leaders to act as trustees to reclaim the funds still held by Woodhouse so they could pay off their bills: Professor Charles Thaddeus Terry of the Columbia Law School, AT & T president Henry B. Thayer, retired Admiral William F. Fullam, and top executives of the Wright, Curtiss, and Hudson aircraft and motorcar companies. To prevent surrendering club property, Woodhouse sued each one, and the M.A.A. as well—seven suits in all! In September 1922 he dragged them, and some thirty lawyers, before the Supreme Court of New York and confounded the proceedings with reams of materials and exhibits. An amazed reporter of the *New York Globe and Commercial Advertiser* (23 November 1922) marveled at the "melodrama" Woodhouse created, acting as his own counsel:

> Mr. Woodhouse ... is calm, suave, pleasant. He jests with his enemies when they will jest with him. He answers any question put to him with the utmost readiness ... [with] explanations, clippings, cancelled checks, photostats of documents—all with a bewildering rapidity. "He's very clever," his opponents admit. I agree with them.

The judge saw through Woodhouse's tactics, however, by rejecting his claims. The court accepted the Aero Club's countersuit that Woodhouse must stop using the name of the Aero Club, return the files, and accept the merger. When Woodhouse could not produce a list of 404 Aero Club members who he claimed supported him, and he kept interrupting the proceedings, the judge held him in "flagrant contempt of court," fined him $100, and threatened to send him to jail. When Woodhouse claimed that the Justice Department held the documents on the aircraft scandal required by the court, the judge found the claim to be false and on examining the papers found them "bold fakes"—like Woodhouse. His bluff finally called, Woodhouse had to turn over all files and dues monies to Aero Club members. It was an expensive victory, the club's legal fees exceeding $40,000, but the new National Aeronautic Association emerged from the ashes of the Aero Club's destruction by Woodhouse.

This restless, plotting mind had not allowed the Aero Club embroglio to distract him from further enterprises. Using the funds obtained from his aviation activities, during 1921–22 Woodhouse purchased options on claims for concessions in Turkey obtained by retired Rear Admiral Colby M. Chester—an early supporter of naval aviation. With these, Woodhouse formed a syndicate that in 1923 purchased all the Chester concessions to build and operate railroads, two seaports, and oil fields in Turkey and was the major shareholder, with Chester as treasurer. When the Department of State refused to guarantee American involvement, Woodhouse convinced the Italian government to invest, no doubt exploiting his native birth. A larger British syndicate tried to take it over, and the dispute dragged on for two years, with Woodhouse allowing Admiral Chester to assume the presidency as major stockholder. As usual for Woodhouse, however, litigation bogged down the whole enterprise for lack of major investors. In the meantime, in 1924 Woodhouse even became embroiled in the Teapot Dome scandal over the navy's oil reserves, claiming to have inside knowledge of British involvement but actually producing nothing of importance.

Ever the self-promoter, Woodhouse conned the publisher of *National Magazine* into writing and publishing an incredible article in the December 1923 issue entitled "Henry Woodhouse—Thinker." In it, the author compared his subject with Leonardo da Vinci, Galileo, Columbus, Confucius, and Ben Franklin as a tireless genius and catalyst of progress in science and technology. The aviation press howled, justly so, but no one could gainsay the fact that Woodhouse had used his distorted "genius" to acquire substantial wealth, social prominence, and a home on Madison Avenue. And then, in November 1924, the New York state supreme court suddenly reversed its decision of a year before that had denied his libel suit against the *New York Evening Post.* It now awarded him one million dollars in damages.

Such occasional successes spurred him on. In 1925 Woodhouse had Admiral Fiske brought before the Supreme Court to testify in his behalf for reinstatement in the Aero Club, without apparent success. At the helm of his spurious Aerial League, he ground out letters of congratulations to famous aviators on their achievements. In 1926 he convinced a prominent philanthropist to underwrite the "world's first air junction" to become Richmond, Virginia's, municipal airport, with himself as president of the development corporation. On 1 January 1929 he oversaw the founding of a similar "Washington Air Junction" near Mt. Vernon to accommodate big rigid dirigibles as well as the first airliners, but which the onset of the Depression apparently scotched. And, of all things, he became involved in musical activities and a collector of artwork and patriotic papers. He amassed 4,500 letters and documents of the Washington and Lee families, which he deposited in the Library of Congress in 1932 (to which he sold them twelve years later).

If anyone could turn the Great Depression to personal advantage, it was Henry Woodhouse. By 1933, he was publishing a magazine called *National Recovery Survey* and billing himself as chairman of the National Recovery Council. And that year he sued a New York bank over his office rent, a hint that the times were rough for him too. In October 1936 he hosted a series of sessions in New York called the World Economic and Monetary Conference and published a statistical survey derived from it entitled *The $150,000,000,000 National Recovery*—his final book. And by 1938, as attention began to shift to the rise of the fascist dictators, he was writing letters to newspapers on world affairs.

The advent of World War II found Woodhouse recalling World War I and tapping conversations with his aged neighbor living at the Waldorf-Astoria Hotel, Admiral Fiske. He targeted the *Proceedings* for new articles, in 1941 submitting an article entitled "Torpedoplanes in World War II." The fact that it appeared in the December 1941 issue, just before Japanese torpedo planes hammered Battleship Row at Pearl Harbor, did nothing to tarnish his image as an expert and prophet. His article on U.S. naval aeronautical policies since 1904 followed in February 1942; another on the sinking of the *Prince of Wales* and *Republic* appeared in April, and still another in June: "The Importance of Naval and Air Task Forces in Global Warfare." All bore the fine hand of Bradley Fiske, who died that April at age 87.

The last record we have of Henry Woodhouse is in 1951 when, at age 70, he donated early American portraits and 500 motion picture and television scripts to the city of Trenton, N.J. But his most positive contribution to his adopted country remained the *Textbook of Naval Aeronautics.*

Here Woodhouse's story ends. No date of his passing has been located, no obituary in the national or aviation press. This is perhaps fitting, since, by his own reckoning, only one person would have been qualified to write it—Woodhouse himself. Anyone else ran the risk of being sued for libel, from the grave.

TEXTBOOK
OF
NAVAL AERONAUTICS

BY
HENRY WOODHOUSE
MEMBER OF THE BOARD OF GOVERNORS OF AERO CLUB OF AMERICA, MEMBER OF NATIONAL AERIAL COAST PATROL COMMISSION, CHAIRMAN OF COMMITTEE OF FLYING EQUIPMENT COOPERATING WITH COMMANDANT OF THIRD NAVAL DISTRICT IN ORGANIZING NAVAL RESERVE FORCES, TRUSTEE AND CHAIRMAN OF COMMITTEE ON AERONAUTICS NATIONAL INSTITUTE OF EFFICIENCY, MEMBER OF THE SOCIETY OF AUTOMOTIVE ENGINEERS, ETC., ETC.

WITH INTRODUCTION BY
REAR ADMIRAL BRADLEY A. FISKE
PRESIDENT OF THE U. S. NAVAL INSTITUTE

NEW YORK
THE CENTURY CO.
1917

INTRODUCTION TO THE ORIGINAL

BRADLEY A. FISKE

The appearance of this book is opportune; because it is essential that the people be told what naval aeronautics can do to help the nation, and that the men who are to fight for us in the air shall be given every opportunity to learn to do it.

The greatest danger confronting the Allies is the German submarine. For combatting the submarine, which was the most modern weapon when the war began, no weapon yet used has been thoroughly effective; with the result that the danger has increased and is still increasing. The most effective weapon has been one more modern than the submarine—the aircraft. If it had been possible to use aircraft in large numbers, the damage done by the submarine would have been materially decreased, and possibly eliminated altogether. For combatting the submarine, therefore, large numbers of aircraft must be employed. For this work, the smaller type of dirigible, usually called "blimp," has been the most effective for coast patrol. It is unfortunate that large dirigibles have not been available; because their long radius of action would have enabled them to patrol the ship-lines of all the seas where the commerce of the Allies goes, and reduce the submarine menace to a minimum.

The Allies can get assistance, however, from aeronautics in other ways than in submarine hunting. They can get assistance in six ways mainly:

First: By taking advantage of the great speed of aircraft and the heights to which they can rise, to get information as to the enemy, his distance, direction, and composition.

Second: By using aircraft to "spot" from great heights; that is, to note how far the projectiles fired by guns missed the target, and thus determine how to correct the range at which the guns were set.

Third: By using aircraft to carry machine guns and rapid-fire guns, and attack the lighter vessels of an enemy's fleet, and make raids on bases.

Fourth: By using aircraft to rise above the water, and see mines and submarines, and make photographs of enemy works and bases.

Fifth: By using aircraft to carry bombs fitted with "delayed action fuses," and drop them near mines, submarines, and shore works.

Sixth: By using aircraft to carry torpedoes, and launch them at the various vessels of an enemy's fleet; using light torpedoes with small charges of high explosive against the lighter vessels, and heavy torpedoes, with charges of from two to four hundred pounds of high explosive, against the heavily armored ships.

Other uses will doubtless develop with the progress of the art.

How great will be the weights which aircraft can eventually carry, and what amount of offensive power they can bring to bear in war, it would be foolish to attempt to prophesy. But aircraft are mechanisms that obey the laws of engineering; they have increased in size thus far exactly as mechanisms in all other branches of engineering have increased, though much more rapidly; and they have carried weights increasingly great, as time has passed: so, the conclusion is unavoidable that the prediction of Marcel Deprez, made in 1883, which came true of electric motors, is applicable to aircraft now: "L'avenir est aux grandes machines" ("The future is to great machines").

Inasmuch as the maximum of military effectiveness is secured when a given power is concentrated in as few and as powerful units as possible; inasmuch as the torpedoplane is the most powerful and mobile weapon now existing; and inasmuch as our enemies are the ablest strategists in the world, comprehend these principles, and will doubtless act in accordance with them, it is of the highest order of urgency that we develop immediately an aeroplane armed with guns adequate for defense against fighting aeroplanes, and capable of carrying a torpedo of the longest range and greatest power. It may be that the side which brings 100 armed torpedo-planes into action the first, will thereby gain the unrestricted command of the sea, and become the victor in the war.

The extended use of chemicals in the present war has started the re-development of a special branch of warfare. For many years before this war, the effective work done by navies and armies against their enemies was almost wholly the mechanical work performed by weapons in striking blows, and thus inflicting mechanical injuries on men and on defenses placed around them. But in the present war, injuries which were chemical and physiological have been inflicted, by the ancient means of flames and gases. To transport swiftly the comparatively light apparatus needed for this class of work, aircraft, even of the sizes of the present day, are admirably adapted.

On many occasions during the past six years, I have called attention to the naval and military possibilities of aeronautics; and on March 24 1916, I pointed out officially to the House Naval Committee that we could improve the national defense more in a short time by aeronautics than by any other means. At the present date of writing, the public has awakened to this fact, and now demands that every effort be made to develop huge air fleets with which to strike Germany over land and sea.

Our people have at least realized the possibilities of aeronautics, and see that aeronautics can be made the most effective agency available, not only to serve the Allies, but even to save the United States. They now see that, while it is well to train a young man to use a musket in an infantry company, more military usefulness can be gotten out of men like mechanics, chauffeurs, and technicians, by putting them in swiftly moving aeroplanes and dirigibles.

It is not sufficient, however, that our people realize the possibiliites of aeronautics in a general way; it is essential that they form an intimate acquaintance with aeronautics in all its branches. It is essential also that a great number of capable young men be instructed as soon as possible in caring for and operating all kinds of aircraft and aeronautical machinery.

To start this urgent work, the most immediate agency is a book like this—clear, correct, and stimulating.

Bradley A. Fiske,
Rear Admiral, U.S. Navy.

THE SECRETARY OF THE NAVY.
WASHINGTON.

June 19, 1917.

My dear Sir:

Many people thought that Tennyson was employing poetic license when he spoke of the "Airy navies grappling in the central blue". We have lived to see that he was a prophet of modern power in warfare. No nation can confidently look for victory because it is mistress of the sea or master of the land. Both may be made impotent by the Nation which commands the air.

Sincerely yours,

Josephus Daniels

Mr. Henry Woodhouse,
407 Union Trust Building,
Washington, D. C.

Congressman Murray Hulbert, co-author, with Senator Morris Sheppard, of the Sheppard-Hulbert Bill, which provides for the creation of a Department of Aeronautics, and who proposed the amendment increasing the aeronautic appropriation by tenfold, in June, 1916:

"Command of the air is the balance of power which decides victories on land and sea. The side which holds command of the air can blind the other side by depriving it of its air scouts and air "spotters," and at the same time, striking its forces on land and sea with bombs, aircraft guns, and torpedoes."

The Right Hon. Arthur J. Balfour: "The time is here when command of the sea will be of no value to Great Britain without corresponding command of the air."

Lord Charles Beresford: "The time is here when the air service of Great Britain will be more vital for her safety than her Army and Navy combined."

Winston Churchill, formerly First Lord of the Admiralty: "Ultimately, and the sooner the better, the air service should be one unified, permanent branch of imperial defense, composed exclusively of men who will not think of themselves as soldiers, sailors, and individuals, but as airmen and servants of an arm which possibly at no distant date may be the dominating arm of war."

Lord Montagu of Beaulieu (1916): "Every nation will before long be forced to create an Air Ministry by the sheer necessity which knows no law, which regards no precedent, and which fears no government. The immense development of aircraft in all directions alone will compel the creation of an air department."

General Petain, the veteran defender of Verdun, commander-in-chief of the French forces: "I see France in the near future with 50,000 aeroplanes."

Lord Montagu of Beaulieu, in the House of Lords (1916): "At the present time the air service is merely auxiliary to the fighting forces of the Navy and Army. I can see a time coming when the air service will be more important than the Army and Navy. We must get into the habit of looking at the air service, not as an auxiliary to the Army and Navy, but as a great service which is an establishment of itself, and to which we shall have to look in future years for the defense of this country."

Lord Beresford: "The new air warfare is going to be of so tremendous a character that it may supersede the Army and Navy. We should be ahead in the air, the same as we are on the water."

Rear Admiral Bradley A. Fiske testified to Congress on March 24, 1916: "Aeronautics is the thing on which we can get to work quicker, and by which we can accomplish more than by anything else."

Alan R. Hawley, President of the Aero Club of America: "Command of the air leads to victory on land and sea."

Rear Admiral Robert E. Peary, Chairman, National Aerial Coast Patrol Commission: "Victory in the present war; the efficiency of our Army and Navy; the protection of our coasts and coastal cities, the safety of Panama Canal; the existence of the nation—all depend mainly on our Air Service. Therefore a Department of Aeronautics is a vital need."

Henry A. Wise Wood stated as early as 1913: "Flight, the final abridger of time and space, incomparably swift and matchlessly direct, suddenly, wholly unexpected, almost unhoped for, it spreads its Wishing Carpet beneath the feet of mankind, so to bear it hither or yon in the twinkling of an eye, all as we dreamed over our childhood's fairy books."

PREFACE

To Train Airmen to Fight the Enemy Over, On, and Under the Water

This text-book on Naval Aeronautics is intended to assist in the training of the thousands of airmen to fight the enemy over, on, and under the water.

Owing to the traditional peaceful disposition of the United States this country has entered the war with a small army and a very limited personnel for the first and second lines of defense, the Navy and the Naval Reserves, and Naval Militia. The most difficult task is now to get and train as soon as possible one hundred thousand officers and men for the first and second lines of defense.

The British Navy has a personnel of 400,000; the United States has about 100,000.

At present Great Britain, France, Russia, Germany and Austria count their aeroplanes by the tens of thousands and dirigibles by the hundred. Their present plans of increase of production provide for increasing the number of aeroplanes to hundreds of thousands. For instance, the British Government is spending $575,000,000 this year for aeronautics and the other countries are spending more or less the same amount. The report of the British Controller of Aeronautical Supplies shows that there are 958 firms engaged in aeronautic work for the British Government. Of these 301 are direct contractors and 657 are subcontractors. The total number of hands employed by the fifty firms of most importance is 66,700.

The British Naval Air Service has over 150,000 officers and men connected with it; our air service has not a twentieth of that number. There are two hundred and twenty naval officers in charge of the administration of the Royal Naval Air Service in London alone. We have not that many officers in the entire air service. The task that we have undertaken may grow as big as the task which Great Britain, France and Italy have performed and are performing, and we must prepare to do our duty in a whole-hearted, substantial way. We must expect to have to keep our first line of defense engaged in preventing successful operations on the part of the enemy's ships, and our second line of defense busy in protecting the merchant ships, transports, the coasts, harbors, and naval stations.

We must expect that as soon as we get ready to ship supplies in large quantities, or troops, the common enemy, the U-boats, and the raiders and mine-layers, will begin their operations in American waters. Then aircraft will be needed in large number in connection with both, the first and second lines of defense. We get an idea of how extensively we may need them from the report which Sir Edward Carson made, on February 21, 1917, to the British Government, in which he stated: that since the commencement of the war the British Navy (with the cooperation of its most efficient aerial coast patrol) had examined 25,874 ships. During the first eighteen days of February, 6076 ships arrived in ports of the United Kingdom, and 5873 ships had cleared from United Kingdom ports. Practically every ship that arrived and every ship that cleared was inspected as it neared the ports, and convoyed by dirigibles or seaplanes. Sir Edward Carson pointed out that from the beginning of the war up to October 30, 1916, the British Navy transported across the seas 8,000,000 troops, 9,420,000 tons of explosives and material, 47,504,000 gallons of gasolene, over a million of sick and wounded, and over a million mules and horses, etc.

Aircraft played an important part in protecting this gigantic accomplishment, but to do it the Allies' naval air services had to be extended to enormous proportions. The British naval air service, for instance, employs scores of seaplane and kite-balloon carriers in connection with the first line of defense, and scores of seaplane stations on the British coasts for the air service of the second line of defense.

PREFACE

There are not less than five thousand British naval aviators, and pilots of naval dirigible balloons and kite-balloon operators—and as many connected with the British land forces.

Naval aircraft in the present war have actually been employed for the following purposes:

1. Attacked ships and submarines at sea with bombs, torpedoes, and guns. (Seaplanes and dirigibles used.)

2. Bombed the enemy's bases and stations. (Land aeroplanes, seaplanes and dirigibles used.)

3. Attacked the enemy's aircraft in the air. (Aeroplanes and seaplanes used.)

4. Served as the eyes and scouts of fleets at sea. (Dirigibles, seaplanes and kite balloons used.)

5. Protected ships at sea and in ports against attacks from hostile submarines and battleships. (Seaplanes and dirigibles used.)

6. Defended and protected naval bases and stations from naval and aerial attacks. (Land aeroplanes, seaplanes, and dirigibles used.)

7. Convoyed troop ships and merchant ships on coastwise trips. (Dirigibles and seaplanes used.)

8. Patrolled the coasts, holding up and inspecting doubtful ships, and convoying them to examining stations and searching coasts for submarine bases. (Dirigibles used.)

9. Prevented hostile aircraft from locating the position and finding the composition and disposition of the fleet, getting the range of ships, naval bases, station, magazines, etc. (Land aeroplanes and seaplanes used.)

10. Located, and assisted trawlers, destroyers and gunners in capturing or destroying hostile submarines. (Seaplanes, dirigibles and kite balloons used.)

11. Cooperated with submarines, guiding them in attacks on ship. (Dirigibles and seaplanes used.)

12. Located mine fields and assisted trawlers in destroying mines. (Dirigibles, seaplanes and kite balloons used.)

13. Served as the "eyes" in planting mines, minimizing the time required for mine planting. (Dirigibles, seaplanes and kite balloons used.)

14. Served as "spotters" in locating the position of the hostile ships and directing gunfire. (Dirigibles, seaplanes and kite balloons used.)

15. Served as carriers of important messages between ships which could not be entrusted to wireless owing to the possibility of the enemy wireless picking up the messages, such as communicating to incoming ships information regarding the location of mines, submarines, and courses, to avoid mistakes and confusion. (Seaplanes and dirigibles used.)

16. Carried out operations over land and sea intended to divert the attention of and mislead the enemy while strategical operations were being carried out by the fleet of squadrons. (Land aeroplanes, seaplanes and dirigibles used.)

17. Have made it possible for commanders to get films of theaters of operation, photographs of the location, composition and disposition of hostile naval forces, and photographic records of condition and of the movements and operations of their own as well as of the hostile naval forces.

As American airmen have not had occasion to do any of these things, they still have everything to learn and the new men who come in to fill the ranks will even have to learn the rudiments of aeronautics.

Realizing this, the author has endeavored to make this a book of reference for the authorities and for every person's library, and a book of instruction for the naval aeronautic student.

CONTENTS

CONTENTS

The photographs are from the private collection of Henry Woodhouse and by courtesy of FLYING, AERIAL AGE, and the ILLUSTRATED LONDON NEWS.

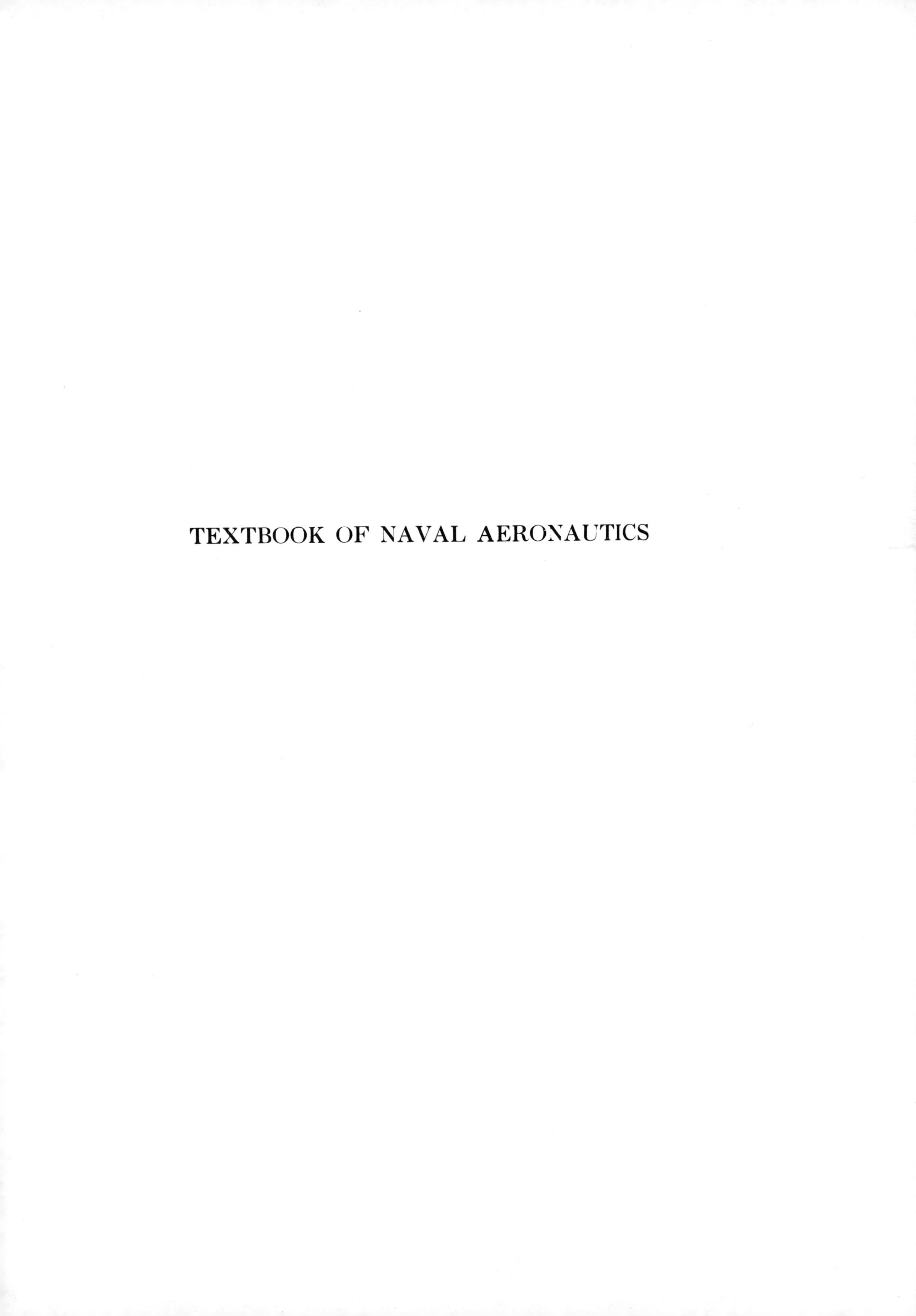

TEXTBOOK OF NAVAL AERONAUTICS

TEXTBOOK OF NAVAL AERONAUTICS

CHAPTER I

AERIAL STRATEGY AND TACTICS

Naval aeronautics as an applied science is in its infancy, but there are definite indications of the course of developments. In a general way, the development of air fleets will be on the same lines as the development of sea fleets; the manœuvering will be governed by the same basic principles of strategy and tactics as are followed for the operation of sea fleets, excepting that the air fleets have a boundless ocean of air, boundless vertically as well as horizontally. An air fleet can fly over or under the enemy's air fleet, and can make its base in a lake in the midst of mountains or in a shallow bay, where no ship of the sea can go. It can operate on land and across promontories, and fly over mountains. In other words, the base of an air fleet can be established almost anywhere, whereas bases for ships of the sea can only be established in a few favorable places where the water and coastal conditions permit. The air fleets can always travel in straight lines, whereas all vessels of the sea must follow the charted channels.

Revolutionary Operations of Naval Aviators Over Land

Bearing in mind the sharp lines of demarcation where, usually, the Navy ceases to operate and the Army begins to operate, and vice versa, we must admit that the operations of naval air fleets in the present war are positively revolutionary, because we find hundreds of cases where naval aviators, flying land aeroplanes, as well as seaplanes, made raids which took them for several hundreds of miles over the enemy's country. In many cases the purpose has been to destroy the enemy's sources of supplies of munitions, as in the case of the raid on the Mauser works, at Obendorf, carried out on October 13, 1916, in which a squadron of British Royal Naval aviators participated. How the raids are carried out will be told hereafter under the heading: "Raiding Operations."

Ships that Navigate Over Mountains!

There have been hundreds of raids by Allied naval aeroplanes overland in the interior of Germany, in the Balkans, Mesopotamia, Asia and Africa. Likewise, there have been numerous raids by the naval aeroplanes of the Central Powers inland on Russian, French and British soil. Instances will be quoted hereafter, under the heading of: "How Far Should Naval Aviators Go Inland?"

There are significant records of flights by naval aviators over mountains. Dirigibles have been flying over mountains for years. For instance, on September 15, 1916, naval aviators flying land machines bombed bases in the interior in Bulgaria, and on September 17, 1916, a seaplane of the Royal Naval Flying Corps bombed a town in Palestine forty-five miles from the coast, crossing mountains several thousand feet high on its outward journey. This was, in effect, a ship being navigated over mountains! As a matter of fact the Italian naval aviator, Angelo Guenzi, on January 10, 1917, flew a hydroaeroplane from the naval base at Gesto Calendre to a height of 18,000 feet, which is higher than some of the highest mountains. Zeppelins are also capable of reaching heights close to 20,000 feet.

It is also interesting to note that oftentimes, in the course of a flight, naval aëroplanes, like dirigibles, pass over parts of England, France, Belgium and Germany.

One of the huge three-motored American seaplanes which are doing such effective work in submarine hunting for Great Britain.

Cooperation Between Army and Navy Air Services

GREAT BRITAIN AND FRANCE HAVE MINISTERS OF THE AIR

The official reports give many instances of army and navy aero squadrons having cooperated in bombing expeditions. Likewise, there are numerous instances of attacks on ships at sea by army aëroplanes.

Perhaps most revolutionary of all was the appointing of Admiral Sir Percy Scott, a naval man, to take charge of the anti-aircraft defenses of London.

All demarcations have been wiped out, and as a matter of fact, both Great Britain and France have now put their air services under a Minister of the Air, who supervises, in a general way, both the army and naval branches of the air service. The supervision deals essentially with getting the equipment and the personnel, and carrying out the broad policy of the Defense Councils. The details of operation are, of course, left to the army and navy authorities. The German air services have always cooperated very closely. A step towards very close cooperation between the U. S. Army and Navy was taken in the early part of March, 1917, when the Joint Board of Aeronautics decided to establish joint training aeronautic stations. A further step towards placing the air services of the United States under a Minister of the Air was taken in May, 1917, when the Aircraft Production Board was appointed.

Functions of the Naval Air Service

IT HAS ALREADY EXTENDED THE FUNCTIONS OF NAVIES

The functions of vessels of war were defined by Admiral Sir Percy Scott some time ago as follows:

DEFENSIVELY

1. To attack ships that come to bombard our ports.
2. To attack ships that come to blockade us.
3. To attack ships carrying a landing party.
4. To attack the enemy's fleet.
5. To attack ships interfering with our commerce.

OFFENSIVELY

1. To bombard an enemy's ports.
2. To blockade the enemy.
3. To convoy a landing party.
4. To attack the enemy's fleet.
5. To attack the enemy's commerce.

The official reports of the employment of seaplanes in the great war show that air fleets have been used to perform every one of the aforesaid functions—and more. Hundreds of aerial attacks on ships and submarines have taken place. Besides attacking ships, convoying ships and landing parties, protecting commerce, bombarding the enemy's ports, and attacking the enemy's commerce, the naval air fleets of the warring nations have done many things, including, as has been pointed out, attacking munition factories far inland, and strategical places, which require flying over mountains. In other words, the naval air services have extended the functions of navies.

Naval aero squadrons equipped with guns have also flown inland and attacked bodies of troops with their guns and bombed railroads far from the coast.

Blockading of Air Fleets Impossible

Blockading of air fleets is, of course, impossible, because they operate on the vertical plane and have roads in every compass direction at every fifty feet skyward up to any height—the altitude record being 26,260 feet.

Scouting may be taken as the equivalent of the work of the cruisers; torpedo launching and bomb dropping, as the equivalent of the work of the destroyers; and the combined damage done by the dropping of several tons of explosives carried by aero squadrons may be taken as the equivalent of the work done by a battleship.

100 BATTLE-PLANES EQUIPPED WITH 3-INCH GUNS AFFORD DEFENSIVE POWER OF 60,000 RIFLES TO PREVENT LANDING OF AN INVADING FORCE

Rear Admiral Bradley A. Fiske, United States Navy, has pointed out that 100 battle-planes carrying 3-inch guns would have a defensive power equivalent to 60,000 rifles. They would have this additional advantage that, whereas 60,000 infantrymen would be hard to transport to any one place, the battle-planes could easily cover a line of 300 miles; in other words, could be mobilized quickly at any one point between New York and the Chesapeake to prevent the landing of an invading force. Admiral Fiske very aptly points out that this extreme mobility of power is unknown in any other arm of our defenses. It is also pointed out that it would take a tremendous length of time to equip and train 60,000 infantrymen, whereas it would take a comparatively short time to get 100 battle-planes with the trained aviators and equipment necessary.

The size and power of aeroplanes is steadily increasing and aeronautic engineers now consider it quite practicable to build aeroplanes that will carry between 30 and 50 tons. This may sound extreme at the date of writing, but it does not sound half as extreme as when in 1908 it was stated that some day an aeroplane would go up to a height of 7000 feet with twenty-one passengers, which is exactly what was done in England in July, 1916. Not later

Photograph taken "Somewhere" in the War Zone, showing a British seaplane mounted with a Davis Non-Recoil Gun, which is now made in three-inch size, weighing less than 500 pounds. A single shot will sink a submarine. A "flight" of five machines mounting such a gun—a "flight" being the smallest tactical unit—can disable the best destroyer.

Courtesy of London Illustrated News

The first Naval Air Raid of the Great War. Cuxhaven, Germany's famous naval base, chief airship base, and mine base was attacked by seven bomb-dropping British naval seaplanes on Christmas day, 1914. (The impregnable position of Cuxhaven, and Heligoland and the Schillig Roads are shown herewith.) The seaplanes were carried to German waters on steamers converted into seaplane carriers which were convoyed by British cruisers.

than 1911 it was considered impossible to fly an aëroplane with two motors. Mr. Edwin Gould offered a prize of $15,000 for a contest governed by very modest conditions, but the prize was not won. We can look forward to amazing and speedy progress in the construction and application of aircraft. Half a dozen nations have between 2000 and 12,000 aviators each, which have been enlisted and trained since the beginning of the Great War. Canada alone, which did not have any aviators at the beginning of the war, sent 600 aviators to England in twenty months.

Aerial Operations Independent of the Fleet

Every raid of the naval Zeppelins has been independent of the German fleet; likewise all the seaplanes raids over land.

In independent aërial operations a number of seaplane carriers are often used. In the Salonika campaign, where there was no danger of attack on seaplane carriers from hostile warships, seaplane carriers were pressed into use in number. The Turkish bases were beyond reach of naval guns and owing to the lack of transportation facilities, the Allied land forces could only advance slowly. There the seaplanes became the most effective weapons for attacking the bases, destroying railroads, trains, supply stations, etc. In July, 1916, a British aviator torpedoed four Turkish vessels.

On May 1st, 1917, a German seaplane torpedoed the British steamer *Gena*. The details of this startling event—which introduces a new method of naval attack—were given in the affidavit signed at Newcastle by the American seaman, Oscar C. Findley to the American Consul. The affidavit reads:

"While I was aboard the British steamer *Gena* in the Channel on May 1, two German seaplanes, 300 feet aloft, passed near by. Without any warning whatever, one dropped a torpedo to the water and the missile sped along the surface and struck the *Gena*. We sank in thirty-five minutes.

A Norwegian steamer which approached us was similarly attacked. We fired while sinking and brought down one seaplane. The other fled. Two Germans on the destroyed plane were picked up by the same trawler that rescued survivors from our ship.

The number and extent of independent aerial operations will increase with the employment of the larger seaplanes being built, which will be

A Zeppelin over the German Fleet. Capable of staying in the air fifty hours, and of traveling at a speed of close to sixty miles an hour, the Zeppelin is the aerial eye and the aerial guide of the German Fleet.

able to carry torpedoes and guns of large enough caliber to sink unarmored ships.

The Air Service in Cooperation with the Fleet, and as an Auxiliary of the Navy

The Great War was only a few months old when the first aerial squadron in the cooperation of the fleet was carried out. It was the expedition against Cuxhaven, Germany's famous naval base, chief airship and mine base, which happened on Christmas Day, 1914. Three steamers converted into seaplane carriers were used to transport the seaplanes and were escorted by cruisers and destroyers. The purpose was to bomb the airship sheds. Seven British seaplanes participated, six of which returned to the seaplane carriers. The pilot of the seventh landed away from the seaplane carriers and was picked up later by a trawler. (See chapter on "Bomb Dropping From Aircraft" for details.)

Numerous Services Rendered by Aircraft as Auxiliaries of Navies

A seaplane from the mother ship *Engadine* was used by the British, and Zeppelins were used by the Germans for reconnoitering in the battle of Jutland on May 30, 1916. But in this case the aircraft were the auxiliaries of the fleet.

This brings us to the numerous services which aircraft can render as auxiliaries of the navy.

Photograph taken during the Jutland Sea Battle, May 31, 1916, which was the first actual naval battle participated in by aircraft.

Seaplanes from one of the many aerial coast patrol stations on the French coasts. In his report to the Chamber of Deputies Admiral Lacaze, the French Minister of Marine, on May 26, 1917, said: "We have organized seaplane stations all around the coasts, so that the zone of action of each station joins that of its neighbor on either side."

Dirigibles, aeroplanes and kite balloons as auxiliaries of navies, have rendered the following services:

1. Attacked ships and submarines at sea with bombs, torpedoes, and guns. (Seaplanes and dirigibles used.)

2. Bombed the enemy's bases and stations. (Land aeroplanes, seaplanes and dirigibles used.)

3. Attacked the enemy's aircraft in the air. (Aeroplanes and seaplanes used.)

4. Served as the eyes and scouts of fleets at sea. (Dirigibles, seaplanes and kite balloons used.)

5. Protected ships at sea and in ports against attacks from hostile submarines and battleships. (Seaplanes and dirigibles used.)

6. Defended and protected naval bases and stations from naval and aerial attacks. (Land aeroplanes, seaplanes, and dirigibles used.)

7. Convoyed troop ships and merchant ships on coastwise trips. (Dirigibles and seaplanes used.)

8. Patrolled the coasts, holding up and inspecting doubtful ships, and convoying them to examining stations and searching coasts for submarine bases. (Dirigibles used.)

9. Prevented hostile aircraft from locating the position and finding the composition and disposition of the fleet, getting the range of ships, naval bases, stations, magazines, etc. (Land aeroplanes and seaplanes used.)

10. Located, and assisted trawlers, destroyers, and gunners in capturing or destroying hostile submarines. (Seaplanes, dirigibles and kite balloons used.)

11. Cooperated with submarines, guiding them in attacks on ships. (Dirigibles and seaplanes used.)

12. Located mine fields and assisted trawlers in destroying mines. (Dirigibles, seaplanes and kite balloons used.)

13. Served as the "eyes in planting mines," minimizing the time required for mine planting. (Dirigibles, seaplanes and kite balloons used.)

14. Served as "spotters" in locating the position of the hostile ships and directing gun-

fire. (Dirigibles, seaplanes and kite balloons used.)

15. Served as carriers of important messages between ships which could not be entrusted to wireless owing to the possibility of the enemy wireless picking up the messages, such as communicating to incoming ships information regarding the location of mines, submarines, and courses, to avoid mistakes and confusion. (Seaplanes and dirigibles used.)

16. Carried out operations over land and sea intended to divert the attention of and mislead the enemy while strategical operations were being carried out by the fleet of squadrons. (Land aeroplanes, seaplanes and dirigibles used.)

17. Have made it possible for commanders to get films of theaters of operation, photographs of the location, composition and disposition of hostile naval forces, and photographic records of condition and of the movements and operations of their own, as well as of the hostile naval forces.

In the United States we are just beginning to realize the importance of aeronautics, and we are just taking steps to develop our air service. The entire country may be said to be cooperating with the Aircraft Production Board and the Army and Navy in developing the air service. About 30,000 applications have been received from young men wishing to join the air service, mostly college men, several hundred of whom are now learning to fly at their own expense, to be ready to meet an emergency. Six units of the aerial coast patrol are under organization, and the members of these units are training at their own expense and have purchased seaplanes, the use of which they have offered to the Government. The same is true in the naval militia. Patriotic people who became interested in aerial preparedness through the efforts of the Aero Club of America have contributed aeroplanes and funds with which to start aviation sections in the naval militia of a number of states.

With this great popular interest, we may expect this country—the country of Langley, the Wrights, Curtiss, and other pioneers—will take giant steps in the development of our much needed air service.

A 200 horse-power twin-motored Curtiss hydroaeroplane, of the type especially adapted for submarine hunting and launching of torpedoes, starting for a flight.

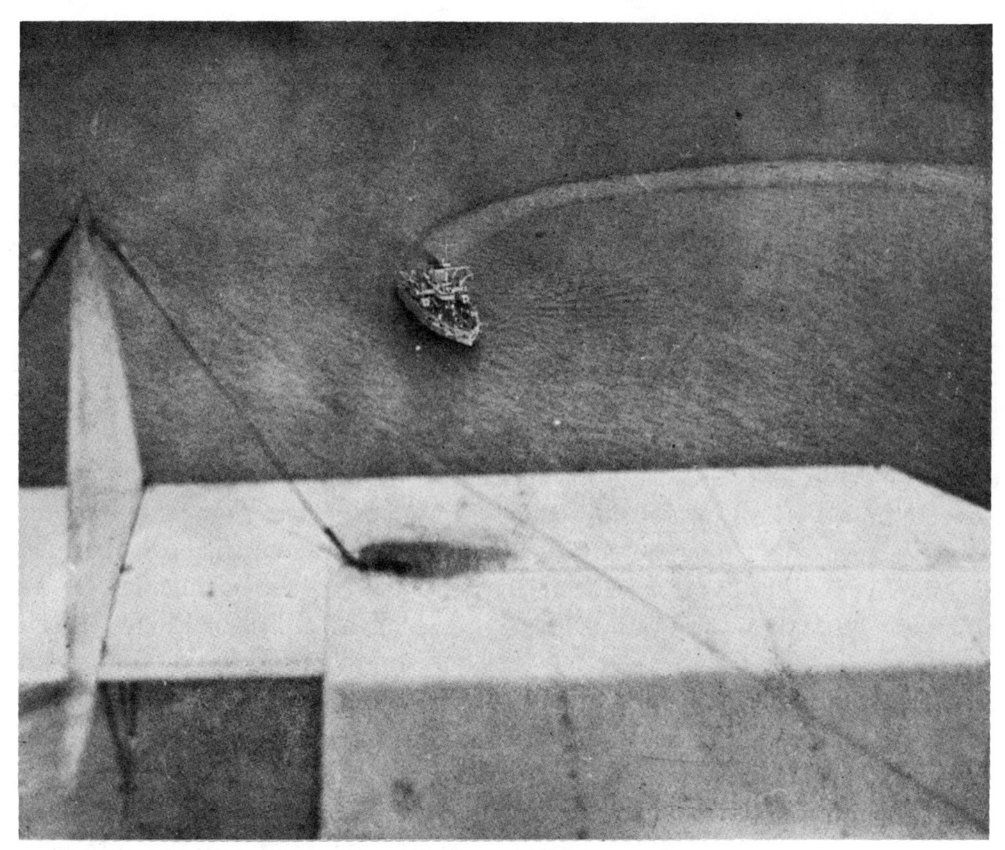

A Russian cruiser photographed from a Curtiss hydr oaeroplane, the tail of which is seen in the photograph.

CHAPTER II

AERIAL ATTACKS ON SHIPS AT SEA

Historic: Aerial attacks on ships, cruisers, destroyers, submarines, and merchantmen at sea began in the early part of the war, as soon as aircraft became numerous enough to permit employing them for offensive operations. A few of the earliest cases of aerial attacks on ships are given herewith:

In 1914–15, a small squadron of Russian seaplanes bombed the German cruisers *Breslau* and *Göeben* which were bombarding the port of Sevastopol. On May 5, 1915, a German naval aero squadron bombed the Russian cruiser *Slava* and a submarine on the Baltic Sea. The French steamer *Harmonie,* on December 11, was attacked by an Austrian submarine. The submersible fired two torpedoes, which were without effect, and then withdrew. The next day the *Harmonie* was attacked by an aeroplane that flew overhead for a quarter of an hour and dropped six bombs, all of which fell into the sea.

German seaplanes have been active particularly in enforcing the "war zone" decree issued against the British Isles. A seaplane cruising over the North Sea attacked with bombs the British steamer *Cordoba* as she was entering Yarmouth Harbor. No damage was done. Seaplanes also dropped bombs on a Dutch and a British vessel, early in 1915, but without hitting them. Zeppelins have been used by Germany for the defense of commerce against attacks by British and Russian submarines. The steamer *Scotia* of Stettin, Prussia, bound from Sweden to Stettin with a cargo of ore, was pursued by a British submarine off Bornholm. In reply to wireless calls for assistance, a Zeppelin suddenly appeared, whereupon the submarine submerged and disappeared. The Turkish army headquarters announced that a hostile monitor, which was firing shells in the direction of Akabah, was silenced by a Turkish aeroplane, which dropped two bombs on the monitor.

An English aeroplane dropped three bombs

A bomb dropped from an aircraft exploding in the water.

from a height of 500 feet on enemy lorries on November 21. The bombs having missed, the machine turned and flew over the lorries again and dropped three bombs in the midst of them. Turning again, the observer directed his machine gun on the enemy from a height of 150 feet. In the course of the first of the raids made by Allied airmen on Bruges, the railway line outside the town was destroyed and a vessel at St. Michel, occupied by Germans, was damaged. In the course of the second attack made upon the port, serious damage was done to three torpedo boats and the steamer *Colchester.* In the third raid a wharf for submarines between Lisseweghe and Zeebrugge was hit. Flight Sub-Lieutenant Ferrand attacked a hostile seaplane November 28, 1915, which was accompanied by three more seaplanes and a destroyer, off the Belgian coast, and brought it down by gunfire when it immediately sank. He then attacked the destroyer, and only abandoned the attack after coming under heavy shell fire both from the destroyer and the shore batteries of Westend.

This is only one of a number of cases of ærial attacks on destroyers off the Belgian coast by British aviators. In several cases the destroyers were reported sunk. There also have been many reports of torpedo boats destroyed in German parts by bombs dropped by Allies' aviators.

The American steamer *Cushing,* which arrived at Rotterdam on April 29, 1916, from Philadelphia with a cargo of petroleum, reported that on the afternoon of April 26, when in latitude 51 degrees, 45 minutes N., and longitude 2 degrees 30 minutes, E., she was attacked by a German aeroplane which threw three bombs at the ship. The first two fell wide, but the third passed exceedingly close to the stern rail and fell into the sea. At the time the ship was flying the American flag and had her name painted on the side in letters six feet long.

It will be remembered that after his early operations in Flanders, Com. Samson took a wing of the R. N. A. S. to Gallipoli, where he and many of his wing were mentioned in despatches by Admiral de Robeck and General Sir Charles Monroe. Later, he was invalided

A bomb dropped from a German aircraft exploding close to a British ship. Photograph taken from a British cruiser during the action off Cuxhaven, December 25, 1914, which was the first naval aerial operation in history.

The Handley Page battleplane. The span of the machine is about 98 feet, length 65 feet, height 20 feet, with seating facilities for five people. It is equipped with two 12-cylinder Rolls-Royce motors of 280 horse-power. It has mountings for 3 Lewis guns. This machine holds all the world's records for large aeroplanes up to a pilot and 20 passengers, which were carried to a height of 7,180 feet, the pilot on this occasion being Mr. Clifford B. Prodger, an American.

home, and, on his recovery, returned to the eastern Mediterranean, where he took command of the *Ben-Ma-Chree.* Despatches from the Egyptian command had previously mentioned this ship as a seaplane carrier, commanded by Squadron Commander (Acting Wing Commander) L'Estrange Malone, now wing commander.

In 1912 the British seaplane carrier *Ben-Ma-Chree* (Wing-Commander C. R. Samson, D. S. O.) was sunk by gunfire in Kastelorizo Harbor (Asia Minor) on January 11. The only casualties were one officer and four men wounded.

An Aerial Attack on a Seaplane Carrier

A vivid picture of a bombing attack upon a ship at sea has been given by Lieutenant François-Bernou, who was aboard the seaplane carrier *Ben-Ma-Chree* in the Salonika campaign. The exact position of the ship has been deleted by the censor for obvious reasons, and a number of photographs taken aboard were skilfully mangled in fear of many valuable details reaching the enemy. Life aboard the mother ship would seem a very novel and thrilling experience to the layman, but familiarity soon breeds not contempt but indifference to the extraordinary activities which form its daily routine. The fascinating spectacle of the aeroplanes rising from the mother ship for their perilous flights of reconnaissance or attack, or their arrival from long aero cruises and the work of swinging them inboard or overboard by powerful cranes, soon became a commonplace. As Lieutenant François-Bernou remarks, these sights, which have never before been witnessed in any war on land or sea, seemed no more unusual than the cranking of an automobile.

The French officer, being a newcomer, was alive to the extraordinary dramatic interest of these stirring days. Many of the flights were made for long distances above the Holy Land, and Lieutenant François-Bernou was impressed by the curious coincidence that the land of miracles should witness this twentieth-century miracle of flight. Day after day the seaplanes ventured forth from the shelter of the mother ship on many daring flights to spy the enemy's positions or direct the deadly fire from the sky upon troops or fortifications far inland, and after raids would return like homing pigeons, bringing in valuable reports.

One of the most dramatic incidents of the life aboard the seaplane carrier *Ben-Ma-Chree* came one day most unexpectedly. An aeroplane which had been out on a scouting trip was suddenly sighted, approaching at top speed, pursued by a German Fokker. The aeroplanes were flying at a high altitude. The British aircraft had managed to elude the enemy, and by a daring volplane landed safely on the water beside the mother ship. Everything was in readiness to retrieve the aeroplane, which was quickly hoisted on board. The German Fokker was not content to give up the chase and continued to fly above the *Ben-Ma-Chree* at a comparatively low altitude, dropping deadly bombs. Such an attack from the sky is extremely daunting. Any one of the bombs, which described black vertical lines against the sky, might bring instant disaster.

The bombing aeroplane succeeded in passing directly above the ship several times. Some of the bombs struck the water so near the vessel that the splash of the waves thrown up by the explosion wet the steamer's bridge. The entire crew stood manfully at their posts. The only hope of escape lay in driving the ship full speed ahead in a series of mad zig-zags, a course, which the aeroplane could not follow. The anti-aircraft guns at last succeeded in driving away the enemy, with what damage could not be known.

After a thrilling experience of this kind, the French officer remarks, the men were almost overcome with sleep, so exhausting had the experience been both on mind and body. The pilot who had been chased in by the German Fokker was, by the way, a very interesting character. In less troublous times he had been a famous jockey, and his thrilling race against time for the mother ship was in a sense a familiar experience. Three days after this experience while on a very daring scouting and bombing trip, a shot from the enemy struck his motor, forcing him to descend, when he was made a prisoner by the Turks.

Weapons and Methods of Attacks on Ships

ATTACKING WITH BOMBS

The weapons employed for attacking ships are bombs, torpedoes, and guns of fairly large caliber. Up to the summer of 1916 all the attacks on ships were made with bombs. In July, 1916, there was registered one case of a British naval aviator who, according to reliable reports, made four flights over the land into the Sea of Marmora in an aeroplane under which a Whitehead torpedo was secured, and sank four Turkish vessels, using 14-inch torpedoes, weighing 731 pounds each. For this service he was given the Distinguished Service Order.

While this was the first case of actual destroying of ships by means of torpedoes dropped by aeroplanes, the idea was by no means new. Rear-Admiral Bradley A. Fiske, U. S. N., patented a device for launching torpedoes from aeroplanes in July, 1912. Captain Alessandro Guidoni of the Royal Italian Navy made experiments for a number of years in dropping weights from aeroplanes with a view of eventually developing a large aero-

A ship makes a perfect target for a seaplane or dirigible. This photograph shows a British destroyer as it looked to the aviator who photographed it from the air.

plane for launching the standard-size torpedoes.

How the Revolutionary Leavitt Torpedo Was Developed

Early in January, 1917, the civilian leaders of the movement to develop our national defenses, after taking stock of the military resources of the United States, came to the conclusion that, owing to our small army and navy and general unpreparedness, the only hope of success on the part of the United States in case of war would be in developing some powerful new instrument which would give us predominance. A committee consisting of Messrs. Alan R. Hawley, Henry A. Wise Wood, Rear Admiral Robert E. Peary, and the writer made an investigation with the purpose of finding one or more new instruments, the value of which would be so great that they would give predominance to the side which employed them. After looking over the field of inventions, the committee came to the conclusion that the torpedoplane patented by Rear-Admiral Bradley A. Fiske in July, 1912, was a revolutionary invention of tremendous possibilities.

The committee then asked Admiral Fiske to deliver an address on the subject, which he did at the Aeronautic Conference held in connection with the First Pan-American Aeronautic Exposition, Grand Central Palace, New York City. Admiral Fiske's address on that occasion is printed elsewhere. It created great interest, and, as a result, a fund was set aside for the purpose of defraying the expenses of the experiments of developing the torpedoplane. Admiral Fiske was asked to be the chairman of a committee to supervise the work of developing the torpedoplane, and he appointed the following as members of his committee: Alan R. Hawley, Henry A. Wise Wood, Rear Admiral Peary, John Hays Hammond, Jr., F. Trubee Davison, Schuyler Skaats Wheeler, Frank M. Leavitt, Lawrence B. Sperry, and the writer.

The committee had the choice between concentrating its efforts in developing large seaplanes and training aviators to drop full-sized Whitehead torpedoes which weigh 2000 pounds, and measure 21 inches in diameter and 17½ feet in length, or to develop a torpedo small enough to be carried by any of the two-passenger flying boats or hydroaeroplanes now in general use. The committee came to the conclusion that, owing to the fact that there were very few aviators in the United States who had any experiences in piloting a large seaplane and, owing to the time that would be required to train men to drop such heavy weights from an aeroplane, it would be best to concentrate efforts in developing a small torpedo, weighing less than 200 pounds, which could be dropped from the average two-passenger seaplane by almost any aviator. This would make it possible in time of war to press into service for launching of torpedoes, almost every civilian naval and military aviator who had had sufficient experience to pilot a machine.

At first it seemed impossible to develop an automobile torpedo weighing less than 200 pounds, having a range of about 1000 yards at a speed of about 25 knots, but the committee was willing to have experiments carried out regardless of the possibility of failure, and three leading experts on torpedoes promptly took up the work and soon advised the committee that such a torpedo could and would be developed. The committee was led to decide by the results of the excellent work of Volunteer Aerial Coast Patrol Unit No. 1, which led hundreds of other college men to interest themselves in aeronautics. It was realized that hundreds of men would follow the example of the members of Unit No. 1, and would make it possible, in case of war, to quickly organize squadrons of aviators equipped with torpedoplanes sufficiently powerful to sink destroyers, transports, and other nonarmored ships. Larger torpedoes will of course, sink armored ships.

John Hays Hammond, Jr., is meeting the difficulty of launching large torpedoes from seaplanes by his revolutionary invention which makes it possible to direct the torpedo to the target from an aëroplane, by wireless.

At the date of writing there are three different torpedoes, one of which, being developed by Mr. Frank H. Leavitt, the expert of the E.

W. Bliss Company, who is responsible for the efficiency of the Whitehead torpedo, is ready for test. The details of this torpedo and the others being developed will probably be made public while this book is on the presses.

Rear Admiral Bradley A. Fiske's paper gives an excellent idea of the revolutionary value of this new development. Since these experiments began, members of the committee have received many expressions from naval authorities stating that the torpedoplane will revolutionize naval warfare.

Memoranda:

CHAPTER III

THE TORPEDOPLANE AND ITS POSSIBILITIES

By Rear-Admiral Bradley A. Fiske, U. S. N.

Messrs. Alan R. Hawley, Henry A. Wise Wood, and Henry Woodhouse, have kindly expressed the thought that the "Torpedoplane" which I patented in July, 1912, is destined to become a dominant instrument in war, giving a marked advantage to the side which employs it and a corresponding disadvantage to the other side. They have also thoughtfully pointed out that, having fallen so much behind other countries in naval preparedness, we can only hope to catch up by including in our naval program the most effective new inventions, and that the "torpedoplane" might, under favorable conditions, make a $20,000 aeroplane a worthy match of a $20,000,000 battle cruiser.

It is said that strategy directs the conduct of war, and that it uses logistics to provide the men, guns, and other details that the plans of strategy demand. This analysis is correct so far as it goes; but it omits the factor that inspires both strategy and logistics, the factor called "invention."

In the United States, and in most other countries, we have come to regard invention as applicable to mechanism only. But to regard invention in that way only is to regard it in a very dim light and to fail to see to how many other things invention is applied to.

With the use of invention in mechanism we

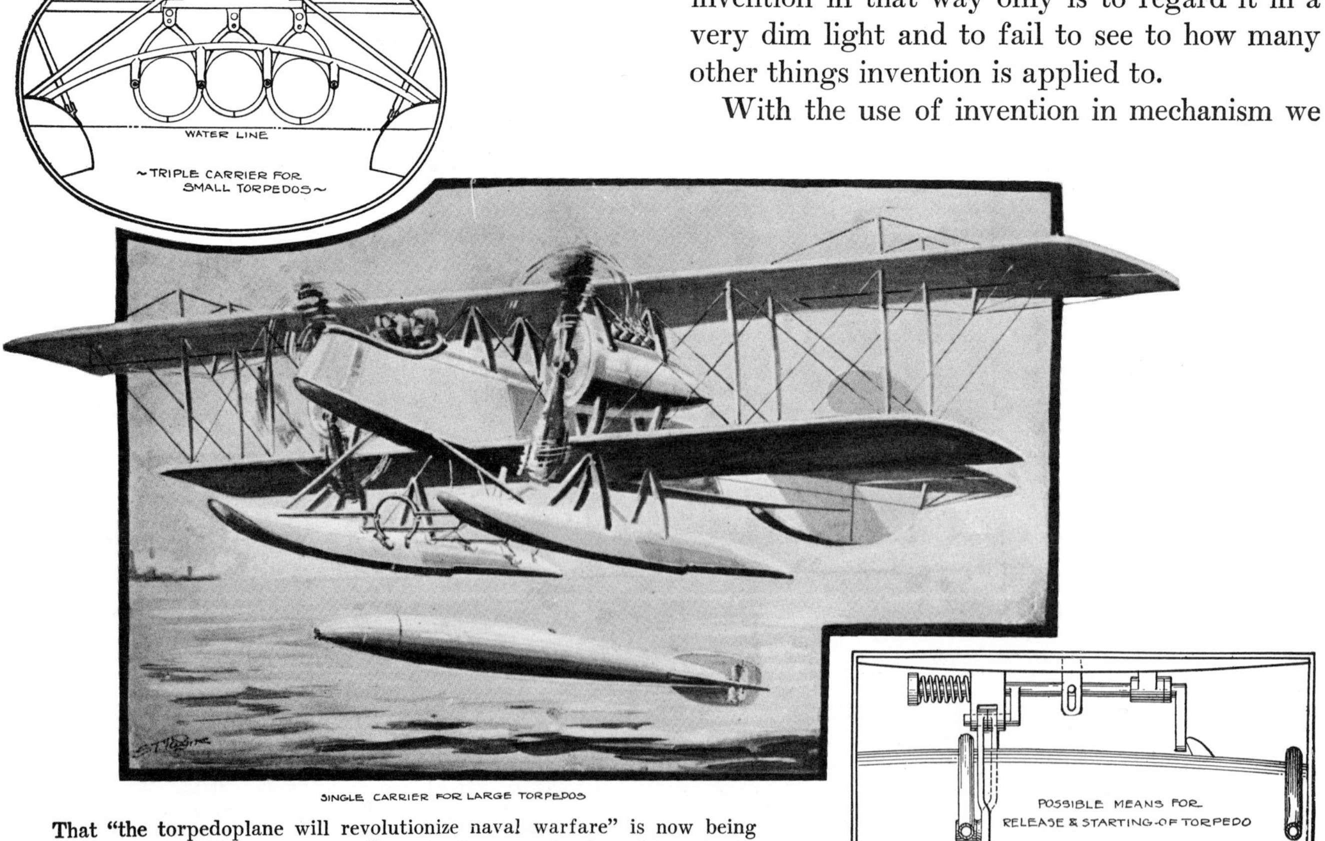

That "the torpedoplane will revolutionize naval warfare" is now being admitted by naval authorities. The simple torpedo launching device developed for Admiral Fiske's torpedoplane is shown herewith. The details of the torpedo itself, which has been developed for national defense under the auspices of a group of patriotic workers for aerial preparedness, will not be made public for obvious reasons.

are very familiar; but it may be pointed out that before the mechanism itself can be invented, the idea of inventing the mechanism must first be invented. Before Ericsson invented the mechanism which we call the *Monitor,* he invented the idea of producing such a thing. Before Alexander started to invade the rest of the world, before Cæsar started for Gaul, before Frederick started for Silesia, before Moltke started for Austria and France, before Washington started for Yorktown, before any policy, or any new line of strategy, or any new enterprise whatever was begun the idea was first conceived by the mind; that is, invented. Shakespeare recognized this truth when he exclaimed, "Oh, for a muse that would ascend the highest heaven of invention!"

The most startling interjection into warfare of a newly invented thing was the Ericsson *Monitor*. Comparatively few of the people living now remember the tumultuous joy that ran through the Northern States, when the news was flashed that the *Monitor* had defeated the *Merrimac* near Hampton Roads, March 9, 1862; and even those of us who are old enough to remember that fact fail to realize what a tremendous menace the ironclad *Merrimac* was. Leaving the Norfolk navy yard on Saturday morning, March 8, she soon rammed and sank the U. S. S. *Cumberland,* which carried more men and guns than she; and, a very few hours afterwards, destroyed the U. S. S. *Congress,* also carrying more men and guns than she. Had the *Merrimac* continued her career as successfully as she began it, she would have destroyed the navy of the Northern States, and brought about the success of the Confederacy. In other words, THE MONITOR SAVED THE UNITED STATES.

The reason why the *Merrimac* and *Monitor* were so successful was because each brought into battle a weapon which the other side did not know how to defend itself against.

That the torpedoplane will become an important factor in naval warfare in the near future, many people have no doubt. It is a scheme whereby the regular Whitehead auto torpedo may be launched from an aeroplane as effectively as it is launched from a destroyer.

As you may know, a destroyer goes toward her enemy at a speed which can rarely be as high as thirty knots an hour, and launches a torpedo from her deck into the water; and by that act of launching throws back a lever on the torpedo, called the starting lever, which causes the propelling mechanism of the torpedo to go ahead full speed. The torpedo, therefore, after reaching the water, goes along in the direction in which it is pointed; and, if it is pointed correctly, it hits its target ship below the water line, and usually sinks or disables her.

The scheme which I submit for your consideration and is herewith illustrated needs little explanation. The aviator approaches his

Torpedo mounted on small scout seaplane of triplane type. Pulling the lever releases the torpedo which is securely held under the seaplane.

target from a great distance and high up in the air; and when, say, six or seven miles away, he volplanes toward the water, runs above the surface of the water a short distance, heading toward his target, and when ready simply pulls a lever. The action of pulling the lever releases the torpedo which is rigidly held under the aeroplane, and at the same time throws back the starting lever, with the result that the torpedo falls in the water in exactly the same way as if it had been dropped from a destroyer, instead of an aeroplane.

I have seen it stated in print several times that Captain Alessandro Guidoni, of the Italian Navy, tried out the scheme two or three years

Model D-1, 300 horse power twin-motored Gallaudet seaplane, built for the United States Navy, which can carry nearly half a ton of explosives or a large gun.

ago, and hit the target nine times out of ten at a distance of 3000 yards. Not having an aeroplane large enough to carry a heavy, long-distance torpedo he used a light short-distance torpedo suitable for the size of the aeroplane.

I received private information from Europe about a year ago that a lieutenant in the British Navy made four flights over the land into the Sea of Marmora in an aeroplane under which a Whitehead torpedo was secured, and sank four Turkish vessels, using 14-inch torpedoes, weighing 731 pounds each. For this service, he was given the Distinguished Service Order. A short time ago, I got a verification of this news from a wholly different source, and I also received further information, which is of absolute reliability, that one of the belligerent countries is taking means to use this plan on a large scale.

I have talked about the scheme to many naval officers and many aviators. The naval officers agree with me that it would be very difficult indeed for the guns of a ship to hit a torpedoplane, for the reason that accurate firing of guns from a rolling ship at an aeroplane, especially if that aeroplane is neither overhead or on the surface of the water, is almost impossible. It is a much more difficult matter than firing at aeroplanes overhead from a stationary platform on land.

The greatest difficulty in firing from a rolling ship at anything near the surface of the water, is to find the range at which to fire; and a rapidly approaching, ill-defined aeroplane makes finding and correcting the range almost impossible. The sudden changes in the height of a torpedoplane as she would swoop down would increase the difficulty tremendously.

Besides, in a contest between a torpedoplane and a ship, in which the torpedoplane seeks to strike the ship below the water, the ship, if she is struck there, is disabled, if not destroyed; while the torpedoplane can be shot full of holes without much damage, unless hit in a vital place.

The aviators tell me that they see no practical difficulties whatever in doing their part of the work.

For an attack on battleships such as might approach our coast, the large size torpedo, weighing over a ton, would be best; and this can be fired successfully from a distance of six sea miles or more. For carrying torpedoes like this, we now have in this country a number of aeroplanes large enough for the task; and this, I think, gives the most ready and practical means of defense that we can provide at present.

But battleships are not the only ships that would be sent against us; the battleships would be accompanied by a vast array of other vessels, which are very important, such as destroyers, colliers, ammunition ships, scout cruisers, and transports. These vessels are lightly built and have thin sides, so that light torpedoes would be thoroughly effective. This would be especially the case for attack on destroyers, because their gun fire is not accurate. Torpedoplanes could, therefore, with comparative safety approach them, and discharge their torpedoes from a distance of a few hundred yards.

In our present state of unpreparedness, it would be a great thing if we could bring out something as revolutionary and effective as the *Monitor* and that could be got ready in the limited time that may be granted us. We cannot hope to catch up to any of the leading powers in ship or submarine-building, as their output is enormous compared to ours and their experience greater. They have outdistanced us in naval preparedness, especially in the air, for some of them have as many as 10,000 aviators, while we have not yet 200 in the Army and Navy combined. We are far behind, but we can catch up if we try, and our national security could be brought up to quite a hopeful condition by establishing say fifty torpedoplanes at each of the ten important naval districts, and on aeroplane mother ships, which would go with the fleet. Such an act would give us a quickly made and inexpensive weapon for defense.

I have talked about this to many naval officers and aviators. They all say that it is a good scheme, and point out that torpedoplanes on either side during the Jutland sea battle would have given tremendous advantage over the other side. Many also say that our navy should adopt this device and immediately put it into service. Hope that it may be developed in America as well as abroad is given by the action of the Aero Club of America and the National Aerial Coast Patrol Commission which are making plans for developing the torpedoplane in a practical way. As these patriotic organizations always carry out everything they undertake, I feel confident that we will have torpedoplanes for the defense of our coasts in the near future.

Since writing the above article, the details of the sinking of the British steamer *Gena,* by a torpedoplane, have been officially announced both by the English and German governments. A dispatch from London dated May 2, states that the admiralty announces that the British steamer *Gena,* 2784 tons was sunk on May 1st by a torpedo discharged from a German seaplane off Aldeburg (Suffolk, England). All hands were saved. Another seaplane concerned in the attack was brought down by the gunfire from the *Gena* and its crew made prisoners. An official announcement from Berlin (via London), dated May 2, says: "A few seaplanes attacked on Tuesday morning enemy merchant ships before the Thames and sank a steamer of about 3000 tons. One of our machines failed to return and is supposed to have been lost."

In a letter to the London "Times" of May 9, 1917, Admiral Sir Reginald Custance says, "Why can we not intercept the submarines off German ports? Because the German High Sea Fleet can issue and destroy any small ships engaged in blocking the exits to their ports. They can do this, because our massed fleet cannot cover our light craft. . . . If the massed (German) fleet is destroyed, the action of the submarine is weakened, since its exit is impeded by the small surface craft and submarines of the victor, which are then free to press in to gun range in the enemy's waters, with mines, nets and every new device."

This seems to mean that, if the German High Sea Fleet could be kept away or destroyed, British small craft could prevent German submarines from coming out. To keep off or destroy the German High Sea Fleet, near the German coast, some device that cannot be sunk by mine or torpedo, but that can deliver a destructive blow is apparently required. The torpedoplane, used in large numbers, is most respectfully suggested.

Twin-motor hydroaeroplane constructed by the United States Navy, equipped with two Curtiss 100 horse-power motors.

CHAPTER IV

ATTACKING SHIPS WITH AIRCRAFT GUNS

By Rear-Admiral Bradley A. Fiske, U. S. N.

[During the first eighteen months of the Great War there were several reports of ships being under fire from guns mounted on dirigibles. The details of these attacks, however, were never made public. In the early summer of 1916, when aeroplanes equipped with machine guns were put in use generally to perform the functions of infantry in flying low and attacking troops, there came a more general employment of guns for attacking ships. The guns were of small caliber. Since then several guns as high as three-inches in caliber have been developed, and large aeroplanes capable of carrying from 1500 to 5000 pounds of useful load have been constructed and successfully tested. Therefore the development to be expected in the near future is the equipping of large aeroplanes with large caliber guns.

A seaplane equipped with a three-inch gun is the most powerful and economic factor in sight for submarine hunting. Flying at a speed of from seventy to eighty miles per hour, an air cruiser represents an extraordinary combination of power and mobility. Rear-Admiral Bradley A. Fiske in a report made to the Aero Club of America in March, 1917, brought out very strikingly the fact that one hundred battleplanes carrying three-inch guns would afford defensive power equivalent to 60,000 rifles. Admiral Fiske's report reads in part as follows:]

A Benoist twin-motored seaplane, equipped with 2 Roberts 100 horse-power motors.

Quickest Way to Prepare Defense of United States Against Invasion Is to Develop Large and Powerful Battleplanes

In view of the backward state of our defenses, especially on the land, I would like to suggest the advisability of considering whether it is not possible to devise some means that is powerful and easily gotten, in order to supplement our present means: something that can do for our army and navy service in a measure what the *Monitor* did in 1862. It was the *Monitor* that cast the deciding vote in our Civil War.

As the submarine threatens to cast the deciding vote in the present war, and as the *Monitor* has been expanded into the dreadnought, it may be profitable to ask what is the inherent cause of the efficacy of those weapons. Clearly it is the same cause as makes any weapon efficacious; that is, concentration of great power in a small space, combined with great mobility and assured control. The combination permits of the application of great force on a given spot at a given time.

In 1911, I published an article called "Naval Power," in the *United States Naval Institute,* in which I pointed out this fact, and suggested what a battleship on land could accomplish, if such a thing could be constructed. The *Popular Science Monthly* republished this suggestion in November, 1915, and a few months later the so-called British tanks appeared, which are small land battleships.

Now the unevenness of the ground is a great obstacle in the way of making land battleships very large and fast, and seems to prevent armies from using units as powerful and swift as navies use. This is unfortunate for two reasons. One reason is that the length of our boundaries on the ocean and on our Northern and Southern frontiers is so great, and the average distance to the boundary from places within the country is so great, that it would be highly advantageous for us to be able to move powerful units at great speed; another reason is, the natural inventiveness of our people would enable them to produce very powerful land battleships, if the difficulties were not absolutely insuperable as they seem to be.

My life in the navy brought me into intimate contact with all the advances in naval construction—from the little *Saratoga* in which I made my first cruise as cadet midshipman to the superdreadnought *Florida,* which was my last flagship. The military value of concentration was, of course, impressed unceasingly upon me, and with it a realization of the fact that the main aim of strategy and tactics is to bring a preponderating force to bear on a given point before the enemy can prevent it. To do this, we need concentration of power in as few units as possible, and ability to move those units as rapidly and certainly as possible. POWER, MOBILITY AND CONTROL ARE THE PRIME AGENCIES OF THE MILITARY ART.

Now, at the present time the unit in all armies is the soldier and his musket. We seem tied down to that slow and feeble little unit. But are we really? The navy seemed tied down to the little sailing frigate; so much so that even after the *Monitor's* achievements in our Civil War, we returned to the sailing frigate. The competition of nations, however, forced us to take up larger units, and now we have the *Pennsylvania.*

Is there no way in which this great inventive and constructive nation can get some more powerful and mobile unit than the soldier and his rifle? Can we not get more defensive usefulness out of the intelligent collegian, technician, or chauffeur than by marching him in a regiment with a little musket in his hand? Is there no device by means of which large units of power can be carried which is not subject to the limitations of speed and size that restrict a land battleship to small dimensions?

Yes, and that device is now being used in Europe, after having been designed and manufactured in the United States. It is called the battleplane. Such a device recently carried twenty-seven passengers, and another, an air cruiser, carried 3500 pounds of crew and equipment. Some of the largest battleplanes are being constructed in the United States, and one of the aeroplane manufacturers states that he can easily build a battleplane capable of carrying and launching a full-size torpedo weighing 2500 pounds.

If battleplanes have a field of usefulness in Europe, where the distances are very small and where the organization, training, and strategical employment of large armies has reached a high state of development, do they not have a much wider field in the United States, where the distances are relatively enormous, and where the organization, training, and strategical employment of large armies are arts almost unknown? Is it not possible that an immediate and strenuous development of battleplanes might save us from invasion, or might enable us to help the Allies effectively, as the *Monitor* saved us from the *Merrimac?*

In case our fleet is defeated in the Atlantic during the next year, we shall not have an army that could stand up against any European army that might land on our shores. But if we had a division of, say one hundred battleplanes near New York, costing about three million dollars, we could certainly prevent the disembarkation, transit in boats to the shore, and landing, of any force of soldiers, especially if the battleplanes were assisted by, say, two hundred small aeroplanes, dropping bombs. Similar divisions at other points, including one at the Panama Canal, could perform similar services, and the great speed of the air craft would enable each division to guard a long extent of coast. A division of one hundred battleplanes could go from New York to the Capes of the Chesapeake in three or four hours.

Twin-motored seaplane constructed by New York Aëro Construction Co.

The size and power of the aeroplane has already gone far beyond the limits set for its possible development by certain engineers only three years ago. The practical difficulties of making it larger still are quite apparent; yet, nevertheless, no theoretical limits to its size and power have yet been accepted by aeronauts. That the aeroplane is now the best single weapon against the submarine, is conceded; that it will rapidly advance in size and power, is the mature belief of many aeronauts. Should we not therefore immediately investigate its capabilities, not only as a scout and accessory, but as a major instrument of warfare; not only for carrying small guns, but guns of as great caliber as—say 3-inch? The energy of 100 3-inch projectiles is equal to that of 60,000 musket bullets, even near the muzzle; and is greater at long ranges.

The Curtiss Model H-12 Flying Boat, constructed for the United States Navy, has a wing span of nearly 93 feet, and a capacity for carrying a useful load of 1556 pounds. The boat is expected to attain a speed of eighty-five miles an hour.

I do not suggest the abolition of the soldier and his musket; but neither do I suggest the abolition of the boat pulled by the oars of rowers. I merely suggest that, as the boat pulled by rowers was superseded for large operations by the sailing ship, and as the sailing ship was superseded by the more mobile steamer with broadside guns, and as this type of warship was superseded by the turret ships, and as the turret ship has been expanded into a superdreadnought, so the soldier and his musket may be superseded for important operations by the immeasurably more powerful and mobile battleplane.

If so, the more quickly we act, the better. "Hindenburg never sleeps."

The *Ark Royal*, the British aeroplane ship which operated with the Allies' fleets in the Great War.

CHAPTER V

AIRCRAFT MOTHER SHIPS

The naval air service is divided into three distinct, separate branches, whose functions are quite different, and which may be designated as: (1) *The Offensive Air Service,* which consists of the squadrons of seaplanes, stationed on seaplane carriers and aeronautic bases, which are used for air raids, independent of the fleet; also of dirigibles, which operate from bases; (2) *The Auxiliary Air Service* of the fleet, including seaplanes and kite balloons, which operate with the fleet, using ships as bases; and (3) *The Aerial Coast Patrol,* which operates from naval stations and naval bases. Aircraft mother ships are, therefore, important.

The report of the Jutland battle established two facts: (1) That the German fleet planned its move on information obtained from Zeppelins as to the whereabouts and composition and disposition of the British naval forces; (2) that the British forces were greatly assisted in their action by a seaplane sent up from the seaplane carrier, the *Engadine.*

Admiral Sir David Beatty's report, dated June 19, 1916, to Admiral Sir John Jellicoe, G.C.B., G.C.V.O., Commander in Chief of the Grand Fleet, reporting the action in the North Sea on May 31, 1916, says:

"From a report from 'Galatea' at 2:25 P. M., it was evident that the enemy force was considerable, and not merely an isolated unit of light cruisers, so at 2:45 P. M., I ordered *Engadine* (Lieut-Commr. C. G. Robinson) to send up a seaplane and scout to N.N.E. This order was carried out very quickly, and by 3:08 P. M. a seaplane, with Flight Lieutenant F. J. Rutland, R. N., as pilot, and Assistant Paymaster G. S. Trewin, N. N., as observer, was well under way; her first reports of the enemy were received by the *Engadine* about 3:30 P. M. Owing to clouds it was necessary to fly low, and in order to identify four enemy light cruisers the seaplane had to fly at a height of 900 feet within 3000 yards of them, the light cruisers opening fire on her with every gun that would bear. This in no way interfered with the clarity of their reports, and both Flight Lieutenant Rutland and Assistant Paymaster Trewin are to be congratulated on their achievement, which indicates that seaplanes under such circumstances are of distinct value.

"The work of *Engadine* appears to have been most praiseworthy throughout, and of great value. Lieut.-Commr. C. G. Robinson deserves great credit for the skilful and seamanlike manner in which he handled his ship. He actually towed *Warrior* for seventy-five miles between 8:40 P. M. May 31, and 7:15 A. M. June 1, and was instrumental in saving the lives of her ship's company."

The seaplane used on the British side in the Jutland battle was a "Short" seaplane, equipped with a 225 horse-power Sunbeam motor. The machine was put overboard and taken back on board the ship by means of a crane, which is the only method so far employed in European navies.

A British Short seaplane being lowered to the water from the deck of a seaplane carrier.

In view of the importance of this naval engagement, and the part played in it by the seaplane, it is well to point out the two main lessons learned through this engagement: (1) that it is absolutely necessary to have seaplane carriers with the fleet; (2) that the seaplane carriers must be capable of manœuvering with the fleet, keeping up with it in speed.

The reports of Vice-Admiral Reginald H. S. Bacon, K.C.B., C.V.O., D.S.O., commanding the Dover Patrol, reporting operations off the Belgian coast between August 22 and November 19, 1915, says: "Throughout these operations attacks have been made on our vessels by the enemy's aircraft, but latterly the vigilance of our Dunkirk Aerodrome, under Wing Commander A. M. Longmore, has considerably curtailed their activity."

Under the heading of "Aerial Attacks on Ships," there will be found elsewhere in this book reports of aerial attacks on ships, which give further facts regarding the important work done by aircraft carried on mother-ships.

The squadron which operated in the Eastern Mediterranean, between the time of the landing on the Gallipoli Peninsula in April, 1915, and the evacuation in December, 1915–January, 1916, had several seaplane mother-ships, and many kite-balloon ships. General Sir Charles Munro, in his report respecting the operations of the Mediterranean Expeditionary Forces, includes among the commendations for services in action the officers and men of the Royal Naval Air Service.

Another amazing report of the activities of the seaplanes connected with the British Expeditionary Forces is found in the Turkish report of May 4, 1916, in which it is stated in connection with the surrender of the British force at Kut, as follows: "They (the British) first threw down sacks of flour from aëroplanes, but Turkish forces put an end to this, shooting down one after another of these old British machines." This was confirmed later in the year by a report from the general officer commanding in Mesopotamia, and is worthy of note that between April 11 and 29, 1916, aeroplanes and seaplanes dropped 18,800 pounds of food into Kut. The amazing part of the feat is that the seaplanes, which are water craft, flew over the desert carrying the food in large quantities.

The report of General Sir John Maxwell, commanding officer in Egypt, recorded that the seaplane carrier *Anne* was torpedoed off Smyrna early in the year, during an armistice, presumably by a German submarine officer who was ignorant of the armistice with the Turks.

C. C. Witmer, the American aviator who trained Russian naval aviators in the beginning of the war tells of the Russian seaplane carriers as follows:

When the need of aerial protection far from the coast became evident, the Russian authorities took the two fast steamers built for the trade between Odessa and Egypt, fitted them with false decks fore and aft for launching and receiving aeroplanes, and sent the two ships with seven aeroplanes each, to afford the aerial protection needed. These steamers were capable of a speed of twenty knots an hour and seven aeroplanes could be snugly accommodated on each. The machines were launched by lowering them to the water with cranes, and taken aboard the same way. After a little practice, this can be done very quickly. I saw seven aeroplanes launched and in flight fourteen minutes after the order was given.

On one occasion, when the Russian Feet bombarded the Bosphorus, six aeroplanes, each equipped with two forty-pound bombs, were launched within fifteen minutes from one of the aeroplane ships. Forty minutes later they commenced to return to the ship for more bombs. They landed on the lee side of the ship, took their loads—a bomb on each side of the machine, connected to the releasing device, and soared aloft.

An official report dated January 11, 1917, stated that the British seaplane carrier *Ben-Ma-Chree* was sunk by gunfire in the Kastelorizo Harbor (Asia Minor). The *Ben-Ma-Chree* was a 2550 ton ship, formerly used as a pleasure steamer between Liverpool and the Isle of Man. She was built in 1908 by Vickers, at Barrow, was a triple-screw steamer, with a speed of twenty-five knots, having accommodations for about 2000 passengers and crew. Kastelorizo is an island to the east of Rhodes, off the Asia Minor coast. Early in 1916 a French detachment landed in the island in connection with some Allied operations against Adalia, and it has been used ever since as a naval base for the Allied squadrons.

Another seaplane carrier, the *Hermes* was sunk by a U-boat in the early part of the war, off the English coast.

The accompanying illustration shows the *Ark Royal,* which operated with the Allied sea fleets at the Dardanelles with two seaplanes on the deck. The nature of its work was described in the following report:

The bombardment of the Dardanelles has been greatly assisted by the cooperation of seaplanes which were sent thither on the British Navy's new hangar ship, the *Ark Royal.*

Numerous reconnaissances were carried out over the Turkish fortifications in order to locate concealed batteries. This work proved to be rather dangerous as the seaplanes had to fly very low so as to get the exact location of the enemy's guns and the Ottomans trained a murderous fire upon the British airmen.

One seaplane, whose pilot was Lieut. Garnett and whose observer was Lieutenant-Commander William-

U.S.S. *Seattle,* an armored cruiser, has been fitted with a runway on her quarterdeck for landing seaplanes, three of which are shown in this picture. The *Seattle* was formerly the *Washington,* but her name was changed last December, as one of the new battleships is to be named for the State.

Ely making the first flight from the deck of the U.S.S. *Birmingham,* November 14, 1910, in a Curtiss biplane.

son, became unstable on March 4 and dived nose on into the sea. Both officers were injured.

Lieut. Douglas, reconnoitering at close quarters in another seaplane, was wounded, but managed to return safely. Seaplane No. 172, commanded by Flight Lieut. Bromat, with Lieut. Brown as observer, was hit twenty-eight times. Seaplane No. 7, Flight Lieut. Kershaw and Petty Officer Merchant being the crew, was hit eight times in locating concealed positions.

The *Ark Royal* convoy to the aeroplanes and seaplanes, is equipped with every appliance for necessary repairs and for maintenance of the numerous aircraft she carries.

There are also shown herewith views of the U. S. armored cruiser *North Carolina,* which has been a seaplane carrier since June, 1915, when it took the place of the U. S. S. *Mississippi,* which was the seaplane carrier of the United States Navy until it was sold to Greece in 1915.

During the overhauling of the *North Carolina,* at the close of 1916, the U. S. armored cruiser *Seattle* became a seaplane carrier. The *North Carolina* 14,500 tons, 23,000 horse-power, twin screws, equipped with 20 guns is of the reserve force Atlantic Fleet; the *Seattle* 14,500 tons, 23,000 horse-power, twin screws, equipped with 20 guns, is also an armored cruiser.

American Aviators First to Fly from and Alight on Deck of Ship

American aviators were the first to alight on and fly from the deck of a ship. On November 14, 1910, Eugene Ely flew from the deck of the U. S. S. *Birmingham,* and on January 18, 1911, flew and landed on the deck of the U. S. S. *Pennsylvania,* at San Francisco, also making the return flight from the ship. These flights were made with Curtiss aeroplanes equipped with wheels.

On January 26, 1911, Mr. Glenn H. Curtiss made the first successful flight ever made with a hydroaeroplane, starting from the water and alighting on the water without accident. Henri Fabre had succeeded in rising from the water on March 28, 1910, near Martigues, France, and in covering a distance of about 1000 feet at a height of about six feet, but met with a mishap in landing. On May 17, 1910, he made a better flight, about one mile, at a height of thirty feet, but on landing, the machine was again wrecked. Mr. Curtiss on February 17, 1911, at San Diego, flew alongside of the U. S. S. *Pennsylvania,* and his hydroaeroplane was hoisted on board by the ship's crane. After the reception accorded to him, the hydroaeroplane was again dropped overboard by the crane, and was flown from there back to the shore.

On November 24, 1911, Lieut. John Rodgers, United States Navy, flew a Burgess-Wright hydroaeroplane at Newport, Rhode Island, rising to a height of 400 feet, circled the U. S. S. *Missouri,* then landed in the lee of the U. S. S. *Ohio,* and was lifted on board by a crane.

The first experiment in starting from the deck of the ship outside of the United States took place on January 10, 1912, when Lieut. C. R. Lawson of the British Army Aviation Section started from H. M. S. *Africa,* anchored in Sheerness Harbor, in a "Short" biplane, equipped with wheels and skids. The machine was hoisted on board by a crane, and the start was made from a platform constructed on the fore part of the ship. On May 8, 1912, when British naval aviators took part in the naval review at Weymouth, England, Commander Charles Rumney Samson made a flight from the platform built on the deck of the battleship *Hibernia* with a Henri Farman biplane, equipped with pontoons and wheels, as the ship was steaming up to Portsmouth.

The French were first to set aside a ship to be used as a seaplane carrier, in 1912. This hangar ship, *La Foudre,* was used for many experiments, employing different types of machines, including a "Voisin Canard" operated by Captain Cayla, a Nieuport, operated by Ensign Delage; and a Curtiss hydroaeroplane, operated by Frank Barra. These experiments took place at St. Raphael.

Since then all the first- and second-class European nations have adopted seaplane carriers and kite-balloon carriers.

Seaplane Carriers vs. Having Seaplanes on Board of Cruisers

At the time of the early experiments in launching aeroplanes from ships in the United States and Great Britain, the world's naval authorities were divided into two camps, one holding that it would be better to make the ship self-sufficient by providing space for launching and landing seaplanes on battleships, with aviators on each ship; the other that it would be better to have regular seaplane carriers, which would supply the entire squadron with an air service.

The results to date would show that the American authorities were more far-sighted. The final decision will depend entirely on the results of a test of the American system in actual naval operations.

It would be illogical to expect a cruiser in action to slow down in order to hoist an aeroplane overboard. It might prove very dangerous to it; but it seems quite possible that the cruiser could launch a seaplane by means of a device, without slowing down, such as the one developed by Captain W. Irving Chambers, United States Navy.

There is another advantage in having each cruiser equipped with its own seaplanes, since it makes each cruiser independent of the seaplane carrier, which may be sunk by the enemy, depriving the entire squadron of the valuable services of aircraft.

Captain Chambers developed a catapult operated by compressed air, and on November 12, 1912, for the first time, launched an aeroplane from a ship in what may be considered a scientific way. The catapult was described by Captain Chambers at the time as follows:

"The catapult is so small that it occupies little space; it can even be mounted for use on top of a turret, it can be transported to any location on the ship, and it can be readily dismounted and stowed away clear of the guns.

"Compressed air is used for the power, as all ships carrying torpedoes are supplied with air compressors. When preparing the apparatus for use, the air is pumped, to a suitable pressure into a receiver, which is connected with a small cylinder conveniently located on deck. The piston of the cylinder has a stroke of about 40 inches, and the piston rod is connected with a small wooden car by means of a wire rope purchase which multiplies the travel of the piston to any desired extent or to any limit fixed by the travel of the car on its tracks.

"The aeroplane, of course, rests upon the car, and, when a flight takes place, both are projected from the tracks together in about one and one-half seconds, the pressure being automatically and gradually accelerated throughout the stroke. The car drops into the water when free from the tracks, and is hauled on board by a rope attached to it."

The device, as used at the Washington Navy Yard, November 12, 1912, was mounted on a float so that the bottom of the hydroaeroplane was not more than two feet above the water. When discharged, the hydroaeroplane gradually rose in a steady, beautiful flight, as soon as it left the tracks, without any tendency to seek the water.

During a previous trial, at Annapolis, the device was mounted rigidly on a wharf. The car and machine were both free to lift from the tracks during any part of the stroke, and after the aeroplane motor had been started full speed, the full pressure of 290 pounds was turned on at once. On this occasion the machine reared at about midstroke, and, as a cross wind was blowing, the right wing was thrown up and a cork-screw dive into the water resulted. Lieut. Ellyson, the aviator who managed the machine on both occasions, and whose iron nerves were relied on to stand the shock, was fully satisfied by this extreme test, that the shock ought not to deter any good aviator. It was also gratifying to note that no part of the machinery or fittings was ruptured or showed any signs of weakness.

When tried at Washington Navy Yard, November 12, the float enabled the apparatus to be pointed toward the wind, which, however, was nearly calm at the time. The car was held down to the tracks by the reverse flanges and extra wheels, and the balanced valve of the cylinder was arranged to be gradually opened to full power by a simple wedge-shaped cam attached to the traveling block on the piston head. The aeroplane was also held down to the car by an iron strap, the ends of which were

Commander C. R. Samson, Royal British Navy, leaving the deck in a Henri Farman biplane.

The French seaplane carrier La Foudre, in 1912, with a Voisin "canard" on board.

tripped automatically at the end of the stroke by studs on the tracks.

Several preliminary tests of the device, with sandbags to represent the weight of the aeroplane, were made before the final test of November 12, and curves of speed and pressure were obtained in each case. These curves were reassuring and demonstrated the possibility of getting, by this method, the curve of velocity to follow any trajectory desirable within practicable limits.

Glenn H. Curtiss introducing the hydroaeroplane to the Navy, 1911.

Another test was made at the Washington, D. C., Navy Yard on December 17, 1912, when a Curtiss flying boat with Lieut. Ellyson, United States Navy, at the wheel, was launched from the catapult mounted on a track. This test was even more satisfactory than the test of November 12, 1912, and demonstrated the thorough practicability of this launching device for launching aeroplanes from ships. In this case it was calculated that a speed of 40 to 42 miles an hour was necessary to support the machine, but after the flight it was found that the recording apparatus showed that the machine had left the track at a speed of 35.6 miles per hour only. In appearance the machine showed a tendency to rise rather than to fall. This demonstrated that it will be possible to shoot off aeroplanes at a lower speed, or possibly on a shorter track, because the tendency to remain in the air would exist for some time, without the accelerating influence of the propellers, owing to the visavisa of the mass which leaves the track at accelerating speed.

P. A. Surg. G. F. Cottle, U. S. N., in the annual sanitary report of U.S.S. *North Carolina,* described the catapult used on the *North Carolina* as follows:

Lieutenant Ellyson shot into the air from the United States Navy's catapult, December 17, 1912.

This apparatus is planned to hurl into space a heavier-than-air flying machine with the aviator seated at the wheel, and to hurl it from the ship's deck at a speed sufficient to allow the machine to fly away from the ship without touching the water. The apparatus is composed of a track, a compressed-air cylinder, a car to run on the track, and a cable connected with the piston of the air cylinder at one end and with the car at the other end. The pilot takes his seat, starts his motor, and when the propeller is spinning at top speed the air is allowed to rush into its cylinder, the cable is pulled upon, and the aeroplane with its pilot is pulled along the track toward the stern of the ship in such a manner that in the distance of 103.25 feet it requires a velocity of forty-five miles an hour. At the end of the track the tripping device releases the aeroplane and by means of its momentum plus the thrust of its rapidly revolving propeller it leaves the car, the track, and the ship, and flies away.

The apparatus must have many more trials before it can be said to be reasonably safe for the pilot, and then must be subjected to tests at sea, with the rolling and pitching of the ship as a factor before it can become a reasonably useful and safe appurtenance of the flying game.

Recovering Seaplanes at Sea

Recovering seaplanes at sea is a much more difficult problem to solve than launching the seaplanes and there is no solution at hand other than hoisting the seaplane by means of the usual boat crane.

The Blériot cable guiding and engaging apparatus proposed by Louis Blériot in 1913. The latch automatically grasps the cable, once the latter is guided into its jaws, and releases it when the cord shown is pulled. The supporting frame is held upright by springs, which enable it to fly back in the event of a too forcible contact with the cable. Having caught the cable while in flight, Pegoud is shown as resting preparatory to another launch, a method proved unpractical.

Photo by American Press Association

Sailors assisting Ely on his departure.

Ely just about to land on the aeroplane platform of the U.S.S. *Pennsylvania,* at San Francisco, January 18, 1911. The platform was 32x127 feet.

Ely landing—showing platform and sand bags on each side, with ropes to check progress.

Experiments at launching hydroaeroplanes conducted at Hammondsport, N. Y. by the naval aviators Lieutenants Ellyson and Tanners in 1912. See "United States Navy Aeronautics."

Curtiss flying boat being launched from the after deck of the U. S. S. *North Carolina.*

Ely's landing on the platform erected upon the quarterdeck of the U. S. S. *Pennsylvania* did not bring a solution. That could only be repeated in calm weather and, as we know, war takes place in all kinds of weather. To make it possible for Ely to alight on the quarterdeck, guide rails were placed along the platform floor, between which the wheels should run, and across the platform were stretched many ropes, a few feet apart, weighted at their ends with bags of sand. When Ely landed these weighted ropes were gathered up in succession by hooks on his machine which brought it to rest within a hundred feet.

In 1913 Louis Blériot, the French inventor and aeroplane manufacturer, conducted experiments intended to show the practicability of recovering seaplanes at sea. The device, consisting of highly suspended cables to which the aviator was to fly and hook onto by means of an automatic clasp connected to the body of the aeroplane, was tried at Buc, France. Pégoud, the first man to loop-the-loop, flew the light Blériot monoplane to the cable, engaged it with the catching apparatus, the latch automatically grasped the cable, and the machine came to a standstill. Then the propeller was again started, the latch-cord pulled to release the machine, which flew off without mishap. This might be repeated under very favorable conditions on board of a ship, but it could not be done under normal conditions, and it does not represent a solution to the problem of recovering seaplanes at sea.

Using large steamers with high freeboard, and turning them broadside to the wind, afford

The flying boat leaving the deck of the *North Carolina.* The catapult makes it possible to secure a launching speed of close to fifty miles an hour in a short run of not more than fifty feet. Thanks to the method of controlling the air throttle, there is no jar or shock from the catapulting.

An early 1912 British method of taking an aëroplane to a ship which is still practical.

a very large area of calm water for landing seaplanes, which can be made calmer by the use of oil, which prevents breaking waves or combers.

Solution Rests with Aircraft Capable of Rising Vertically from Deck of Ship

The solution rests with the aircraft capable of rising vertically from the deck of the ship. This suggests the helicopter—and brings forth the problems of making the helicopter efficient.

Recently, the writer had the pleasure of meeting in New York the Danish inventor, Ellahammer, who made one of the earliest flights ever made, in 1906, and to see the photograph of a remarkable aircraft invented by him which promises to do everything that an helicopter should do, but without the objectionable helicopter features. This would make it possible for the seaplane to rise vertically from the deck and alight in the same way on its return. Experiments should be conducted as soon as possible along this line.

Submarines as Seaplane Carriers

The use of submarines as seaplane carriers is a possibility. According to reports, Germany is building submarines especially for this purpose. The progress in submarine construction has been amazing, and further progress must be anticipated. The relative dimensions of the *U-7* and the *U-53*, both built in 1916, show extraordinary developments. The *U-1* was 139 feet long, displacement, 240 tons; speed (surface), 11 knots; speed (submerged), 9 knots;

One of the kite-balloon mother-ships of the Allies.

The kite balloon and balloon ship employed by the Italians in the Tripolitania campaign.

cruising radius, 700 miles; torpedo tubes, 1 forward; torpedoes, 3 18-inch; only one periscope, and no guns on deck. The *U-53* was 213 feet 3 inches long; displacement, 800 tons; speed (surface), 18 knots; speed (submerged), 10 knots; cruising radius, 10,000 miles; torpedo tubes, 2 forward and two aft; torpedoes, 10; three periscopes, and two guns on deck.

Some of the latest United States submarines, now under construction, are 250 feet long, of 1200 tons displacement, with a radius of 8000 miles, and a speed of 20 knots on the surface. The seaplane carrier *Engadine,* which supplied the seaplane that gave such great assistance to the British squadron during the Jutland battle, is of less than 1000 tons displacement. A submarine of 1000 tons displacement has, of course, much less room for seaplanes than a ship of the same size. But even the *U-53* could carry several small, fast seaplanes, such as are used for bombing raids. Larger submarines may be built that can carry a number of seaplanes easily. A fleet of submarine seaplane carriers would operate very much like the fleets

The aft-deck of a British balloon-ship, showing the cylinders of gas for inflating the balloon ranged in rows.

of seaplane carriers operated at Salonika, mention of which is made in another chapter.

Kite-Balloon Carriers

The employment of kite balloons for observation and spotting the fall of shots has become general. The old-time kites, which were flown from ships—when weather permitted—have been entirely replaced by the kite balloon, which is steadier and easier to operate.

Kite-balloon ships have formed part of the Allies' squadrons throughout the war, and hundreds of kites have been used by both sides, off the coasts as observation posts, and for guarding the approaches of ship lanes, harbors, and naval stations.

The Allies had but few kite balloons at the beginning of the war. Their value was recognized by the Germans for many years previous

An observation balloon partly inflated on board of a British balloon ship. The foreground shows the winch used for winding in the balloon after its work is finished.

A kite balloon being inflated on the balloon-ship, *H. M. Canning.* The white "fence" protects the balloon from the wind.

An observation balloon ascending from the "hangar" in the balloon-ship's main deck, where the balloon is stowed, off the coast of Flanders.

The observation balloon in action. It is tethered to the ship by cable. The two officers in the basket report their observations by telephone and other ways.

Directing the firing of Allies' ships that shell the enemy's coast fortifications at Zeebrugge.

to the war, and they had many kite balloons in service when the war opened. The Allies had but few kite balloons, and had at first to press into service spherical balloons, which were, however, soon replaced by kite balloons. British and French private ships were likewise pressed into use as balloon ships, and stationed outside of harbors, where the observers kept watch for enemy ships and aircraft and helped to locate mines and submarines.

In the accompanying illustrations views are shown of two of the British kite-balloon ships. One illustrates a ship with a balloon on board, with a "fence" to protect it from side winds; another, the balloon being inflated and the winch used to take the balloon down; a third, the hydrogen tubes which supply the hydrogen to inflate the balloon. As shown herewith, there are two types of balloon ships: (1) One on which the kite balloon is inflated on the aft deck; (2) another where the balloon is kept and inflated in a hold in the aft deck. The latter is the better because the balloon can be inflated in the hold, where it is covered with canvas and kept in readiness, to be sent up at the opportune moment; whereas, in the former, the inflating can only be done under normal conditions. On ships which carry both seaplanes and a balloon, the latter is kept in the hold of the aft deck. Kite balloons were first placed on board of United States ships the *Nevada* and the *Oklahoma* in 1916. Both these ships are of 27,500 tons displacement, twin screw, 26,500 horse-power, equipped with 31 guns. They took on board Goodyear type kite balloons, 80 feet long, 25,000 cubic feet ca-

Observation balloon ascending from a ship operating with the Italian fleet.

How British and French private boats were used as balloon ships and stationed outside of harbors, where the observers kept watch for enemy ships and aircraft and helped in locating mines and submarines. This shows one of the small spherical balloons used in the beginning of the war.

pacity. These are inflated with hydrogen carried in cylinders containing 200 cubic feet each, hundreds of which are carried on board of the ships.

There is a disadvantage in carrying balloons on board a battleship, because the ship must slow down to send up the balloon. Since European navies do not have balloons on battleships it has not yet been determined whether this disadvantage is compensated by the promptness of service made possible by having the balloon on board.

Rail and truck of the Catapault on the U.S.S. *North Carolina* for launching seaplanes.

The white arrow shows exactly what the periscope of a submarine looks like to an aviator about 3500 feet up. This photograph also shows the submarine mothership, the U. S. S. *Columbia,* second class cruiser, 7350 tons displacement, 18,509 horse-power.

CHAPTER VI

SUBMARINE HUNTING BY AIRCRAFT

The submarine menace can be checked by present-day aircraft.

In any discussion of what can be done against the submarine, it must first be stated whether we mean the protection of ships at sea or on coastwise trips. Nothing could protect the sea lanes so well as large dirigibles, capable, as the Zeppelins are, of cruising for 3000 miles without stopping.

Unfortunately, no country outside of Germany has large dirigibles for use for this purpose. If we had such airships, they could be used to patrol the ship lanes daily. No submarine would be safe, no matter where at sea, if large dirigibles were thus patroling, because large dirigibles carry guns of sufficient caliber to sink a submarine with a single shot. Likewise, the observers from a dirigible, as in the case of an observer from an aeroplane, can see a submarine miles away, when a man from a ship cannot detect it, and as the airship travels many times faster than the submarine and the submarine could not easily detect the approach of the airship, the submarine would stand no chance.

Unfortunately, the Allies are not now in a position to patrol these sea lanes with a large number of airships, although there is a possibility that Great Britain will put some into service within a few months.

The submarine menace can be checked by present-day aircraft—seaplanes, small dirigibles, and kite balloons. We are now building large seaplanes which are capable of carrying fuel for continuous flights of over fifteen hours and capable of gaining a speed of over seventy-five miles an hour. American manufacturers have supplied quite a number of large seaplanes of this type to England, and, as Mr. C. G. Gray, the editor of the "London Aeroplane," has pointed out, "If America is seriously perturbed about the facts of American shipping and American citizens traveling by sea, it should not be a difficult matter for America to rig up in a very short space of time quite a fleet of aeroplane carriers suitable for handling these big seaplanes."

According to the figures of the French Minister of Marine the submarines sank ships aggregating 2,085,380 tons in 1916, and 3,000,000 tons in the first four months of 1917. This is far more tonnage than the United States has, and as tonnage can be produced but slowly, every means which affords protection to our ships must be employed.

On sighting the submarine under water the aviator summons the cruisers and trawlers by wireless. If the submarine comes to the surface, and there is danger of its escaping before it can be netted, the aviator bombs it.

Historic

Some months before the beginning of the Great War, the British submarine *A-7* was lost near Plymouth, and an aeroplane was employed, among other means, to find it. The aeroplane proved to be the most efficient means for finding the submarine; it found it with such promptness as to give an idea of the possibilities of employing aeroplanes for submarine hunting. But, at the time, the efficiency of the submarine itself as a naval weapon was doubted by many naval men, who could not believe that a submarine could be constructed that would have a cruising range of over a thousand miles. Even the most far-seeing naval experts could hardly appreciate the advent of a submarine like the *U-53,* which would have a cruising radius of 10,000 miles. It will be remembered that Sir Percy Scott, the British naval authority, created a sensation a few months prior to the war when he stated that submarines and aircraft would revolutionize naval warfare, and urged that Great Britain concentrate its efforts on building fleets of submarines and aircraft. He also pointed out that the aircraft would be the most powerful weapon to be used against submarines. Admiral Sir Percy Scott's prophecy was very generally laughed at in naval circles the world over.

The submarine's periscope and its wake, which is easily visible from an aircraft.

The Great War was only a few months old when the revolutionary value of both submarines and aircraft became evident. But the submarines and aircraft available were not sufficient in number to permit any of the nations to employ them for offensive purposes. They were all used for defensive purposes within a small radius of their respective bases. As the number of both submarines and aircraft increased, their operations extended more and more, and as the submarine menace grew, the nations had to meet it, and found that the aircraft was the best weapon for hunting submarines. To best understand the tremendous task which the British Navy had to perform and how important the protection afforded by aircraft became, one must read the report which Sir Edward Carson, the First Lord of the Ad-

In reproducing this photograph the "Illustrated London News" says: "This photograph, from a French source, gives an excellent view of the car of one of the small naval dirigibles used for scouting and observation, as it appears in flight. The head of the pilot can be seen near the front of the machine, with the observer sitting behind him to the left. The whirring blades of the propellor can be faintly discerned just above the right wheel. The car is a slight modification of that of an aeroplane. The little scouting airships were first introduced by our own Navy, and have been found very useful in tracking submarines, which can be seen when under water in clear, calm weather. A dirigible can itself attack a submarine by dropping bombs upon it. Provided as it is with wireless apparatus, a scouting airship, on sighting a submarine, at once communicates with the patrol-boats." The Allies have hundreds of such dirigibles in use.

miralty, made to the British Government on February 21, 1917. He reported that since the commencement of the war the British Navy (with the cooperation of its most efficient aerial coast patrol) had examined 25,874 ships. During the first eighteen days of February, 6076 ships arrived in ports of the United Kingdom, and 5873 ships had cleared from United Kingdom ports. Practically every ship that arrived and every ship that cleared was inspected as it neared the ports, and convoyed by dirigibles or seaplanes. Sir Edward Carson pointed out that from the beginning of the war up to October 30, 1916, the British Navy transported across the seas 8,000,000 troops; 9,420,000 tons of explosives and material, 47,504,000 gallons of gasoline, over a million of sick and wounded, and over a million mules and horses, etc.

Little had been done by the navies of the world to develop naval aeronautics prior to the war. The German Navy had concentrated on developing its naval Zeppelins, and both the German and the British navies only really began to give serious consideration to naval aeronautics in 1913. When the war started, the British Navy had less than 100 seaplanes, and but very few dirigibles available for service. France had been concentrating her efforts in developing the army branch of the air service, but had done very little in naval aeronautics, outside of the few experiments made at San Raphael. That was also true of Italy. In fact, when the war started, naval aeronautics was in a period of experimentation. Until then most navy people, trained to face the crushing force of the elements, looked at the frail aeroplane askance and asked for the supreme test, seaworthiness, before admitting it as a naval auxiliary. Without seaworthiness they could not see any use for the aeroplane, and accordingly postponed the organization of naval aeronautics.

When it became necessary to build up a system of protection against submarines, the warring nations pressed into service thousands of small vessels, destroyers, trawlers, and submarine chasers, and as fast as they could obtain them they put into service seaplanes and dirigibles, to cooperate with the ships in locating and capturing and destroying hostile submarines; and convoying ships, protecting them from submarine attacks.

The fame of the aircraft which convoyed every troop- and supply-ship which crossed the Channel from the beginning of the Great War and of those that convoyed ships on coast-wise trips, is world-wide.

The first attack on a submarine base was reported on March 24, 1915. British naval aviators bombed Cockeriel's Ship Yard and wharves at Hoboken near Antwerp, destroying two submarines. The British Admiralty on that occasion issued the following statement:

The following has been received from Wing Commander Longmore: "I have to report that a successful air attack was carried out this morning by five machines of the Dunkirk Squadron on the German submarines being constructed at Hoboken, near Antwerp.

"Two of the pilots had to return owing to thick weather, but Squadron Commander Iver T. Courtney, and Flight Lieut. H. Rosher, reached their objective

Photograph of a bomb dropped by an Allies' dirigible on a submerging German submarine.

Watching for submarines: A kite balloon anchored to a ship watching for submarines on the North Sea. The observer in the basket can spot his quarry many miles away and summon seaplanes, destroyers, and trawlers by wireless to deal with the U-Boats.

and after planing down to 1,000 feet, dropped four bombs each on the submarines.

"It is believed that considerable damage has been done to both the works and two submarines. The works were observed to be on fire. In all, five submarines were observed on the slip.

"Flight Lieut. B. Crossley-Meates was obliged by engine trouble to descend in Holland. Owing to the mist, the two pilots experienced considerable difficulty in finding their way, and they were subjected to a heavy gun fire while delivering their attacks."

The first report of an attack on submarines by an aircraft was issued by the German Admiralty on May 4, 1915. It stated that on May 3 a German naval dirigible fought several British submarines in the North Sea and dropped bombs on them, sinking one. The submarines, the report stated, fired on the dirigible without success.

On May 31, 1915, the German Admiralty announced the sinking of a Russian submarine by bombs dropped by German naval aviators near Gotland.

On July 1, 1915, the following despatch from Rome told of the sinking of the Austrian submarine *U-11* in the Adriatic by a French aviator:

The Minister of Marine states the action took place on Thursday. The *U-11* was lying lazily on the surface and apparently failed to notice the aviator as he circled overhead. With a sudden swoop the aeroplane shot downward to within forty-five feet of the submarine's deck. By this time it was too late for the underseas craft to submerge. Three bombs were dropped, all of which struck the submarine near the turret and exploded.

The submarine sank almost instantly and did not reappear, although wreckage was afterwards found about the scene. The *U-11* was one of the newest of the Austrian submersibles and displaced about 860 tons. She is supposed to have had aboard a crew of twenty-five men.

A later report stated that this submarine was destroyed by French Naval Sub-Lieutenant Rouillet of the French seaplane squadron, operating in the Adriatic with the Italian naval forces.

On July 27, 1915, it was reported that a German submarine headed for a British transport laden with troops and ammunition was put to flight by an Allied aeroplane in the Dardanelles. The aviators saw the submarine preparing to launch a torpedo and gave the alarm. Pending the arrival of the destroyer, the aeroplane dropped bombs at the submarine. Although none of the bombs took effect, they forced the submarine to submerge. Soon after the periscope reappeared on the surface, and the aviator dropped two more bombs. The submarine submerged and did not reappear.

On August 19, 1915, the Turkish war office stated that an Allied submarine had been sunk in the Dardanelles by a Turkish aeroplane.

In an official note issued on August 26 by the British press bureau, the following report of the destruction of a German submarine by a British aviator was given:

The Secretary of the Admiralty announces that Squadron Commander Arthur W. Bigsworth, R. N., destroyed single-handed, a German submarine this morning by bombs dropped from an aeroplane. The submarine was observed to be completely wrecked and sank off Ostend.

"It is not the practice of the Admiralty to publish statements regarding the losses of German submarines, important as they have been, in cases where the enemy had not other sources of information as to the time and place at which these losses have occurred.

"In the case referred to above, however, the brilliant feat of Squadron Commander Bigsworth was performed in the immediate neighborhood of the coast in occupation of the enemy, and the position of the sunken submarine has been located by a German destroyer."

On November 28, 1915, a brief official report stated that a French aviator had destroyed a German submarine off the Belgian coast by dropping bombs on it. This report was confirmed through the awarding of the Victoria Cross to Lieutenant Viney and the recommendation for the Legion of Honor of Lieutenant de Sincay. The official mention of the conferring of the honors reads as follows:

For his services on November 28, 1915, when accompanied by le lieutenant en second de Sincay as observer, he destroyed a German submarine off the Belgian coast by bombs dropped from an aeroplane. Le

A submarine emerging off the Belgian coast, detected by the observers from one of the kite balloons anchored off the coasts to watch for them.

Lieutenant en second Colley Saint-Paul Comte de Sincay, attached to No. 1 Wing, R. N. A. S., to be an Honorary Companion of the Distinguished Service Order—for his services in connection with the destruction of a German submarine by bombs dropped from an aëroplane on November 28, 1915.

The story of the sinking of the German submarine is told by Lieutenants Viney and de Sincay on their return to Paris from Dunkirk a few days later as follows:

It was noon on Sunday. We had left half an hour before on a French biplane, to look for submarines, which were reported near by. We rose 10,000 feet, and had been cruising about for some time, when we saw two submarines five miles off shore, west of Nieuport.

It was an ideal spot for our purpose. The sea was shallow, giving the submarines little chance of escape. By plunging in wide spirals, we descended on one of the boats which, being above a sand bank, could not dive. She made desperate efforts to get away, steering in wild zigzags.

We realized we could not get her, and so turned our attention to the other boat. Apparently it was more difficult to handle her, for despite all endeavors, she failed to get out of the circle we traced as we pounded down on her.

We came down to about 300 feet above the sea. When we were certain of not missing, we let go the first bomb, and had the satisfaction of seeing we had made a hit. Even with the naked eye we could observe that serious damage had been inflicted on the deck of the boat.

We circled around twice more over the doomed submarine. A second bomb did the rest of the work. She broke in half and sank.

We did not wait to see more. Moments were precious. We had to get back to Dunkirk as quickly as possible, for the submarines were sure to have given warning and we were liable to find our retreat cut off by the enemy's aeroplanes if we lingered.

Numerous other reports of attacks on submarines and sinking of submarines were made public in 1916, mostly successes of the Allied aviators. A detailed report of how an Austrian seaplane sank the French submarine *Foucault* and two Austrian aeroplanes rescued the twenty-nine officers and men of the submarine was told in the "Tageblatt." The Austrian Admiralty's statement read as follows:

An Austro-Hungarian naval aeroplane in the Southern Adriatic sank, by means of bombs, the French submarine *Foucault*. The aeroplane's pilot was Lieutenant Celezeny and the observer was Lieutenant von Klimburg. The entire crew of the submarine, comprising two officers and twenty-seven men, many of whom were in a drowning condition, were rescued and made prisoners by the naval aeroplane mentioned and by another piloted by Lieutenant Komjovee, with Cadet Severa as observer.

Half an hour later the imprisoned crew was taken over by a torpedo boat, while the two officers were transported to land on the naval aeroplanes.

The craft sunk was the submarine *Foucault*, built in 1912 at Cherbourg. She was 167 feet long, 16.3 beam, with a speed of 12.5 knots above water and 8 knots submerged. She was equipped with 6 torpedo tubes.

In the story of the rescue of the twenty-nine men, the officers and crew of the *Foucault*, the "Tageblatt" stated that the sea was rough at

A French seaplane starting on a submarine patrol at Dunkirk.

the time and there was the danger that the Austrian aeronauts would be captured by hostile warships, as well as that the aeroplanes, overloaded by taking on board so many men, might collapse. Nevertheless, the Austrian aviators told the men from the French submarines to swim to the seaplanes and take hold of them. The commander and second officer of the submarines were allowed to climb into the pilots' seats. The aviators signaled for help, and half an hour later a torpedo boat arrived and took on board the men from the submarine.

Many other submarines were also captured or destroyed through the cooperation of aircraft. The policy has been to capture the submarines whenever possible. The report of one of the latest cases where two submarines were enmeshed as the result of the cooperation between aircraft and trawlers was reported by Captain E. L. Smith of the American steamer *Alaskan,* which arrived at Newport on March 19, 1917, from La Pallice, France. The U-boats were detected beneath the surface by a patrol seaplane. The aviator signaled for trawlers and circled about directing the placing of nets. Soon these were drawn completely about the unsuspecting submersibles, which were brought to the surface. They were lying side by side in the harbor of La Pallice when the *Alaskan* sailed.

There are also many instances where submarines about to attack ships were chased by aircraft.

How a French aeroplane drove off a U-boat shelling a British freighter that was in flames on March 19, 1917, is told by Captain D. S. Ramsdale, commander of the *Eastgate,* which arrived from La Pallice, as follows:

We left New York on December 20 bound for La Pallice, France, with a general cargo, principally of twine and gasolene in barrels, consigned to the French Government. On December 26, when well off the Newfoundland Banks, a fire was discovered in the coal in the 'thartship coal bunker. It quickly spread and fed on the twine stored in hold No. 2, located forward.

Things began to look rather bad, as the seas were rolling high. We tried turning the hose on the flames, but without success. I then ordered the hatches battened down and ran a steam pipe forward, hoping to smother the flames.

We were in a tight place, as our gasolene was stored in hold No. 1, and if the flames ever succeeded in reaching the barrels—well, it was n't a thought to improve any one's sleep.

After three days and nights of fighting the fire with steam, our coal supply ran low, since it was impossible to reach the coal without getting in the 'thartship hold and shoveling the coal down.

It was impossible to remain in the 'tharthold more than three or four minutes at a time, but we had to have that coal. Accordingly, every member of the crew took turns entering our "little furnace" and shoveling as long as we could before the smoke became overpowering. Oh yes; I shoveled too.

On January 3 we were advised by naval authorities by signals to put into a certain bay, as there was danger ahead, but I signaled back that my vessel was afire and that I intended to continue on to La-Pallice.

We were well in sight of land the following day when a shot whistled across our bows. I had no gun or wireless to call for assistance, and as the submarine was not in sight, decided to make a run for it.

A few minutes later the submarine appeared about two miles off our starboard quarter and bore down on us at great speed. Ten shots struck the *Eastgate.*

One shot tore through the skylight above the engine room and burst just above the boilers. The shrapnel caught our third engineer and rendered him unconscious. Another shell found our starboard lifeboat and blew it to atoms.

Meanwhile I had ordered the crew to take to the port lifeboat. Just as we were rowing away I heard the roar of an aeroplane, and looking up saw a fast French battleplane approaching.

The aeroplane dropped down to within 500 feet above the undersea craft. At the same time two destroyers summonded by the aviator came up from opposite directions at top speed. The submarine had only one avenue of escape from the two destroyers and the aeroplane. She submerged with all possible speed.

A few minutes later we rowed back to the *Eastgate* and headed for La Pallice. We arived that night and quenched the flames by flooding the holds.

Methods of and Weapons for Aerial Attack on Submarines

Attack by Seaplanes and Dirigibles

In a recent report, Sir Edward Carson, the First Lord of the Admiralty, gave instances of the sinking of submarines with bombs dropped by the small coast patrol dirigibles and seaplanes. The method employed by both the dirigibles and the seaplanes is similar. Hundreds of these aircraft are employed to cooperate with destroyers, trawlers, and submarine chasers in capturing or destroying hostile submarines and searching coasts for submarine bases. The usual evidence of the submarine's presence is the wake of the periscope. This wake cannot easily be seen from ships, but can always be clearly seen from aeroplanes. For one thing, the aviator is not troubled by the refraction of the rays of light, which interfere with the vision of the person on a ship. For another thing, the aviator, flying at a height of from 1000 to 5000 feet, has a range of vision of many miles, and the whitish wake of the periscope is clearly visible against the dark surface of the waters, even in cases where the sea is fairly rough and white caps are showing.

In clear weather an aviator from a height of between 1000 and 3000 feet can also see a submarine under water. In clear weather and clear water, he can see the submarine even when it is down to a depth of 100 feet. In less clear water, the submarine can be seen at a depth of 20 to 30 feet.

The present-day submarines are so large that

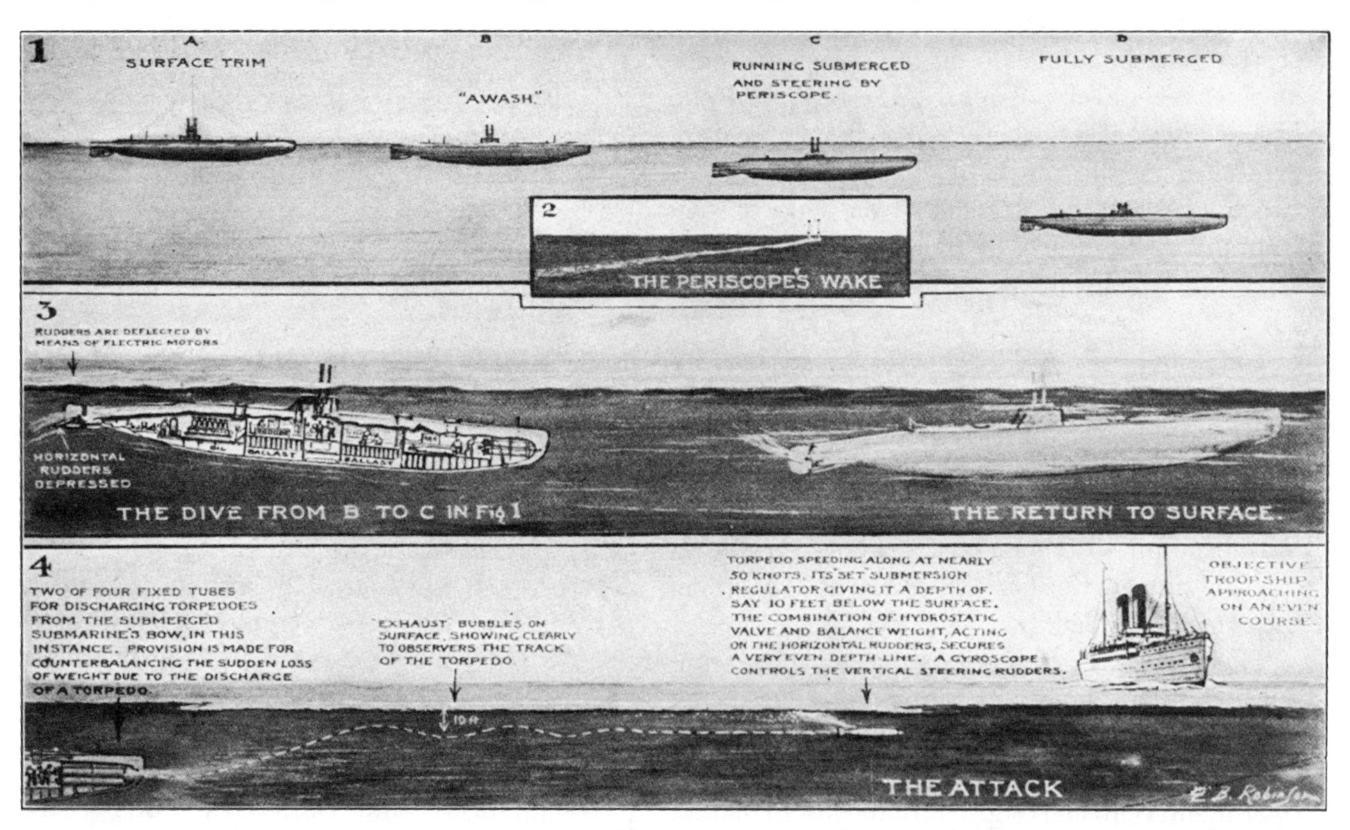

DIAGRAMS ILLUSTRATING THE SUBMARINE'S METHOD OF DIVING, RETURNING TO THE SURFACE, AND ATTACKING BY TORPEDO

When a submarine commander observes an enemy vessel he submerges his boat whilst still at a distance from his target, and then approaches to a position within firing-range of, say 2000 yards (Fig. 5). If the sea be rough, and it is consequently difficult to observe small objects on its surface, he keeps the enemy under continual observation by means of his periscope, a vertical tube projecting above the surface of the water fitted with an arrangement of lenses whose design enables them to project the picture within the field of their object-glass on to a suitable lens under observation inside the vessel. If, however, it is inadvisable to show even a periscope, the object is approached in a series of "porpoise dives," observations being taken when the periscope is above the surface (Fig. 5).—[Drawn by W. B. Robinson, for the "Illustrated London News."]

they not only can be detected through the contrast which they make against their surroundings in the water, but also through the foamy wake at the stern, which is clearly visible by contrast.

The *U-53* is 213 feet, 3 inches long, and later ones are even larger. Such submarines present a very large target, and whereas their speed submerged is between 10 and 15 knots at most, the seaplanes, which go at a speed of up to 90 miles an hour, and even the small dirigibles, with a speed of only about 35 miles an hour, have an advantage over the submarines—and the latter steer clear of places where it is known that aircraft are employed for submarine hunting. Seaplanes and coast patrol dirigibles are employed in daily patrols to search for submarine bases and for submarines that may be lying in wait for ships. If a submarine is seen under water, the aircraft, whether seaplane or dirigible, being equipped with wireless and bombs, first send a wireless summoning destroyers, trawlers, and submarine chasers. Whenever possible an opportunity is given to the trawlers or the ships which operate the nets to come up to the submarine and enmesh it in the huge net. That saves the submarine, and the crew is made prisoners. If, however, the submarine comes to the surface and there is a possibility of its escaping, the aircraft makes its attack by dropping the bombs. While it is difficult for the submarine to see the aircraft, and it takes several minutes to submerge, it has also been difficult in the past for untrained aviators who had to be pressed into this service to hit the submarines. While the above reports show that many submarines were destroyed, it is also known that many more escaped, because the pilots dropped several bombs without hitting them.

But now aircraft guns of up to 3-inch caliber are being turned out, which will make it possible for the pilots to shoot at the submarines, cutting down the difficulty of hitting to about one quarter, because it is easier to sight the target with a gun, and the average man finds it natural to shoot a gun, and more difficult to drop a bomb with precision.

The best success is obtained, of course, through bombing a submarine from a height of from 300 to 500 feet. Then the target is not missed so easily. The submarine, unless it happens to be on the surface, with its guns ready to fire, stands very little chance of fighting back, because to do so it must come entirely on the surface, and the hatches where the anti-aircraft guns are must be opened, the gun must be aimed, etc. That requires time, and gives an opportunity to the aviator to drop bombs and anything else he may have available on the submarine and gunner. If a submarine finds itself in danger and submerges, it leaves an oily patch, which is clearly visible from the air, although far less visible from a ship. As the submarine can only make a speed of between 10 and 15 knots, and usually comes up to the surface at intervals of between 50 to 100 miles, it is comparatively easy to keep a watch on that particular submarine—although, of course, not so easy to capture it or destroy it. But as the aircraft and the trawlers and submarine chasers are watching for it, warning is given and the submarine cannot make a surprise attack upon ships.

The "deep-sea vision" afforded by the seaplane is an even more valuable asset for fighting a submarine than is its superior speed. Deep sea vision enables the seaplane to detect a submerged submarine to a depth of about 100 feet in very transparent waters such as are found in the Mediterranean or in the Caribbean Sea; in the northern Atlantic the visible depth is more limited. It generally varies according to the color of the sea bottom and of the sky.

The possibility of detecting submarines by means of a seaplane reconnaissance has obviously an immense importance for the safety of a fleet, for it eliminates to a great extent the deadliest danger ships of the line have to cope with in time of war. Whereas the submarine cannot launch a torpedo without getting its bearings, i.e., without showing its periscope above the water, it is an easy matter for a seaplane to follow the course of a submerged submarine and attack it with bombs at the very moment the periscope pops out of the sea.

Painting Submarines to Make Them Less Visible

Exhaustive experiments have been made in painting submarines so as to make them less visible when they are under water. But while they can be made less visible through painting them in colors that blend well with the water and sea bottom, it is hardly possible to so paint the large fleet submarines and the ocean-going super-submarines, which are used for long cruises, in such a way that they will blend with the water and conditions existing in different places in which they are cruising. The smaller coastal type submarine, which is used entirely for coast defense, and does not have to go on long distance cruises, can be painted more effectively to blend with the color of the water in which it operates. But nothing can be done to eliminate the foamy wake of the periscope, and if the periscope were eliminated entirely, then the submarine would have to come close to the surface and it would make a better target for the aeroplane.

To Distinguish Hostile Submarines from Our Own

There are many submarines still operating, because the aviators could not distinguish the hostile submarines from their own, and could not afford to take any chances in destroying them. Marking submarines does not afford a solution, because the enemy can adopt the markings and carry on its work of destruction under disguise.

Considering that when a periscope shows the pilot has to decide how to act, and that unless the aircraft is flying low, it is hard to distinguish the features of the submarine from a height, one can well understand why even naval men in different countries have found it hard to tell whether a given submarine was one of their own or the enemy's. In this respect, it was found that naval men who acted as observers were no better in detecting submarines than the aviators who had had practically no experience in naval work. The aviator was used to judging things from the air, whereas the naval man, with little experience in flying, found it hard to define the things he saw. Of course, this does not last long; and after a score or so of flights, the naval observer becomes accustomed to flying, just as the aviator becomes accustomed to distinguishing naval craft.

The only way to prevent mistakes and not to let hostile submarines get away is for the commanders to give the aerial submarine hunters information regarding the movements of friendly submarines operating in the locality.

Kite Balloons as Lookouts for Submarines

Hundreds of kite balloons have been used as lookouts for submarines in the Great War. These balloons are sent up from barges or kite-balloon ships, and are sent up to a height of from 1000 to 2000 feet, where they stay throughout the day, the observers scanning the surface of the water, looking out for submarines. When they see a submarine or a doubtful ship, they summon the seaplanes, destroyers, and submarine chasers by wireless. The employment of kite balloons as lookouts releases dirigibles and ships from continuous patrol of different localities which are equally well protected through the work of the observers in the kite balloons.

The First Aerial Submarine Hunt in American History

The first aerial submarine hunt in American history took place during March, 1917. On Monday, March 26, the keeper of the lighthouse at Quogue, Long Island, New York, reported to Commissioner Putnam, of the Bureau of Lighthouses, Department of Commerce, Washington, D. C., that there was evidence that two U-boats were "lying in toward the Sound."

The supposed U-boats had been sighted at the Montauk Point entrance of the Sound at about six o'clock that afternoon, headed into Long Island Sound.

Remembering the exploits of the *U-53*, which had done everything that had theretofore been pronounced as impossible for a submarine to do, including crossing the Atlantic and sinking half dozen ships in succession, off the

American coasts, then disappearing from sight, the authorities, naval and aeronautic, had to take steps to ascertain the truth of the report. While the tendency of some people in such a case was to laugh incredulously, the authorities realized that unless something was done immediately there might be a repetition of the work of the *U-53*. But, unfortunately, the navy had not yet established an aeronautic base in the East, and the score of seaplanes owned by members of the Aerial Coast Patrol and members of the Aero Club of America were in Florida, where the members of the Aerial Coast Patrol were training. The Naval Militia Aviation Section was not in a position to assist because the two seaplanes presented to it by patriotic people were worn out by the training of last summer and the Navy Department had not yet supplied the Militia with machines. The nearest seaplanes available were at Pensacola, Florida.

So the distinction of doing the first aerial coast patrolling over our coast line, to hunt for submarines, went to the civilian aviators who later became part of the Aerial Reserve Squadron at Governor's Island, and to the civilian instructors and aerial reservists connected with the Mineola (L. I.) Army Aviation School.

The following morning four fliers rose from the Mineola field in a forty-mile an hour gale and rain and a bad fog. They were detailed to patrol the Long Island coast from Oyster Bay to Montauk Point, while Governor's Island aviators watched over the shore from the island to Oyster Bay. The Governor's Island aviators were. First-Lieut. H. H. Salmon (Aerial Reserve Corps), with Edwin M. Post, Jr., as observer, and First-Lieut. Wm. P. Willetts (A. R. C.), with observer. The Mineola aviators were Capt. A. W. Briggs (British Royál Flying Corps), pilot; Lieut. H. F. Wehrle, formerly of the West Virginia National Guard, now of the Aerial Reserve Corps, as observer; Leonard W. Bonney, instructor; Alan S. Adams, observer; Bertrand B. Acosta, instructor; Douglas E. Manning, observer; A. Livingston Allen, instructor; Harmon C. Norton, observer. Two of the men, Acosta and Briggs, were out for three days. They did not return to their headquarters, merely landing when they were forced to.

Considering that there was a gale blowing and the weather was foggy, and it was their first experience, it was quite a difficult task. But it was well done by both groups. The Mineola aviators had the hardest task. Acosta and Allen got to Port Jefferson in the first evening, Bonney to Southold, and Captain Briggs and Lieutenant Wehrle to Springs, on Gardiner's Bay, after having searched the bays and inlets around Big Gull Island, Little Gull Island, Gardiner's Island, and Gardiner's Bay. The total distance covered by Captain Briggs and Lieutenant Wehrle in a driving rain-storm, was 124 miles.

The machines went out between five and eleven miles at sea, the inlets and bays were searched, vessels plotted, compass directions and time when located were given. But the submarines were not found.

The machines were not equipped with wireless, and there was not a wireless receiving station in operation to receive their message if one had been sent. But there was a cruiser and other vessels which could have been summoned if the submarines had been found.

The submarine hunt lasted three days, after which the Navy Department issued the following statement:

"The Navy Department has chased down the rumor that two strange submarines were sighted off Montauk Point at 6 o'clock on the evening of March 26, headed into Long Island Sound.

"These supposed submarines were two patrol motor boats returning from a trial trip. The builder has stated that these boats passed Montauk Point at the time stated and that one was trailing the other, which was in accordance with the report of submarines sighted.

"The builder also stated that he has been told that his boats looked so much like submarines that there was danger of their being mistaken for such.

"The Navy Department has expressed its gratification at the prompt, efficient and timely assistance of the army in detailing its aeroplanes for search duty, on which they were constantly engaged for three days.

"This incident emphasizes the need of hydroaeroplanes for naval scouting purposes."

French System of Patrol Against U-Boats

The French system of patrol against U-boats, including the employment of seaplanes from possibly hundreds of seaplane stations, were described briefly on May 26, 1917, in the public session of the Chamber of Deputies, by Admiral Lacaze, the Minister of Marine, who gave an interesting outline of the means of defense France had adopted against the undersea boats.

"I see no reason why I should not speak of these methods in public," said Admiral Lacaze. "It would be childish to think they are unknown to the enemy. They consist of a system of patrol boats, of arming merchantmen with guns, and fitting them with wireless; of seaplanes, nets, mines, smoke-raising devices, and dragnets.

"I sought to get patrol boats built here and buy them abroad. I scoured the world over with missions, covering the ground from America to North Cape, from the Cape of Good Hope to Japan, but England had been beforehand. When I entered the Ministry I found 243 patrols. Now we have 552." (A Socialist voice: "It is formidable.")

"I do not say it is formidable," continued the Minister, "nor even sufficient, and I have drawn up a scheme which will increase the figure to 900. I continue to buy in London, the world's center for shipping. I am obliged to do so because our shipyards had been almost completely abandoned; because, as a result of that short-war theory which weighed so regrettably upon all decisions taken at the outset of the war, the yards had been transformed into war material factories to meet the pressing need of the national defense. We have now got back most of the arsenals and a number of private yards, together with skilled workmen.

"The guns we mount on the patrol boats have been referred to disdainfully, but you cannot put ten-centimeter guns on a small vessel. A patrol boat on guard, armed with 95-centimeter guns, met two submarines armed with 105-millimeter guns, sank one and put the other to flight.

"We have 1,200 dragnets as well as 170,500 curtain nets and 5,000 twenty-foot float nets, which indicate the presence of submarines. We have special bombs for submarines and apparatus to throw them.

"We have organized seaplane posts all around the coasts, so that the zone of action of each post joins that of its neighbor on either side. By October all merchantmen and patrollers will be fitted with wireless and all merchantmen supplied with guns of as heavy caliber as possible, for which measures programs have been drawn up even beyond what was thought possible."

Memoranda:

Photograph of a coast patrol dirigible taken from a flying boat of Harold D. Kantner at the Italian Naval Station of Taranto. In the foreground may be seen the submerged beds of shell fish, the Cocchi.

CHAPTER VII

LOCATING SUBMERGED MINES WITH AIRCRAFT

One of the most important uses to which aircraft have been put since the beginning of the Great War has been the locating of submerged mines. Hostile submarines and steamers masquerading under false colors lay mines whenever the opportunity presents itself, and no ship lane is immune from them; ships are in danger of being sunk unless lanes are properly patroled by aircraft, and subsequently, if necessary, swept by mine sweepers. Aircraft are employed extensively to direct the planting of mines and locating mine fields, and mines set adrift by storms or other causes.

The size of mines is different in different countries; and they are spherical or cylindrical in shape. In the United States the average contact mine is about forty inches in diameter, and contains about 100 pounds of TNT, gun cotton, or other explosives. They run from that size up to seven feet in diameter.

Mines are usually submerged to a depth of ten feet, and are usually arranged in clusters of four, the total number of mines in a mine field being unlimited. Whether in clusters or in fields, the electric contact mines protecting harbors and stations are controlled from a shore station, each mine being connected to a submerged connecting box, which is connected to this station by cable, through which the electric current is flashed to explode the mines. The officers in charge of the mine field and operating from the stations on shore have detailed maps of the mined areas, showing the exact location of the mines. The maps are plotted in squares, and when a hostile ship approaches a mine field, the observers at this station, who get information either through seeing the ship through telescopes or from a kite balloon or a dirigible or a seaplane, follow the ship's movements and notify the officer in charge of the electric switchboard to fire the mines of a given square, through which the ship or ships are passing.

All mines controlled from the shore can be made contact mines, that is, mines that will explode when touching a ship or resistant body. The officers in charge are warned of the presence of the ship by signal or bell worked electrically by the contact, and if the vessel is hostile, a fact which is determined by the officer, the mine is fired.

Other contact mines used extensively are only anchored but not connected to the shore. These are exploded by electric contact caused by coming in contact with the ship or a resistant dy, or by the extreme tipping of the mine, which completes an electrical circuit, which explodes the mine.

The latter mines are the most dangerous, because they explode on contact, and are a danger to one's own vessels as well as to the enemy's vessels. All harbors and channels and important approaches are protected by mine fields, the number of fields and location of mines being known, of course, only to the authorities.

This information cannot be divulged, for obvious reasons, to the commanders of steamers, transports, and other ships coming into or clearing from ports. These ships must, therefore, be guided in and out of ports, and through the clear channels between mine fields, which is often done by dirigibles.

The anchored mines are apt to be set adrift by storms, and aircraft are used extensively in locating them. Nothing else can enable mine planters to plant or locate mine fields so quickly as the aërial observer, who from a height of 500 feet or more in a dirigible or observation balloon, can plot the location of the mines in a few minutes, whereas it would take days by employing any other method.

The observer from the basket of the kite balloon cooperating with the mine planter or mine sweeper advises the officer in charge of the locations of mines, which are clearly visible to him,

Planting a submarine mine.

but owing to the refraction, cannot be seen from the ship.

Dirigibles are especially valuable for locating submerged mines. The first employment of dirigibles for this work was the result of the Italian dirigible *P-4* finding by accident, in October, 1914, while on a cruise over the Adriatic, a number of Austrian floating mines. Thereafter, it was found that the small coast patrol dirigible, which is capable of flying very low and of almost standing still over a spot, is the best means of locating submerged mines. Upon locating them, the observer summons the mine sweepers, or if it is important that the mine be destroyed immediately, he destroys it by firing at it or by trolling a weight, attached to a wire, against it under the water. Seaplanes are also used for locating submerged mines, but less satisfactorily, owing to the fact that they cannot travel slowly or stand over a given spot. The same purpose, but less efficiently, is accomplished by the seaplane circling over a given spot until it has detected the mines that may be submerged in that spot.

Aircraft can also be used for experiments in painting mines in different colors, which makes them less visible from the air. The color must, of course, change with the color of the background where the mine is located. In clear water and when the sun strikes in a way to create a shadow of the mine, this ruse is of little avail, but in other conditions the locating of mines is made very difficult by painting them.

CHAPTER VIII

NAVAL ANTI-AIRCRAFT DEFENSES

War with Germany brings the possibility of aerial raids on American cities by German aircraft, which may rise from ships at sea or from temporary bases. Such a possibility was discussed and admitted in Congress in February, 1917, and as a result, the number of anti-aircraft guns for the defense of naval stations was doubled. Since then reports have appeared that Germany is building ten large submarines for the special purpose of carrying seaplanes for raiding purposes. As Mr. Alan R. Hawley, the president of the Aero Club of America, has pointed out, under the present conditions, until our anti-aircraft defenses are developed and anti-aircraft gunners have experience, the worst that could happen to enemy aviators who found it necessary to land after dropping bombs on American cities is that they would be made prisoners.

Much of the work of operating the anti-aircraft defenses will have to be carried on by the Naval Reserve Flying Forces, how much will be shown by the following definition of what the naval aircraft defenses comprise.

Naval anti-aircraft defenses are divided into three classes, as follows:

(1) *The Flying Defenses,* consisting of fighting aircraft, including fast-armed, fighting seaplanes and armed dirigibles, which on receipt of wireless messages from ships at sea, or through other means, that aircraft are on their way to attack, fly out to sea and endeavor to shoot down the hostile aircraft or force them to retreat.

The purpose of anti-aircraft defenses is to prevent the enemy's aircraft from reaching our shores and inflicting damage. The real victory does not consist in repulsing the enemy or even in destroying them after they have reached our shores, because if they succeed in reaching our shores we are forced to fire in the sky and bear the responsibility of the damage done by the fall of our own shots, and the fall of the enemy's bomb-laden aircraft, which, if hit, may do extensive damage.

(2) *The Floating Defenses,* consisting of anti-aircraft guns mounted on different types of ships, stationed out at sea as far as possible from the shore, so that in case the aircraft is not hit, wireless messages can be sent to other anti-aircraft ships to be on the watch-out, and to summon aircraft to fight them. Closer to the shore fast motor boats with anti-aircraft gu of smaller caliber can be of great value.

(3) *Shore Anti-Aircraft Gun Defenses,* consisting of anti-aircraft guns, searchlights, "listening towers," equipped with huge microphones to magnify the sound of the motors of the approaching aircraft, and range-finders and other instruments for gaging distances.

The 3-inch anti-aircraft gun mounted on the U. S. S. *Pennsylvania.*

Naval Anti-Aircraft Guns

The Sixty-fourth Congress at its second session, which ended March 4, 1917, allowed $3,800,000 for anti-aircraft guns for the defense of naval stations, to be available until 1920. The year before, provision was made for 134 anti-aircraft guns to mount on ships.

These guns, which are illustrated herewith, are 3-inch, 50-caliber guns, firing a 13-pound shell or shrapnel. The movement is so arranged that the gun can fire from 10-inch depression to 90-inch elevation, and at any angle of train. The gun is semi-automatic, allowing a rate of fire of more than 20 shots a minute under favorable circumstances.

These guns have been mounted in many United States ships, including the U. S. S.

The anti-aircraft gun on board of a British ship and the bluejacket who operated it when it brought down the Zeppelin "L 20" near Salonica. This official photograph shows the telescopic sighting-apparatus of the gun and part of its range-finding mechanism, the laying wheel and the firing key and pistol-grip, which are held by the gun-layer's right hand. See opposite page for photo of the gun and the control officer.

Pennsylvania, the flagship of the Atlantic fleet, a photograph of which is shown herewith, and ships of the *Texas* class, as well as destroyers of the *Davis* class.

At the beginning of the war, the German Navy was using 4.1-inch guns manufactured by the Krupps, each of 45-caliber, firing projectiles weighing 34 pounds, with a muzzle velocity of 2630 feet to a height of 12,000 feet. The rate of fire was 15 shots per minute. Since the beginning of the war reliable reports state that other guns have been put in use which are capable of firing as high as 28,000 feet, although the practical limit of vision may be placed at between 10,000 and 12,000 feet.

F. W. Lancaster, the British aeronautic authority, states that:

"Anti-aircraft firing is very inaccurate, hence numbers are employed to compensate.

"The German guns are: 71 mm., firing 20 to 25 shots per minute, shell weighing 11 pounds, vertical range 19,000 feet; speed of motor car 60 km. per hour, climbs grade of 1 on 5, with 12mm. armor; submarine anti-aircraft gun, 3.5-inch caliber, 20 to 25 shots per minute, muzzle velocity 2500 feet per second, weight of projectiles, 20 pounds; Krupp, 6-inch, firing 35-pound projectile discharging smoke trail, is mainly used by coast defenses and by the navy.

"The Ehrhardt factory at Dusseldorf makes various anti-aircraft guns, of 2-inch to 6-inch caliber.

"The most remarkable of these is of 2.6-inch caliber, weight of projectile 9 pounds, elevation 75 inches, muzzle velocity 2000 feet per second, vertical range 17,500 feet. Three classes of projectiles—shrapnel, smoke shrapnel, and 'balloon grenade' are used. This gun is mounted in an armored car, weighing complete about seven tons.

"The Skoda works, Pilzen, Austria, makes a 1.5-inch gun with a muzzle velocity of 3000 feet per second, weighing complete about 1350 pounds. It can be carried on any high-powered motor car."

The French naval aircraft gun, illustrated herewith, shows that the French guns are very similar to the 3-inch anti-aircraft guns.

Some British and German submarines have been equipped with disappearing guns, which fit into special hatches. As soon as the submarine has reached the surface, the gun hatch opens and automatically places the gun in firing position. While it is true the submarine faces danger in fighting a seaplane than is the case vice versa, there have occurred in the Great War several encounters of this kind, and on April 30, 1916, a German submarine shot down a British seaplane.

The anti-aircraft guns mounted on British and French war ships have been responsible for bringing down a number of Zeppelins, and a large number of aeroplanes. On May 5, 1916, two Zeppelins were brought down, one by a

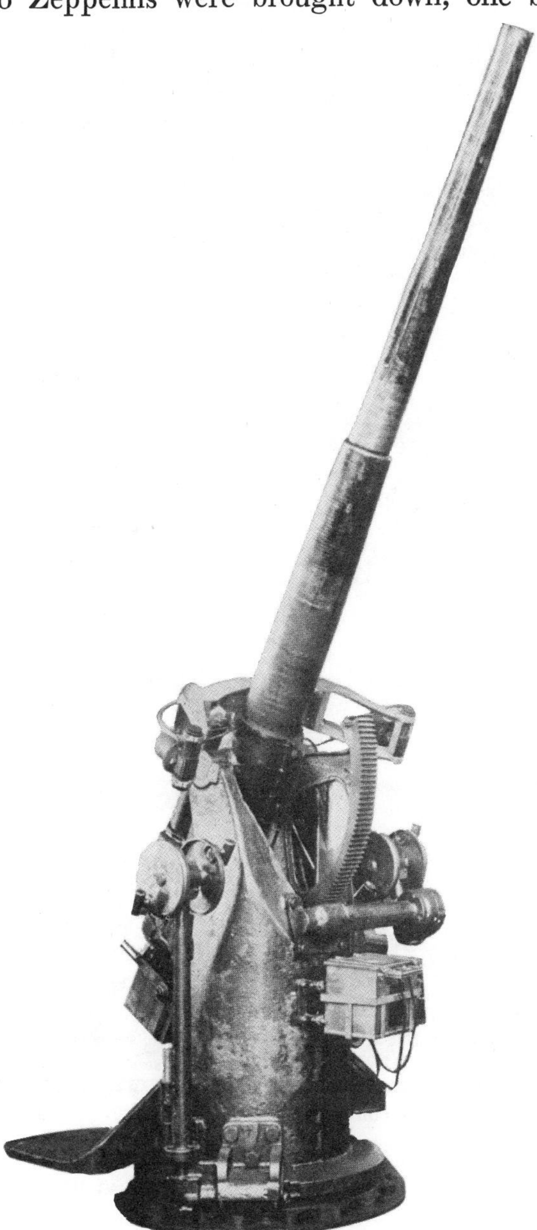

One of the 3-inch anti-aircraft guns of the U. S. Navy.

German motorboats equipped with anti-aircraft guns on the Vistula pursuing a Russian seaplane. From a drawing by C. Barber, Copyright Illustrirte Zeitung.

Hundreds of such motorboats equipped with anti-aircraft guns are needed to organize the anti-aircraft defenses of the United States.

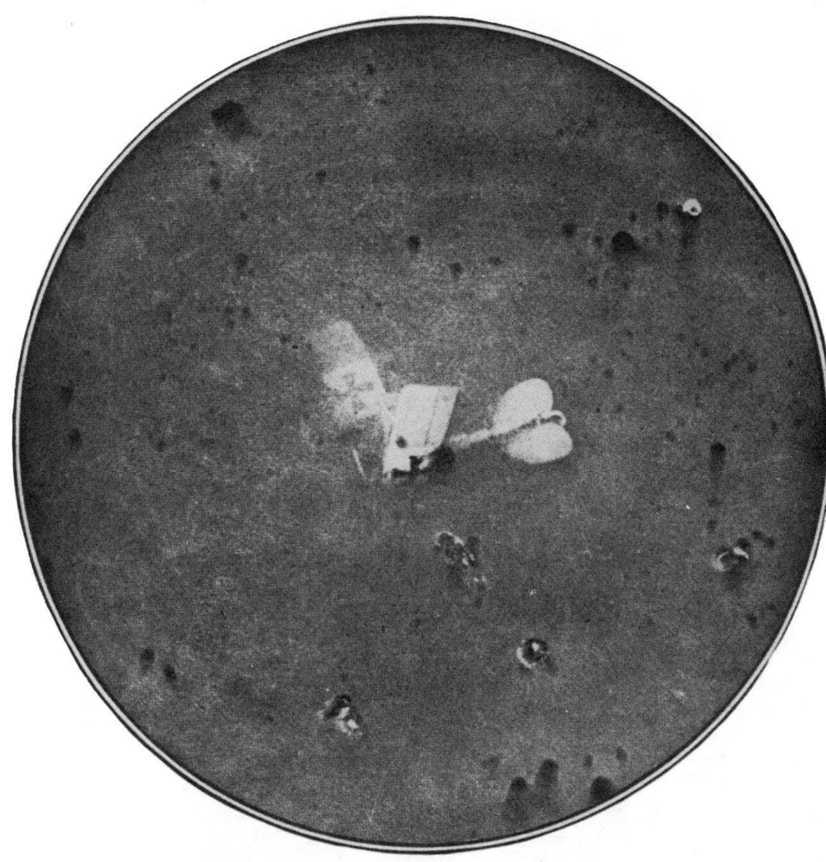

Photo of a German aeroplane which attempted to drop bombs on the residence in La Panne occupied by King Albert I and the Queen of the Belgians, and was brought down by a Belgian aviator and fell into the sea.

British light cruiser off the coast of Schleswig, and another at Salonika, shot down by the anti-aircraft guns mounted on the French battleship *Patrie*.

There is also shown herewith an illustration taken from the "Illustrirte Zeitung," showing motor boats on the Vistula pursuing a Russian seaplane, shooting at it with small caliber anti-aircraft guns.

Jurisdiction Over Naval Anti-Aircraft Defenses

While the introduction of aircraft in military operations has practically removed the lines of demarcation, and we find an overlapping co-operation between the air services of armies and navies, it is best, for our purpose, to adopt the old rule that a navy's duties begin outside of the three-mile limit, but it must protect and defend the navy yards, naval stations, and magazines.

As Senator Swanson stated on the floor of the United States Senate on March 1, 1917, "The defense of New York, the defense of other cities, and all inside of the three-mile limit, is left to the army, except the navy yards and magazines."

When the appropriation for 224 3-inch anti-aircraft guns for the defense of naval stations came before the House of Representatives on February 6, 1917, and a point of order was made by Congressman J. J. Fitzgerald, that "it is a part of the coast defense to provide anti-aircraft guns for these naval stations and not within the jurisdiction of the Committee on Naval Affairs," the Chair ruled that "the defense of naval stations within those stations" are "within the jurisdiction of the Committee on Naval Affairs.

Necessity often changes laws. The necessity of defending London from Zeppelin attacks made the authorities set aside prece-

Lieutenant of the Royal Navy who, as control officer, regulated the firing and A. B., who laid and fired the anti-aircraft gun which brought down the "L 20." The officer is holding one of the shells, the A. B. holding the cartridge-case of the gun.

One of Uncle Sam's new anti-aircraft guns mounted on torpedo destroyer *Davis.*

dents and appointed Sir Percy Scott, the naval authority, in charge of the anti-aircraft defenses of London, a position which he held until February, 1916. Then, on February 16, the entire anti-aircraft defenses were put in charge of Field Marshal Viscount French.

British Anti-Aircraft Defenses

Following is part of Viscount French's illuminating official report on Home Defense to the Secretary of State for War, dated December 31, 1916:

"At the date of my assumption of command, the question of the anti-aircraft defenses of the country was under consideration.

"On February 19 it was decided that the London defenses should be handed over to me, and on February 26 it was further decided that I should be responsible for the whole of the anti-aircraft land defenses of the United Kingdom. Previous to this I had given considerable attention to the subject of anti-aircraft defense, and I submitted a scheme for consideration, which was approved and has been carried out.

"During the winter there was little hostile activity in this direction, but since I assumed charge of these defenses enemy airships and aeroplanes have invaded the country whenever conditions have admitted. The numbers of airships taking part in a raid have varied considerably. On April 3 only one was engaged, while in the raid of September 2–3 not less than twelve ships are believed to have taken part. In all, nineteen raids have been made by aeroplanes. The damage done has been comparatively small, and nothing of any military importance has been effected.

"Taken as a whole, the defensive measures have been successful. In very few cases have the enemy reached their objective. They have been turned, driven off, seriously damaged by gunfire, and attacked with great success by aeroplanes. Seven have been brought down, either as the result of gunfire or aeroplane attack, or of both combined.

"The work of the Royal Flying Corps and the Gun and Light Detachments, including the Royal Naval Anti-aircraft Corps, has been arduous, and has shown consistent improvement; the guns and lights have been effectively handled, and the pilots of the Royal Flying Corps have shown both skill and daring. All are deserving of high praise.

"Close cooperation with the Navy has been maintained and the R. N. A. S., by their constant and arduous patrol work on the coast and overseas, have shared in successful attacks on the enemy."

It is seen by this report that there is the closest of cooperation between the British Army and naval air services in maintaining the anti-aircraft defenses. The following part of the official report of the committee which investigated the Royal Flying Corps in 1916, gives further information regarding the respective share of the two services in the anti-aircraft defenses of Great Britain:

"A good deal of confusion has arisen upon the subject as to whether the air services of the Army or the Navy are responsible for home defense, or whether the responsibility is divided. The truth is that the Navy was entirely responsible till the middle of February last. Since that date the responsibility has been divided. The Navy is responsible until hostile aircraft reach our shores. From that time the Army is responsible. It is hardly necessary to state that if a naval machine was attacking hostile aircraft it would not cease to do so because the aircraft crossed the boundary line (highwater mark), nor would an army machine cease to pursue hostile air craft when it passed over the line seawards.

"The Royal Flying Corps is not responsible for anti-aircraft guns. It has no control over them. Nor has it any responsibility for or control over the searchlights which work in connection with those guns. The Royal Flying Corps now has, however, its own searchlights where-ever home defense machines are maintained.

"The defense of the London area is under the immediate control of the Commander-in-Chief for Home Defense. In other areas, subject to his general control, it is under that

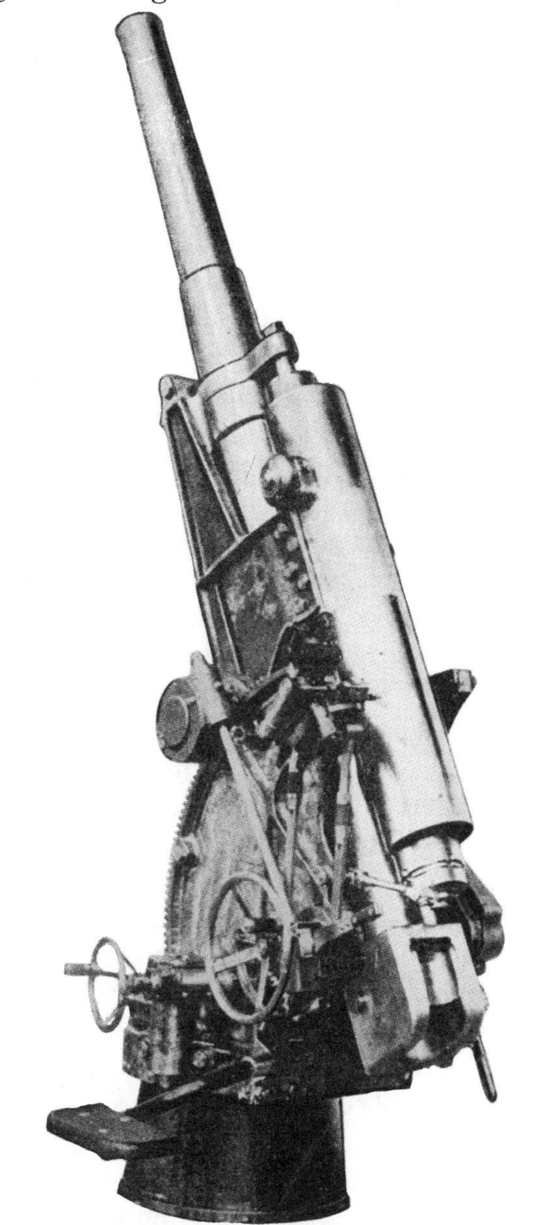

A French naval anti-aircraft gun used during the 1914–1915 campaign.

An anti-aircraft gun mounted on a German submarine, which disappears in a special hatch when the submarine submerges.

of the army officers in command of the particular anti-aircraft defense areas. Those areas are not co-terminous with the districts commanded by the officers in charge of home defense from attacks other than by aircraft.

"It seems desirable to mention that, while the Navy was still solely responsible for home defense, Lord Kitchener issued an order that army aeroplanes were to render all possible assistance, an order which was very willingly obeyed.

"The Navy have aerodromes for their own purposes along the coast, and we think it reasonable to assume, although we have no knowledge on the subject, that, now that the Army is responsible for home defense, from the coast inward, a similar order has been issued to the Navy.

"Having regard to the limitations imposed by the number of aeroplanes, pilots, and night landing places as yet available, we do not know that, so far as the Royal Flying Corps is concerned, anything more can be done.

"It ought, we think, to be generally known that home-defense machines and pilots are not now stationed at every aerodrome. It must be supposed that, because aeroplanes are seen flying freely day by day from a given aerodrome, there are necessarily any aeroplanes kept at that aerodrome fit for night flying or any night flying pilots there to fly them. Home-defense machines, with their pilots, are now grouped at various centers, a plan which, after careful consideration, we approve."

In this report, which covers a dozen large pages in small print, is found the following report of the investigation of the complaint that no machines went up on the occasion of the raid at Dover by a German seaplane over Dover on or about January 31, 1916:

"It appears that Dover is a naval war station, and that the Royal Flying Corps merely has a training and mobilizing ground there. On the date in question it happened that General Henderson was inspecting there. He had just gone into the messroom when he heard the anti-aircraft guns firing. The only portion of the Royal Flying Corps at Dover at the time was a half-completed squadron ready to go abroad. Directly the guns were heard, the pilot on duty ascended in pursuit of the German seaplane, and was immediately followed by two naval machines, and these again by another army machine. It appears that the machines went up in the opposite direction to Dover, so were not seen by the inhabitants of that town. The day was rather misty and the

A Lewis aeroplane gun mounted to fire through geared down hollow propeller shaft of French 180 horsepower Hispano-Suiza motor on fighting aeroplanes.

An anti-aircraft gun mounted on an ammunition barge on the Tigris, part of the British forces operating in Mesopotamia.

German seaplane 8000 feet up, so that the British machines were unable to catch it. Another allegation was that the anti-aircraft guns fired at the British aeroplanes, and there is evidence which points to some rounds having been fired at one of the naval machines."

This brings up both the fact that there is close cooperation between the two British air services, and that there is danger of shooting at one's aeroplanes under certain conditions.

Aircraft Brought Down in 1916

According to official reports, during 1916 the British destroyed 247 and brought down in a damaged condition 142 German aeroplanes, and destroyed eight Zeppelins, four of which were brought down by aviators and four by anti-aircraft guns. The French destroyed 417 and drove down in a damaged condition 195 machines. Twenty Zeppelin raids on England, and six on France, were reported in 1916. Hundreds of raids by German aeroplanes were reported, but only eleven cases where the German aviators succeeded in passing the three lines on anti-aircraft defenses, and dropped bombs on English soil.

The official number of aircraft brought down by the anti-aircraft defenses of the Central Powers are not available at date of writing.

Following is the list of the Zeppelins brought down by the British and French anti-aircraft defenses in 1916:

February 21—LZ77 brought down by French artillery at Revigny.

March 31—L15 hit by gunfire and fell into the sea at the mouth of the Thames.

May 4—L7 destroyed by the fire of British light cruisers off the coast of Schleswig-Holstein.

May 5—LZ85 destroyed by the fire of the Allied Fleets at Saloniki.

September 2—A Schutte-Lanz destroyed by Lieutenant W. Leefe Robinson, R. F. C., at Cuffley.

September 23—L32 brought down in flames in Essex by Second Lieutenant F. Sowrey, R. F. C.

September 23—L33 landed in a comparatively undamaged condition in Essex.

October 1—L31 brought down in flames at Potter's Bar by Second Lieutenant W. J. Tempest, R. F. C.

November 27—A Zeppelin brought down off the northeast coast by Second Lieutenant I. V. Pyott, R. F. C.

November 28—A Zeppelin brought down off the east coast by Flight Sub-Lieutenant E. L. Pulling, Flight Lieutenant G. W. R. Fane, of the R. N. A. S.

Efficient Anti-Aircraft Defense in 1916 Bring Reduced Aircraft Insurance Rates

Efficient anti-aircraft defenses in 1916 resulted in the British Government cutting down the insurance against enemy aircraft. This

The armored cars of the British Royal Naval Air Service.

The Austrian seaplane shown in this photo, attacked the French cruiser *La Savoie,* which was embarking Serbian troops, at Valona, in the Adriatic, and was brought down by the fire of French gunners.

first change in the British Government rate for insurance against enemy aircraft and bombardment risks since the scheme was inaugurated on July 19, 1915, was announced on February 13. It takes the form of a discount of 50 per cent. in respect of all rates, subject to the minimum premium payable in respect of any one insurance not being reduced below 2s. Two shillings is the present rate for insuring private houses and their contents and buildings in which no trade is carried on, for £100, so that the rate for insuring all private property of the value of £200 or more is halved to 1s. per cent. The existing rates under the scheme are as follows:

	s.	*d.*
All other buildings and their rents	3	0
Farming stocks (live and dead)	3	0
Contents of all buildings, other than in private houses and in premises specified below	5	0
Merchandise at docks and public wharves, in carriers' and canal warehouses and yards, in public mercantile storage warehouses, and in transit by rail; timber in the open; mineral oil tanks and stores (wholesale)	7	6

The Zeppelin "L 15" photographed off the Kentish Coast just before it disappeared beneath the waves, having been hit by the anti aircraft guns, during the night of March 31–April 1st, 1916.

British naval anti-aircraft gun crew getting ready to fire at German and Turkish aeroplanes at Salonika.

The new terms, which represent half of the above rates, went into effect until March 1. All the government rates are for twelve calendar months, except as regards the property named in the last paragraph above. For these risks policies are issued for six months at three fourths of the annual premium, for three months at one half the annual rate, and for one month at one fourth of the annual premium.

The damages on which insurance was paid by the British Government went up to tens of millions in 1915; they are reported to have been less in proportion in 1916.

Our anti-aircraft defenses are far from being sufficient to meet an emergency, and it will be necessary to make up for this deficiency by organizing numerous squadrons of fighting seaplanes and mounting anti-aircraft guns on the new submarine chasers and fast motor boats. The task of the Reserves will be to prevent aircraft from reaching the shores after the aircraft have passed the anti-aircraft guns of the battleships, which will be the first line of defense. There should be large fleets of yachts, submarine chasers, and motor boats ready to meet any emergency.

Above all, practice in shooting at flying kites should take place as soon as possible, so that the anti-aircraft gunners may have experience, and may be ready to protect our coasts from aircraft attacks.

Memoranda:

CHAPTER IX

THE AERIAL DEFENSES NEEDED FOR THE THIRTEEN NAVAL DISTRICTS OF THE UNITED STATES AND TWO INSULAR NAVAL DISTRICTS

If it requires hundreds of aeroplanes, dirigibles, and observation balloons to patrol the coasts of Great Britain, which aggregate about 1500 miles of coast line, we will need thousands to patrol our 2500 miles of coast line in the United States and almost as much in our possessions—and to protect the millions of dollars worth of supplies which are to be sent to Europe and the ships which carry them.

The thirteen naval districts in the United States and their headquarters are as follows: (1) Eastport, Maine, to include Chatham, Massachusetts; headquarters, Boston. (2) Chatham, to include New London, Connecticut; headquarters, Narragansett Bay Naval Station. (3) New London, to include Barnegat, New Jersey; headquarters, New York. (4) Barnegat, to include Assateague, Virginia; headquarters, Philadelphia. (5) Assateague, to include New River Inlet, North Carolina; headquarters, Norfolk. (6) New River Inlet, to include St. John's River, Florida; headquarters, Charleston. (7) St. John's River, Florida, to include Tampa, Florida; headquarters, Key West. (8) Tampa, Florida, to include Rio Grande; headquarters, New Orleans. (9) Lake Michigan, headquarters, Naval Training Station, Great Lakes. (10) Lakes Erie and Ontario; headquarters, Naval Training Station, Great Lakes. (11) Lakes Huron and Superior; headquarters, Naval Training Station, Great Lakes. (12) Southern Boundary to Latitude 42°N.; headquarters, San Francisco. (13) Latitude 42°N., to Northern Boundary; headquarters, Port Townsend, Washington.

A very comprehensive plan of aërial coast patrol suitable for the fifteen naval districts was submitted to Rear-Admiral Nathaniel R. Usher, the commandant of the third naval district by the Aero Club of America's Board cooperating with the commandant in the organization of the Naval Reserve Forces, of which the writer is secretary.

This report, which was made a congressional document, is reproduced herewith in part:

New York, April 2d, 1917.

TO: *Rear Admiral Nathaniel R. Usher, Commandant, Third Naval District.*

FROM: *Alan R. Hawley, Chairman, Advisory Committee on Aeronautics, cooperating with the Commandant in the organization of the Naval Reserve Forces.*

SUBJECT: *Aerial Defenses Needed for the Third Noval District.*

AERIAL DEFENSES NEEDED FOR THE THIRD NAVAL DISTRICT

(Extending from New London to Barnegat)

Duties.—The air service of the Third Naval District has the following duties to perform. All of these have been done in the Great War:

(a) To locate, and assist destroyers, trawlers and submarine chasers in capturing or destroying hostile submarines (both seaplanes and dirigibles are needed).

(b) To locate submerged mines and assist trawlers in destroying mines. (Seaplanes, dirigibles and observation balloons needed.)

(c) Searching the coasts for submarine bases. (Seaplanes and dirigibles needed.)

(d) To convoy troop and merchant ships on coastwise trips. (Dirigibles best adapted for this work.)

(e) To patrol the coasts, holding up and inspecting doubtful ships and convoying them to examining stations. (Dirigibles best adapted for this work.)

(f) Attacking hostile ships and submarines that may show up near the coasts, with torpedoes, bombs and guns. (Large torpedoplanes and large seaplanes mounting guns best adapted.)

(g) Protecting ships at sea and in ports against attack from hostile submarines and battleships. (Seaplanes and dirigibles needed.)

(h) Communicating to incoming ships information

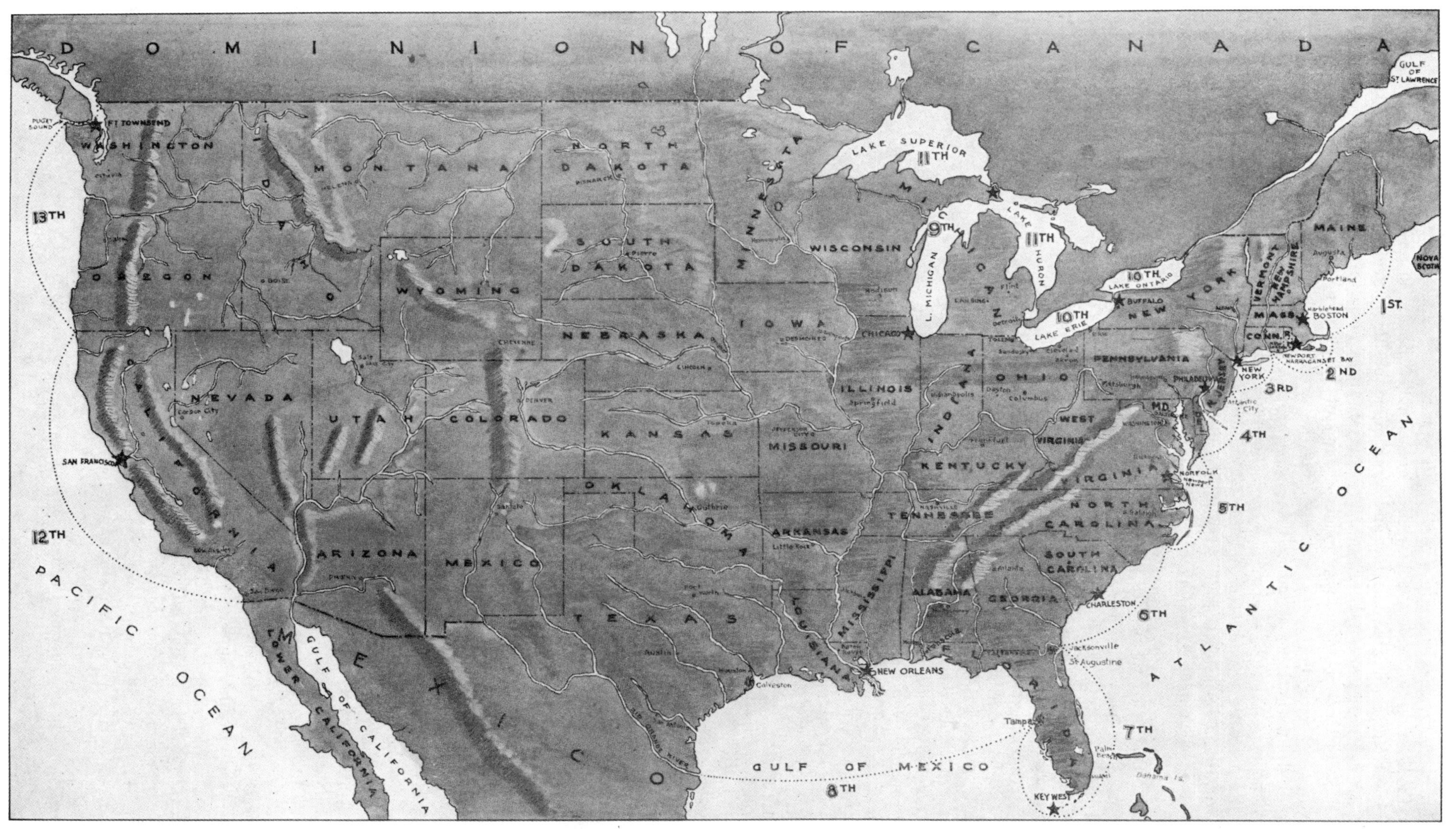

This map shows the 13 naval districts of the United States. The districts, and their headquarters are as follows: (1) Eastport, Maine, to include Chatham, Massachusetts; headquarters, Boston. (2) Chatham, to include New London, Connecticut; headquarters, Narragansett Bay Naval Station. (3) New London, to include Barnegat, New Jersey; headquarters, New York. (4) Barnegat, to include Assateague, Virginia; headquarters, Philadelphia. (5) Assateague, to include New River Inlet, North Crolina; headquarters, Norfolk. (6) New River Inlet, to include St. Johns River, Florida; headquarters, Charleston. (7) St. Johns River, Florida, to include Tampa, Florida; headquarters, Key West. (8) Tampa, Florida, to include Rio Grande; headquarters, New Orleans. (9) Lake Michigan; headquarters, Naval Training Station, Great Lakes. (10) Lakes, Erie and Ontario; headquarters, Naval Training Station, Great Lakes. (11) Lakes, Huron and Superior; headquarters, Naval Training Station, Great Lakes. (12) Southern Boundary to Latitude 42 N.; headquarters, San Francisco. (13) Latitude 42 N., to Northern Boundary; headquarters, Port Townsend, Wash. Great Britain alone has 107 aeronautic stations; France and Germany have about 150 each.

regarding the location of mines, submarines and the courses to follow to avoid disasters and confusion. (Seaplanes and dirigibles needed.)

(i) Serving as the "eyes" of mine planters, minimizing the time required for mine planting. (Dirigibles and observation balloons best adapted for this work.)

(j) Defending and protecting naval bases and stations from naval and aerial attacks. (Armed air cruisers and combat planes needed.)

Besides the above, the Naval Air Service in other countries has been used for many other purposes, but the Air Service of the Third Naval District need not concern itself with the other purposes, which are to be performed by the aviators connected with the fleet.

Divisions.—The territory comprised in the Third Naval District should be divided into divisions to be served by aeronautic stations established in each division, so as not to weaken the efficiency of the service by sending aircraft on too extended cruises. The lack of large dirigibles capable of long cruises necessitates costly and inefficient makeshifts, and it is necessary to follow as closely as possible the example of Great Britain and establish the aeronautic stations close enough to get utmost efficiency out of each type of aircraft available. Great Britain has an aeronautic station at about every twenty miles along her coasts, numbering 107 in all. The aircraft available for the Air Service of the Third Naval District within sixty days will be seaplanes equipped with two or three motors, which may be entrusted with flights of about eighty miles out to sea and return, at a speed of eighty miles an hour; and seaplanes equipped with a single motor which may, if supported by water-craft, be entrusted with flights extending twenty miles out to sea and return at a speed of 60 miles an hour; and observation balloons which can be put on board ships or barges and can be used from whatever positions these ships occupy.

The limit on continuous flight out to sea is placed here because while many aviators have made flights of several hundred miles along the coasts, very few of our aviators have had experience in actual aerial navigation over water. Until they have gained this experience their flights seaward should be limited in distance and with every seaplane sent out on patrol duty there should be sent out one or two boats with observers, whose duty is to keep track of the flight with powerful glasses, and be ready to rush to the assistance of the aviators.

After ninety days it will be possible to get larger air cruisers, some of which are under construction; and small dirigibles of the coast patrol type, sixteen of which were ordered by the Navy Department on March 12th.

Location of Divisions.—In considering the location of divisions, there must also be considered the necessity of establishing one of the divisions as far as possible out at sea where the aircraft may have the opportunity of detecting hostile submarines while they are on the surface. On approaching the land, hostile submarines logically submerge so as not to be seen. Therefore, a division should be established at Montauk Point. On the other hand, the divisions having charge of keeping channels clear of mines and submarines must be located as near as possible to the channels. For instance, the New York Division, having as one of its duties to keep the channel clear up to the 50th fathom curve, which is about 85 miles from Sandy Hook, must include in its equipment large multiple motored seaplanes, capable of long distance cruises.

For efficiency, ten aeronautic stations should be established in the Third Naval District, to be located approximately as follows:

(1) Sandy Hook (Aeronautic Base).
(2) Montauk Point (Aeronautic Base).
(3) Bay Shore (Station being established by the New York Naval Militia).
(4) Port Washington (Already established by the America-TransOceanic Co., Offices, 280 Madison Avenue, N. Y.).
(5) Amityville (One hangar and workshop already established by the Sperry Gyroscope Co., Manhattan Bridge Plaza, Brooklyn, N. Y.).
(6) Ocean Beach (New Jersey).
(7) Seaside Park (Barnegat Bay).
(8) Rockaway or Manhattan Beach or Massapequa, Great South Bay, L. I.
(9) New Haven (Connecticut).
(10) Southampton.

Equipment for Stations.—The flying equipment of each station varies according to the duties of that station. In a general way, until dirigibles and observation balloons are obtained, the territory to be covered by each station will be covered by units consisting of one aviator and one observer, having at their disposal three aeroplanes, one of which must always be in flying order. Each of these units is expected to fly about 200 miles each day when the weather permits. Whenever the distance to be covered is so great that a unit only covers it once in the course of its 200-mile flight, as in the case of the unit which will have to protect the channel down to the 50th fathom curve and back, starting from the Sandy Hook base, there will be required a sufficient number of units to make it possible for a unit to start every half hour beginning with daylight and ending at sundown.

As at present the daylight lasts about twelve hours, there would be required 24 units to patrol the channel course, a unit starting every half hour. That would necessitate having 24 units with three machines to each unit or 72 machines in all. The same is true at the Montauk Point station, which would have to patrol part of Block Island Sound and go as far out at sea as possible to look out for hostile submarines. The smaller stations, the duties of which are essentially to search the bays for submarines and convoy coastwise shipping, would need a smaller number of units, unless the call for aerial patrol and convoying was heavy.

The equipment required for each station depends on the importance of the station from a strategic standpoint. In some cases, the station can consist of only a sufficient number of hangars to house the aeroplanes, with a workshop, storage for gasoline, oil, etc., and the necessary housing for the officers and men. In other cases, the stations must have hangars for aeroplanes, dirigibles and observation balloons; motor and machine shops; hydrogen plant, magazines, erecting shops, stores, an aerologic station, wireless station, listening towers, searchlights and anti-aircraft guns for the protection of the station. Also provision for aeroplane mother ships, kite balloon ships, and mine laying ships, to cooperate with the aeronautic station; and the necessary watercraft.

Until dirigibles and observation balloons can be obtained, the entire work must be done by aeroplanes.

A single dirigible of the Zeppelin type could do the work of patrolling the channel from Sandy Hook to the 50th fathom curve, which is 85 miles out at sea, better than the 72 seaplanes hereinbefore mentioned. But no number of small scouting dirigibles could do that same work, excepting in the best of weather conditions because the small dirigibles would be carried away by or could not travel against the average wind to be met along the channel.

Under fair weather conditions there could be placed four or six observation balloons along the channel, anchored on barges or suitable ships. Slow moving ships with the observation balloons could, under normal conditions, do the work of the 72 aeroplanes. On sighting a hostile submarine, or mines, the observers would wireless the information to the shore station and summon cruisers, air cruisers, submarine chasers or the trawlers in charge of mine sweeping.

There should be in addition to the stations at least one aeronautic base in the naval district. It may be stated that all the personnel required for the air service of the district has to be trained, there being practically no trained personnel available. The personnel should be trained at the aeronautic base.

The Sandy Hook and Montauk Point stations should be most complete, their equipment including the seventy-two aeroplanes required to maintain a steady patrol for twelve hours daily and at least two dirigibles and two observation balloons.

The aviators, dirigible pilots, observation balloon operators and observers for the three kinds of aircraft would be trained partly at these stations and partly at two other stations which, while not so extensive in general equipment, would have extensive facilities for instruction. The equipment needed for the last mentioned two stations would consist of about 24 seaplanes to be used for coast patrol, and about 18 aeroplanes and one dirigible and one observation balloon, respectively, for training.

The number of torpedoplanes and of large seaplanes mounting three-inch guns needed for the aerial defenses of the Third Naval District is not estimated herewith, because the number required will depend entirely on how extensively shipping in the Third Naval District is subject to attack from hostile ships and submarines.

Large seaplanes equipped with three-inch guns would be powerful factors of offense and defense also as they can sink destroyers, submarines, transports, etc. The large seaplanes required for this purpose are obtainable, one having been delivered to the Navy recently. Both the torpedoplane and the seaplane equipped with a three-inch gun represent an extraordinary combination of mobility and power, which combination promises to revolutionize naval warfare. Their great speed and their ability to fly in a straight line over all natural obstructions, make it possible to mobilize their power at any point from Barnegat to Montauk Point, within two hours.

The other six stations would require about twelve seaplanes each. The above is, to some extent, based on the British and French experience. At the beginning of the War, Great Britain had only 18 aeronautic stations. To-day she has 107, one fifth of which are large aeronautic bases. France has about 150 aeronautic stations.

There is practically no trained personnel available, but it will be possible to get, to start, at least twenty civilian aviators, professional and amateurs, who have had some experience in marine flying, although no experience in actual naval operations or in the operation of twin motored aeroplanes. There can also be had about twenty students who are about to complete their preliminary course in the operation of single motored seaplanes. About two hundred more students, mostly college men who have joined the Aerial Coast Patrol units, will be under training in the Third Naval District within two months, several large aviation training camps being established at private expense. These are part of the hundreds of college men who wanted to join the Naval Reserve Flying Corps, but could not, because the Navy Department has not the aeronautic training schools at which to train them.

The entire personnel of the Air Service will comprise for the Third Naval District about 150 aviators and aviation instructors; thirty dirigible balloon pilots and thirty observation balloon operators. Also as many observers as there are pilots and operators. There will be required an average of one chief mechanic and three assistant mechanics to each aviator. For each dirigible in operation there must be a crew of mechanics and a company of enlisted men to act in docking the dirigible. To each observation balloon there is required a crew of mechanics and a company of men.

How Far Should Naval Airmen Carry Their Operations Over Land?

How far should naval airmen carry their operations over land is a question that began to be asked in the early days when the first Zeppelins, which were built for naval work and housed in a floating hangar, were first put in operation. No definite answer has been given so far, although the matter has been brought up in a number of instances in Europe in connection with questions of defining the responsibilities of the land and the naval air services. The committee which investigated the administration of the British Royal Flying Corps in its official report to the British Government established a line of demarkation to divide the respective duties of the land and naval aviators as follows:

"A good deal of confusion has arisen upon the subject as to whether the air services of the Army or the Navy are responsible for home defense, or whether the responsibility is divided. The truth is that the Navy was entirely responsible till the middle of February last. Since that date the responsibility has been divided. The Navy is responsible until hostile air-craft reach our shores. From that time the Army is responsible. It is hardly necessary to state that if a naval machine was attacking hostile aircraft it would not cease to do so because the aircraft crossed the boundary line (high-water mark), nor would an army machine cease to pursue hostile air craft when it passed over the line seawards."

But any line of demarkation is subject to be taken exception to, as often as necessary, as was shown by the fact that throughout the present war naval aviators have been used for raids over land, flying land aëroplanes. It will be remembered likewise that the United States Navy aviators flew over land during our occupation of Vera Cruz, in 1914. They actually flew in hydroaeroplanes and flying boats over the Mexican hills and mountains. (See chapters on "Aerial Strategy and Tactics," "United States Navy Aeronautics.")

CHAPTER X

ADMINISTRATION OF A NAVAL AERONAUTIC STATION

A naval aeronautic station may be an aeronautic base or merely a station from which seaplanes, dirigibles, and observation balloons are operated. The equipment required for each station depends on the importance of the station from a strategic standpoint. In some cases, the station can consist of only a sufficient number of hangars to house the aeroplanes, with a workshop, storage for gasoline, oil, etc., and the necessary housing for the officers and men. In other cases, the stations must have hangars for aeroplanes, dirigibles, and observation balloons; motor and machine shops; hydrogen plant, magazines, erecting shops, stores, an aerologic station, wireless station, listening towers, searchlights, and anti-aircraft guns for the protection of the station as well as provision for seaplane carriers, kite-balloon ships, and mine laying ships, to cooperate with the aeronautic station; and the necessary watercraft. The accompanying drawing shows the plan for a complete naval base, with all the important buildings and departments of a naval aeronautic base. On the following page will be found the photograph of a smaller seaplane station.

An excellent illustration of the completeness of the organizations of naval aero stations may be gained from the telephone directory of the United States Government Station at Pensacola, Florida, which includes extensions to the commandant; captain of yard; inside superintendent; radio station; planning division, aeronautic secretary; yard section; sentry, office building; labor board; flying school; watch tower; pay officer; hangars; supply department; accounting department; supply storehouse; supply purchasing section; aeroplane erecting shop; motor erecting shop, storeroom; truck house; joiner shop; machine shop; storehouse, yard division; power house; quarters of captain of watch; pumping station; dispensary and sick quarters; drafting room; commanding officer, marines; post quartermaster; ship's 'phone, crib wharf; wet basin; seamen's barracks, and bachelor officers' mess.

Regulations for the United States Navy Aeronautic Station, Pensacola, Florida

While the regulations for the administrations of a naval aeronautic base or station necessarily vary in detail, the fundamental regulations are essentially the same. The following regulations for the United States station at Pensacola may be applied:

Ships at Anchor or Berthed at the Yard

Berths at the piers will be assigned by the Captain of the Yard. Ship's boats will be secured at places designated by the Captain of the Yard.

The crew of the ships will not be permitted to wander around the yard or shops. In case of work being done in the shops by the enlisted force of a ship, only those directly connected with the work will enter the shop, and some responsible person will account for all tools used.

Yard regulations will be supplied to ships on arrival at the yard and should be returned to the Captain of the Yard before departure. Attention is invited to the fire bill. In case of fire, ships will send fire details in accordance with their ship organization bills.

Liberty parties will be formed on the dock and marched to the main gate. Liberty men will return through this gate.

Ships will keep the pier clean where the ship is berthed. Garbage cans will be provided in which all garbage will be placed. These cans will be collected by yard force at regular hours.

Ships at anchor will furnish the Captain of the Yard with a copy of their boat schedule, and will have their mail orderly call at office of Inside Superintendent on regular trips.

Long wharf and the basin will be used by ship's boats. The center wharf is reserved for boats used by the Flying School.

Ships at piers will dump ashes in places designated by the Captain of the Yard. Boxes will not be thrown

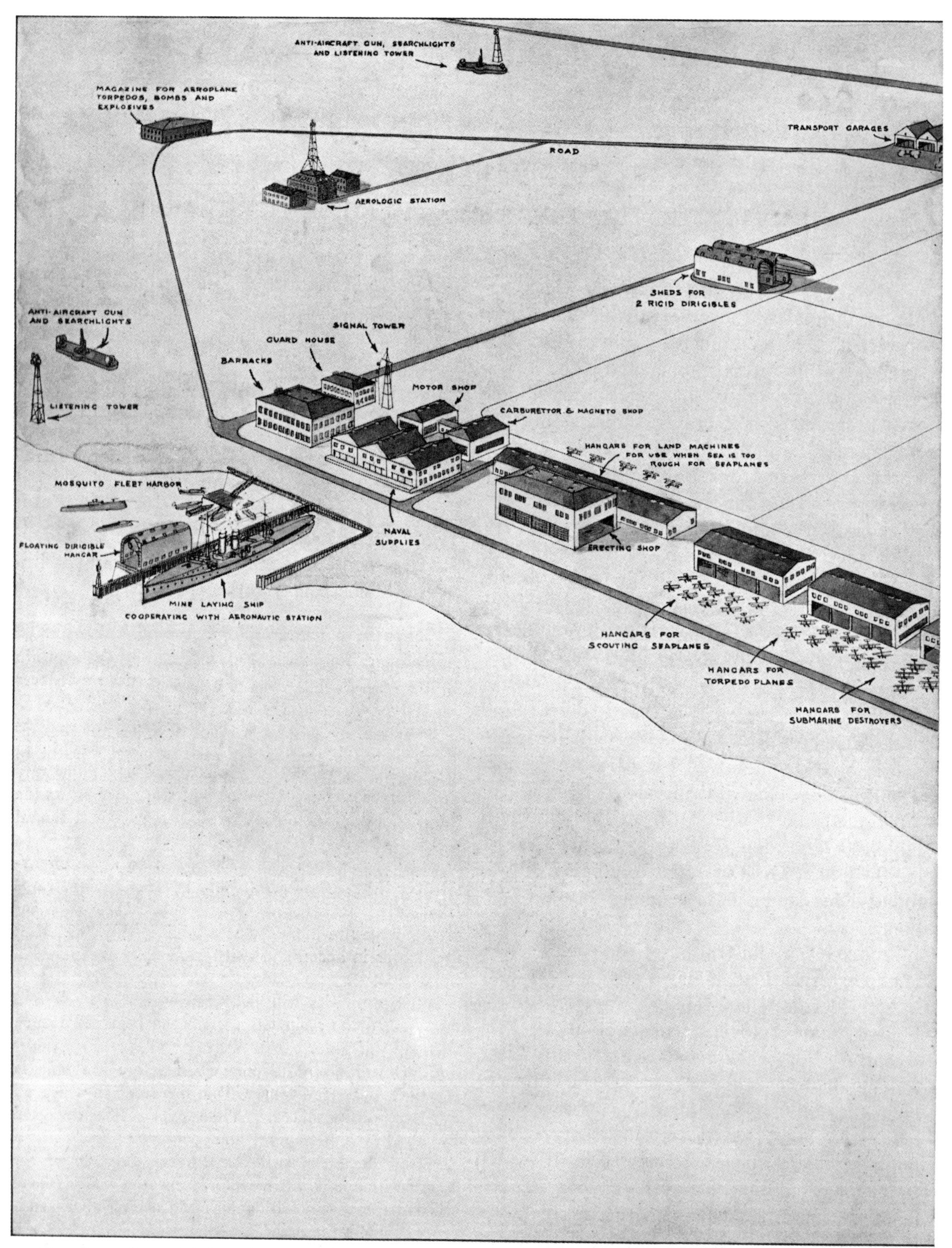

Plan of a well appointed naval aeronautic center. Establishing completely equipped aeronautic stations will preclude the necessity of continuous adding, which is so inefficient and wasteful.

No provision is made here for extra large cruisers because there has not yet been decided how such air cruisers can best be housed, being somewhat too large to be taken to a hangar on shore. The solution will probably rest with building large hangars on the water edge.

It will be noted that provision has been made for four anti-aircraft guns, placed so as to almost form a square, for the protection of the station from aircraft attacks. There are also four Listening Towers and four searchlights provided. The dirigible sheds are of the revolving type, so that the airships can go out no matter which way the wind blows.

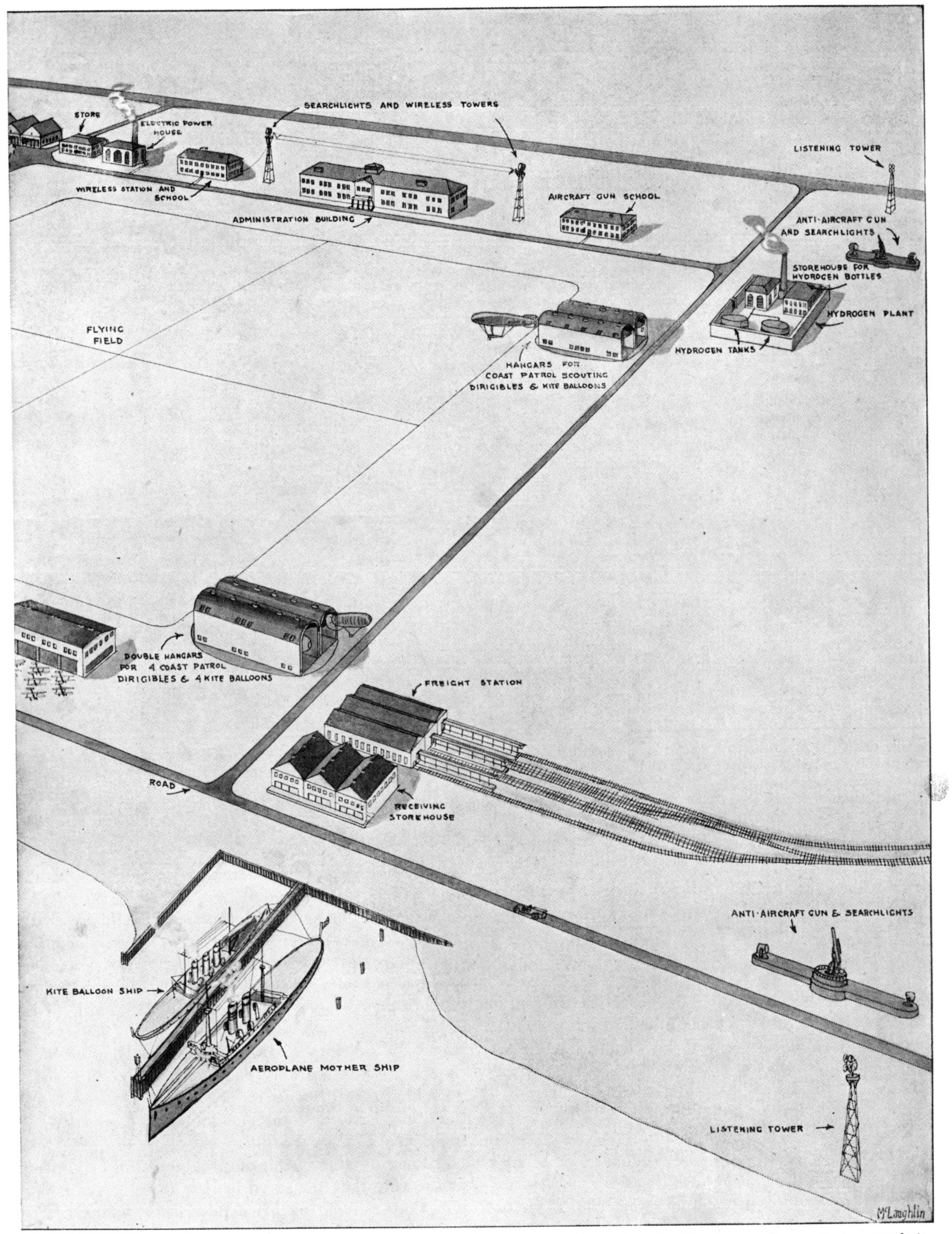

Provision is made for a hangar for a squadron of land aeroplanes, to provide against the times wnen the sea is too rough to permit launching seaplanes, but the atmospheric conditions are suitable for land machines. Also for the defense of the station from aircraft attacks. Land aeroplanes are used to a great extent by naval aviators in all countries.

Provision is made for a mine layer to cooperate with the station, so that the aviators and the dirigible and kite balloon operators can practice in locating mines and assist in planting mines.

The seaplane carrier and the kite balloonship are absolutely necessary to have ready for the fleet to train personnel for the mother ships. A seaplane carrier and a balloon ship should be allowed for every eight battleships. The aerologic station is a most essential thing and it should be completely equipped with the latest instruments.

One of the 107 British seaplane stations which supply the daily aerial coast patrol for the protection of British shipping. This aerial photograph, taken from one of the seaplanes, shows one of the smaller stations. One of the seaplanes is shown in the water, "taxying" to the runway.

The seaplanes from these numerous stations keep a constant watch for U-boats and protect ships from U-boat attacks.

overboard as they are liable to damage the pontoons of the hydroaeroplanes.

Ships in passing the yard will slow down to such speed that their bow or stern wave will not damage the hangar runways and boats at the docks.

Automobiles

Officers and enlisted men attached to the station or to ships at the station, and others residing on the naval reservation, owners of automobiles or motorcycles, may obtain Aero Station License by application to the Captain of the Yard.

An annual fee prescribed by the Captain of the Yard will be charged for such license, payable on purchase of machine and renewed January 1st of each year. A record of all licenses issued will be kept by the Captain of the Yard and no automobile or motorcycle will be allowed to operate on the reservation without a license. The proceeds of such licenses will be used for the upkeep of the roads leading to the navy yard.

The owner of an automobile or motorcycle license will be held responsible at all times, irrespective of who may be driving, for the proper observation of traffic and station regulations under the penalty of having such license revoked.

No enlisted man or yard workman shall do any work necessitating the use of government tools, machines, etc., on private automobiles, motor boats, etc., nor shall automobiles be taken inside the Machine Shops without permission of the Captain of the Yard. Enlisted men may work on automobiles, motor boats, etc., owned by individuals attached to the station on holidays and after working hours, but in no case shall such work entail the use of government material.

Parking space for automobiles will be designated by the Captain of the Yard.

All motor vehicles entering the yard will be notified at the gate that the speed limit is 12 miles per hour; also that running on brick pavements and sidewalks and the use of muffler cut-outs is prohibited.

Fire-Regulations

The signal for fire will be the rapid ringing of the fire bell at the North West Gate and near flagstaff, also blowing of the whistle followed by a pause then a number of strokes or toots designating the district. The reservation and yard is divided into districts as follows:

No. 1—Navy Yard—North Avenue to North Wall —1 toot on whistle and 1 stroke on bell.

No. 2—Navy Yard—South Avenue to North Avenue—2 toots on whistle and 2 strokes on bell.

No. 3.—Navy Yard—Water front to South Avenue—3 toots on whistle and 3 strokes on bell.

No. 4—Naval reservation—Warrington—4 toots on whistle and 4 strokes on bell.

No. 5—Naval reservation—Woolsey—5 toots on whistle and 5 strokes on bell.

The fire main and plugs are situated as follows: From pumping station near old hospital down Canal Street, Warrington to Navy Yard. In navy yard along West Avenue, North Avenue, East Avenue, South Avenue and Center Avenue. A line also cuts off from West Avenue near G. S. Storehouse to Center Avenue near Power House. A line branches off from North Avenue near alley East of Commandant's house to the Magazine along Magazine Street in Woolsey.

The fire hose and reels are situated as follows:

No. 1—Near flagstaff, North Avenue.

No. 2—Opposite Building No. 9, East Avenue.

No. 3—Near hydrogen plant, South Avenue.

No. 4—Near West Gate, South Avenue.

No. 5—Warrington on Canal Street, near corner of Newton Avenue.

No. 6—Woolsey on corner Magazine Street and Howard Stret.

General Fire Quarters

1. When a fire signal is sounded, the telephone operator will notify the pumping station and tell man in charge to start pumps, will ascertain place of fire, notify Commandant, Captain of the Yard, Officer of the Day, Flying School Office, Erecting Shop, Commanding Officer of Marines, Seamen's Barracks. All messages in regard to the fire will take the precedence to other messages.

2. In case of fire in the yard all gates will be closed and no one not connected with the station will be allowed to enter until secure is sounded.

3. In case of fire in or near the hangars or buildings where hydraeroplanes are stowed, the hydroaeroplanes will be run out into water and canvas hangars knocked down to smother the fire or keep it from spreading.

4. All boats will be manned and gotten ready for use as directed.

5. All men not detailed on a hose reel will equip themselves with buckets and axes and proceed to scene of fire, officers in charge of such details will provide for such details.

6. The Captain of the Yard will be in general charge at scene of fire. The officer of the Day will be his assistant. In the absence of the Captain of the Yard, the Officer of the Day will take charge until the different details have arrived, when the senior line officer present will take charge.

7. Naval ships in the harbor will send details in accordance with their ship organization bills, and the officer in charge of details will report to officer in charge at scene of fire.

8. The recall will be three blasts on the steam whistle.

Fire in Yard—Day

The enlisted men of the Flying School under the officer in charge will man hose reels Nos. 1 and 4 and proceed to scene of fire reporting to Captain of the Yard. The enlisted men of the Erecting Shop and Machine Shop under the senior officer of the two shops will man hose reel No. 3 and proceed to scene of fire reporting to Captain of Yard. The marines will man hose reel No. 2, proceed to scene of fire and report to Captain of Yard.

Fire in Yard—Night

The seaman watch will man hose reel No. 4 and proceed to fire. Other enlisted men in the barracks will be divided into details by the Captain of the Watch, one detail to hose reel No. 1, one detail to hose reel No. 3, and other details equipped with buckets and axes. The marines will man hose reel No. 2.

Fire Outside Navy Yard—Day

A detail from Flying School will man hose reel No. 4, a detail will be equipped with buckets and axes and sent to scene of fire. The Marines will man hose reel No. 2 and proceed to scene of fire. The citizens of Warrington and Woolsey will man hose reel Nos. 5 and 6.

Towing a seaplane back to its station after a flight.

A United States Naval Aeronautic Station "Somewhere in America."

Night

The watch will man hose reel No. 4, a detail of enlisted men in the barracks will be equipped with buckets and axes. Otherwise fire bell same as day.

Washwomen

Washwomen will be permitted to receive clothes on Mondays only and deliver them on Thursdays and Saturdays only. They will remain outside buildings 18 and 25, and on no account will enter the buildings or loiter around the entrances.

Lunches may be sold in the yard by applying for permission to Captain of Yard. The time of such traffic will be limited to the noon hour, mid-day to 1 P.M., and only authorized articles will be sold. The dealers of such articles may sell their goods in front of the shops but are not to enter any building and must clean up any paper or food stuff left over. Uncooked food will not be sold.

Visitors

Visitors will not be permitted to enter any ship or hangar except when accompanied by an officer at any other than noon hours. No visitors will be permitted under any circumstances to enter any shop at any time without a written permission. Officers in charge of shops will make necessary arrangements to enforce this regulation, and in all cases the permission must be shown and initialed in each shop visited.

Shops will not be opened to visitors on Sundays and Holidays except by special permission from the Commandant. Visitors will not be allowed to question any of the workmen. All officers and enlisted men should assist in preventing unauthorized persons from entering or wandering around the yard. Visitors will not be allowed to bring kodaks in the yard or to take any pictures. All officers and enlisted men will keep a careful watch for visitors taking unauthorized pictures.

Fishing

In the future no written passes with permission to fish will be accepted by the marines on duty at the gates. Twenty-five brass checks will be furnished the sentry at the Main Gate, and each colored person desiring to fish will be given a check which will serve as his pass. This check must be turned in at the Main Gate upon leaving the yard. Twenty-five checks per day will be issued for the use of whites. All colored fishermen will be required to fish from the East wharf. The wharf to the Westward of the wet basin will be reserved for the use of whites. All persons are warned that the docks must be left clean and neat and free from all scraps of unused bait, paper, etc. Persons guilty of neglect of this order will be denied fishing privileges for a period commensurate with the extent of their untidiness. Fishing hours are from 8 A.M. to sunset.

Shop Regulations

1. Ships will be kept clean and in good order at all times.

2. Smoking, except in offices, is prohibited.

3. No open light, such as blow torch, acetylene light, etc., shall be left unattended, and no open light of any sort will be brought near engine assembling or testing stands. Fire extinguishers will be kept ready for use at all times.

4. In case of fire notify the Captain of Yard immediately.

5. All tools shall be kept in the tool room and drawn only by check. Tools shall be returned after using. In case a tool is broken in use, it shall be turned into the tool room and attention called to its condition. The tool room keeper shall keep a list of all tools broken or lost, and by whom broken or lost, which list will be submitted to the office on Friday morning.

6. No shop machines will be used by outside men without authority from the Planning Division.

7. No work will be done in the shops that is not covered by a written work order, or, in an emergency, by a verbal work order.

8. Shops will be closed by the janitors after working hours and the keys turned in to Captain of the Watch. The janitors will get the keys and open up the shops by 8 A.M. No unauthorized person will be allowed to have keys to any shop, and doors and windows will be kept locked except when authorized work is being done in a shop.

Regulations for Enlisted Personnel Uniforms

Enlisted men will wear regulation uniforms. During working hours regulation dungarees may be worn and full piece bathing suits may be worn by those handling aeroplanes in the water. Liberty parties will wear the prescribed uniform in entering and leaving the yard. Civilian clothes will not be permitted to be kept in the yard.

Officers assigned to office duty will wear the prescribed service uniform or civilian clothes. Officers in the Flying School and in the Shops may wear flying uniform when engaged in these duties. All officers will keep a complete service uniform in the yard; lockers will be provided for those officers not living in the yard.

Liberty

No liberty list will be made. After working hours all enlisted men not on watch or otherwise restricted may go on liberty. Liberty is up at 7:50 A.M. of the following morning unless extended by special permission from the Captain of the Yard. No liberty granted during working hours except by special permission. A list of those restricted from liberty will be published in the seamen's quarters and at Main Gate.

Barracks

The chief master at arms will have full and absolute charge of Building 25 and the surroundings and will be held responsible for both the sanitation and discipline. All subordinates will carry out his orders promptly, absolutely and rigidly.

Routine

WEEK DAYS

6:30 A.M.	Reveille.
7:00 A.M.	Breakfast.
7:50 A.M.	Muster in and go to work.
8:00 A.M.	Turn to.
11:45 A.M.	Knock off.
12:15 P.M.	Dinner.
12:55 P.M.	Muster.
1:00 P.M.	Turn to.
4:15 P.M.	Knock off.

One of the Curtiss school machines, which is the most popular type here.

5:00 P.M. Supper.
9:00 P.M. All unnecessary lights out, except standing lights.
10:00 P.M. All lights out, except standing lights.

SUNDAYS AND HOLIDAYS
7:00 A.M. Reveille.
7:30 A.M. Breakfast.
8:30 A.M. Muster.
Otherwise same as week days.

Bedding will be aired on Friday by turning the bedding over the foot of the bed and rolling the mattress back at the head of the bed.

Reading Room.—The reading room will be used as a place to keep magazines and to write letters. No loud conversations will be permitted.

Returning at Night.—Men who have been on liberty and return at night will be quiet and not disturb others who have turned in.

Visitors.—Visitors will not be permitted in Building 25 except by permission of the Captain of the Yard, or, in his absence, by the Officer of the Day.

Telephone.—The telephone will not be used for personal conversation.

Turning Out.—All hands will turn out promptly at 6:30 A.M., except on Sunday.

Inspection.—The quarters will be inspected daily by the Officer of the Day.

Late Meals.—Whenever a man or party of men will be delayed for a meal, they will endeavor to notify the C. M. A. A. or Commissary Steward at least half an hour before meal time of the delay and how long they will be delayed.

The Commissary Steward will have charge of the mess gear and mess cooks will sign receipts for same and be responsible.

Field Day.—Field day for general cleaning will be held every Friday.

Mess Tables.—Mess tables and mess benches will be scrubbed and dried in the sun every Thursday.

The Watch

The watch shall consist of two chief petty officers, one of whom shall be a line petty officer, and a designated number of lower ratings. The watch list will be made out in the Captain of the Yard's Office and shall contain a sufficient number of men for sentry duty and to handle the yard fire equipment. The watch list will be published at the 7:50 A.M. muster and the details to sentry post made. No changes shall be made in the watch list without permission from the Captain of the Yard. The tour of duty shall be for 24 hours; during the day the watch may proceed with their regular duties but are subject to a call.

The number of posts and duties of the sentries shall be as prescribed by the Captain of the Yard.

The Captain of the Watch shall notify the watch what sentry duties they have, see that the sentries are posted at 4:30 P.M., shall personally inspect the electric flash light in the watchtower, shall make a thorough inspection of the water front after working hours and shall make at least two inspections of the yard and sentry posts after eight o'clock at night. He shall keep himself informed as to the whereabouts of the Officer of the Day. He shall be familiar with the fire regulations and shall divide the watch in accordance with fire bill.

The janitors of each building and shop will lock up their respective shop after working hours and turn in the keys to the Captain of the Watch. In case of work being done in a shop after working hours the key will be turned in and the Captain of the Watch will lock up when work is completed. The Captain of the Watch will make inspection of all buildings after receiving the keys to see that they are properly secured. The janitors will get the keys before 8 A.M. and open up the shops. The keys of the offices in Building 45 will be kept by the sentry on duty in that building. The watchman on duty will have a set of keys to all buildings. All other keys will be turned in to the Captain of the Yard.

Sentry watches other than marine posts will be stood as follows:

WEEK DAYS	SUNDAYS
4:30 P.M. to 8 P.M.	Regular four hour watch
8 P.M. to midnight.	from midnight Saturday
midnight to 2 A.M.	to 8 A.M. Monday.
2. A.M. to 4 A.M.	Saturdays: During months
4 A.M. to 6 A.M.	of half holidays watch
6 A.M. to 8 A.M.	starts at noon.

The men on watch will be relieved by their reliefs for meals.

The Log

The log book and necessary instruments will be kept in the Radio office. The radio man on watch will fill in the columns. The Officer of the Day will see that the columns are properly kept and after conferring with the Captain of the Yard will write in the remarks. The smooth log will be written up by the yeoman in the Captain of the Yard's office.

Duties of Boat Officer

1. All student naval aviators not assigned to permanent detail in the shops will do duty as boat officer.

2. The tour of duty will start at 8 A.M. on flying days and finish with the completion of flying on that day.

3. The tour of duty will be taken in order of rank. In case a tour of duty as Officer of the Day and as Boat Officer should occur on the same day for same officer, the duty as Officer of the Day takes precedence and duty as boat officer falls on next officer in rank.

4. The Boat Officer is not to leave the dock during

flying hours under any circumstances without a relief.

5. In case he is due for flying he will get the next available officer on detail to relieve him for the necessary time.

6. He will see that speed boat is kept ready and equipped during flying hours. He will keep in touch with the watch tower and go out in the speed boat when called in case of accident.

7. He will report to Officer in Charge of Flying School at 8 A.M., number of boats available for duty.

DUTIES OF NAVAL OFFICER OF THE DAY

1. All commissioned line officers of the station other than those doing duty as heads of departments or flying school instructors will do duty as Officer of the Day, taking turns in order of rank.

2. Tour of duty will be for 24 hours beginning at 8 A.M.

3. The new officer of the day will see that the boats are ready for duty, the speed boats tested out, and other boats started to their stations.

4. He will see that his name plate is posted at Sentry Box, Quarters A, Captain of the Yard's Office and Flying School.

5. He will report to the Captain of the Yard at about 8:30 A.M. for new orders or instructions.

6. He will see that the proper watch detail is made and will let the Captain of the Watch, Commandant's Orderly and Radio Station know where he can be found during the day by telephone.

7. He will be present when mast is held at 11 A.M.

8. He will be present at the 12:50 P.M. and 7:50 A.M. musters, will receive reports and publish orders. All absentees will be reported to the Captain of the Yard.

9. He shall be responsible for the keeping of the flying school log and making the proper entries therein.

10. At 4:30 P.M. he shall make an inspection of the different sentry posts and see that sentries are posted and understand their orders. At 4 P.M. he will report to the Captain of the Yard for orders.

11. He shall see that boats are secured after flying orders.

12. He shall make frequent inspections of the shops during the day and see that no unauthorized persons enter the shops.

13. He shall make arrangements with the C. M. A. A. about inspection of provisions.

14. He will inspect the crew's mess hall before at least one meal a day.

15. He may proceed with his regular duties at the station but must not leave the yard without permission of Captain of the Yard or without a relief.

16. He will make two or more inspections of the yard and sentries during the night, one of which shall be between midnight and 4 A. M.

17. He may get relieved by any other commissioned officer of the station.

18. He must be familiar with the yard regulations, particularly as to fire.

19. Night quarters will be provided for those officers of the day not living in the yard.

20. A desk will be provided for the convenience of the officer of the day in the Captain of the Yard's office.

21. He will not alter or change in any way the orders of instructions of the Marine Sentries.

1. The officer in charge of the Flying School will submit a list of aeroplanes available for use of naval aviators daily to the commandant. These aeroplanes may be used at any time by naval aviators during flying hours, 8 P.M. to 11:45 P.M. and 1 to 4 P.M. Naval aviators will apply to the senior instructor present at the Flying School, who will inform him what aëroplanes are available and ready for use.

2. The course of instruction and required qualifications of personnel for the air service of the navy will be such as is specified in the circular letter issued semi-annually by the Navy Department.

3. In the future the speed boats will be used only for the purpose for which they were bought, i.e., for necessary rescue work. At the direction of the boat officer they may be used in cases of emergencies to prevent aeroplanes drifting ashore or to prevent a collision. Flying at the station will be suspended whenever it happens that there is not at least one speed boat in first class condition equipped for rescue work. During flying hours a hospital apprentice with first aid kit will be continuously on duty at speed boat wharf for an immediate call to go out in the speed boat on duty for the day or to go to any hangar or shop.

CHAPTER XI

SAFETY ORDERS AND REGULATIONS PERTAINING TO THE FLYING SCHOOL AT UNITED STATES NAVAL AERONAUTIC STATION

4. The following revised safety orders for pilots and assistant pilots will be complied with; all flight officers will report in writing to the commandant that they have read and understand these orders.

Procedure Before Flight

(1) Obtain a flight order card signed by the proper authority, and be certain of the exact meaning of the orders thereon.

(2) Note direction of wind and character of gusts; if any previous flights have been made during the day, ascertain if there are any unusual atmospheric conditions. The course to be used will be indicated by flag signals displayed on the observation tower (or near the Flying School beach) a blue flag indicates "Northerly Course," a red flag "Southerly Course." In using the northerly course the circuit of Pensacola Bay is made, making right hand turns; southerly course, vice versa.

(3) Note the number of aeroplanes out with special reference to the Rules of Road, Air, and Beach. Ascertain the course being used.

(4) See that radiator is filled, that oil level is correct, and that ample amounts of oil and gasoline are on board; test oil feed and see that gasoline pump is working.

(5) Inspect thoroughly and test all controls and their leads, and see that they are "hooked up" properly. This is most important.

(6) Receive report from chief mechanician that aeroplane has been inspected in accordance with the prescribed inspection routine and that it is in good condition and is in all respects ready for flight.

(7) See that no loose tools or other articles have been left in or on any part of the aeroplane.

(8) See that motor is warmed up, running properly, and that carburetor adjustments are correct.

(9) See that throttle connections are in good order and that the adjustment is correct for the released position.

(10) Put on and secure the prescribed safety jacket and helmet.

(11) See that the assistant pilot, student or passenger has his safety jacket and helmet on properly.

(12) Examine safety straps, see that they are in good condition and that the releasing device is in good order; adjust straps and put them on.

(13) See that the assistant pilot, student or passenger is properly secured in his seat.

(14) If carrying a passenger, caution him as to interfering with the controls and the foot throttle, and as to remaining secure in his seat.

(15) Do not start from runway until chief mechanician has signalled "All Clear."

The Curtiss seaplane in which Captain Francis T. Evans, U. S. Marine Corps, looped-the-loop on February 14, 1917, and in which Lieut. Edward O. McDonald, U. S. N., duplicated the feat the following day. In the photograph are seen various members of the State Militia detailed for instruction at Pensacola.

(16) The pilot shall inform the assistant pilot of the purpose of the flight.

(17) Make sure that pontoons contain no water and that plugs and hand hole plates are secured.

(18) On aeroplanes equipped with Christenson self starters:

- (a) See that valve from air flash to distributor is open.
- (b) See that air gauge registers the required amount.
- (c) See that starting lever is at the neutral point before starting on flight.

(19) See that air pressure on the fire extinguisher line is up to the required amount.

(20) See that the valve to fire extinguisher air charging line is closed, that the main stop valve is open, and that the supply valve is working freely.

(21) See that gasoline cut-off valve is working freely.

(22) See that shorting button functions properly.

(23) Before starting a scouting flight obtain a compass error card for your compass; see all instruments in proper working order, properly calibrated and set. Barograph reading "zero"; inspect signalling apparatus and see that it is in perfect working order. See armaments in proper shape and ammunition secured; bomb-dropping device properly adjusted and equipped.

Procedure During Flight

(1) All pilots shall familiarize themselves with the prescribed Rules of the Road, Air and Beach.

(2) Upon taking the air, if motor does not develop its proper reserve power, or if there is anything unusual in the action of the aeroplane, land at once. Investigate and correct it, returning to the runway if necessary.

(3) In horizontal flight use only enough power and no less than is required for the normal angle of incidence, except when under orders to make a high or low speed test.

(4) If motor develops any unusual sound while on the water, return to the runway at once and investigate; if sound becomes more pronounced while proceeding to the runway, cut off motor and signal for a tow.

(5) If motor develops any unusual sound while in the air, throttle down and come into a glide; if sound becomes more pronounced, cut off; after landing proceed as prescribed in preceding paragraph.

(6) In flight, when motor misses or dies, come into a glide instantly.

(7) In all glides use approximately the safe angle of incidence prescribed for the type of aeroplane concerned; always hold this angle until time to flatten out for landing.

(8) At all times, either on the water or in the air, note as much and as continuously as possible the behavior of the structure of the aeroplane; in event of dectecting any break, looseness, or defect of any part, or any irregularity of action, return to runway at once and correct.

(9) In the event of fire, turn off pet cocks in gaso-

ROUTINE FLIGHT REPORT.

Began: ....hrs., ...min., ...M.
Ended: ....hrs., ...min., ...M.
Elapsed time: ....hrs., ...min.
Motor stopped: ....hrs., ...min.
1 Operating Time:..........hrs., ...min.

Pilot's weight: ....lbs.
Ass't Pilot's weight: ....lbs.
Av. weight, fuel: ....lbs.
Av. weight, oil: ....lbs.
Extras: ....lbs.
2 Load: ....lbs.

(Began: ....gals.
(Ended: ....gals.
3 FUEL
(Expended: ....gals.
(Av. carried:............gals.
(Began: ....qts.
(Ended: ....qts.

4 OIL
(Expended: ....qts.
(Av. carried: ....qts.

6 Weather: ....
5 Wind Direction: ....**Av. velocity:** ....**M. P. H.**
5 Gusts: ....
7 Sea: ....
8 Altitude: Max.: ....**ft.; Av.:** ....**ft.**

Remarks:

(Sig.).............. ....**U. S.**............**Pilot**

In the office, a continuous record of each student's flying is filed. The facsimile of the flight report card is shown herewith.

A French seaplane in the harbor at Dunkirk just about to be launched.

line line as soon as possible and open cock on extinguisher line—land.

(10) Except when orders on the flight card direct otherwise, all flights will be restricted to the limits prescribed in Beach and Air Rules.

(11) Courses over land or over shallow water must not be made unless an altitude has been attained that will give an ample margin for gliding to deep water in event of failure of the motor.

(12) Except in an emergency, glides will not be started in a direction towards land, or from an altitude less than 500 feet.

(13) In event of landing, the following signals shall be made by either occupant of the aeroplane facing the lookout station and standing in a position unmasked by the motor:

(a) Signal: Waving one arm from vertical to horizontal position—meaning: "Cannot return under power; require tow."

(b) Signal: Waving both arms from vertical to horizontal position—meaning: "Emergency; send boat as quickly as possible."

The above signals will be repeated at intervals until a boat is seen approaching.

(14) The general recall is a large rectangular canvas flag, checkerboard red and white, rolled down on south end of machine shop roof; when this is shown, all aeroplanes out, whether on the water or in the air, will at once return to their runways.

(15) After turning over the controls to the assistant pilot or to a student, direct him as to course and altitudes and be prepared to resume control instantly. When carrying a student, conform at all times to the prescribed system of flying instruction.

(16) When making a glide from high altitudes, cut in the power momentarily at intervals in the approved manner for the type motor used, to prevent its choking up with oil and to provide that power will continue available; in event of feeling indisposed by the change in barometric pressure, cut in power and hold the altitude for a few minutes then resume the glide.

(17) Side slip spirals are prohibited.

(18) In gusty weather or in doubtful air conditions, when possible, land under power.

(19) Execute no turning manœuver before reach-

A French seaplane of the flying boat type preparatory to its flight. See chapter on "Submarine Hunting from Aircraft" for report of Admiral Lacaze, the French Minister of Marine, regarding the extensive employment of seaplanes for coast patrol work.

ing an altitude of at least 300 feet, except where local conditions make turning safer than continuing a straight climbing path. The spirit of this order is that the pilot shall endeavor to make a straight flight into the wind to an altitude of at least 300 feet before commencing a turn.

(20) Do not attempt any unusual performance unless proper authority has been obtained.

(21) Under all circumstances in the air avoid attitudes that either in climbing, horizontal flight, are on the verge of stalling.

(22) Returning from flight approach the runway at slow speed; be ready to cut out the spark. Remember that the lives of the men waiting to handle your machine may depend on your presence of mind.

Procedure After Flight

(1) Fill out the back of flight order card and send it to the office it was issued from; if flight order requires a flight report make it out at once and send it to the commandant's office.

(2) If any part of the aeroplane or its equipment has developed defects or irregularities of operation fill out data required on a trouble report blank, giving the details of the trouble, and turn in to the officer in charge of your aeroplane division.

(3) If the aeroplane is ordered out with a different pilot, inform him of any peculiarities of its handling, and of any irregularities that he may meet in the local atmospheric conditions.

(4) If the aeroplane is ordered into its hangar, the officer to whom it is assigned, unless absent with proper authority, will see that the crew cleans and secures properly.

(6) When any overhauling, repairs, alterations, renewals, or adjustments are to be made in any part of an aeroplane or its flying equipment, the officer to whom it is assigned shall personally inspect the progress of the work; at its completion he shall fill out the work report (N. O. A. No. 4).

Procedure in General

(1) The pilot's responsibility and authority, irrespective of his rank relative to the assistant pilot, student, or passenger, begins when the aeroplane is being inspected in preparation for the flight, and does not end until the aeroplane is in its hangar, or is formally turned over on its runway to another pilot or to the officer to whom it is assigned.

(2) Any pilot that has any reason to believe that his physical or mental condition, on account of fatigue or any other reason, is not entirely up to its usual standard, before undertaking a flight must report the fact to the officer signing the flight card.

(3) Student ~~naval~~ aviators flying alone will not in any given week fly at 500 feet higher than the highest attained by them in the previous week.

(4) Student naval aviators who have already attained to altitude of 3000 feet may fly at this altitude if they so desire, but will not increase it from week to week by steps greater than 500 feet.

(5) Student naval aviators are strictly enjoined from attempting to make rapid climb until they are so authorized—this to prevent danger of stalling.

(6) No student naval aviators, except those who have qualified under the supervision of the officer in charge of the Flying School, will attempt to make whole or partial spirals unless forced to do so.

(7) In addition to contents of previous orders, student naval aviators will in the future, until further orders, be restricted in their flying as follows:

(a) Flights shall be limited as prescribed by officer in charge of Flying School.

(b) A ratio of climb of 200 feet per minute shall not be exceeded for the first 1000 feet; in other words that altitude, 1000 feet, shall not be attained in less than 5 minutes. A noticeably steep angle of climb shall never be used.

Competition and the spirit of rivalry among students is discouraged. Careful and conservative flying is desired. Accidents are either caused from carelessness, recklessness, lack of information, pride on the part of the aviator, and, to a far less degree, by mechanical defects in the construction of the aeroplane.

Safety Orders for Mechanicians—Procedure Before Flight

(1) The chief mechanician will make a careful inspection, in the approved manner, of all accessible parts of the aeroplane's power plant, controls with their leads, and instruments and equipment; also note quantities of oil, fuel, and circulating water on board, and shall see that pontoons are free from water and that drain plugs and hand hole plates are properly secured.

(2) The second mechanician shall make a careful inspection, in the approved manner, of all accessible parts of the aeroplane's structure.

(3) Any unusual condition noted in the above inspection shall be referred to the pilot about to make the flight, no matter how unimportant that condition may seem.

(4) Special care shall be taken that all loose tools or other loose articles are removed from all parts of the aeroplane before motor is started.

(5) When inspections are completed and all is found satisfactory, the chief mechanician shall report "Ready" to the pilot about to make the flight, notifying him as to quantities of fuel, oil, and water on board, condition of starter, fire extinguisher and other equipment.

(6) After the motor has been started and all men of the crew are stationed as directed by the chief mech-

anician, he shall hold up his right hand as a signal to the pilot "All Clear."

(7) No person shall be permitted under any circumstances to pass under the lower plane or to stand in line of the propeller blades after the motor has been started or while there is a possibility of its being started.

(8) Before aeroplane leaves runway note if pilot and passenger have adjusted their safety straps, and if either one has not done so invite his attention to the fact.

Procedure During Flight

(1) The chief mechanician shall detail one of his crew to the watch tower to keep his aeroplane in sight until it is on the water returning to the runway, in case there is no regular tower watch.

(2) The other members of the crew shall remain in the vicinity of the hangar ready to receive their aeroplane upon its return to the runway.

(3) In aeroplanes that are capable of carrying a passenger, the chief mechanician shall be taken on the first flight each day for observation of its behavior in the air, with special reference to the power plant operation. The other members of the crew shall be given flights as often as convenient.

Procedure After Flight

(1) Upon return of the aeroplane to the runway, the inspections specified in paragraphs 1 and 2 of PROCEDURE BEFORE FLIGHT shall be repeated. In addition, the pontoon shall be carefully examined for indications of leaks or damages that may result in leaks.

(2) Fuel, oil, and water shall be replenished as required.

(3) Minor repairs and readjustments as directed by the officer assigned to the aeroplane shall be made and recorded.

(4) See that the valve stems, push rods and controls are well lubricated, especially after a long flight.

(5) Upon completion of the aeroplane's last flight for the day, in addition to the above, the aeroplane shall be carefully cleaned and dried down; oil, grease, and finger marks removed from fabric and varnished woodwork with soap and fresh water or with the approved cleaning compound—care being taken to wash off all soap; treat bare metal parts with oil or vaseline as specified, being careful not to get any of it on the fabric or other parts where not required. Put on motor cover and see that vents in oil and fuel system are closed; open hand hole plates and allow interior of pontoon to dry.

(6) After each 5 hours' flight drain the oil out of the crank case and give the motor the kerosene treatment as specified for its type; put in fresh supply of oil.

Morning Routine

(1) Dust off all parts of aeroplane thoroughly before removing motor cover and vent covers.

(2) Check up valve timing and interrupter gap, readjusting to specified clearance if necessary.

(3) Check up propeller lock nut adjustment and readjust if necessary.

(4) Make a careful examination for indications of leaks over night in fuel, oil, and water service.

(5) Make careful examination for development of rust on wires and fittings; where this is found the rust shall be scraped off to bare metal and a coating of vaseline applied.

(6) Supply aeroplane with fuel, oil, and water as required.

(7) Carry out the provisions of paragraphs 1, 2, 3, and 4 of PROCEDURE BEFORE FLIGHT. In addition, the chief mechanician shall repeat the inspection of the second mechanician.

(8) Oil valve stems, push rods, and controls.

(9) See all clear for turning over motor; start motor and allow it to idle for at least five minutes and until cylinders are warmed up, then very gradually increase to full speed; note operation of motor, readjusting carburetor to specified adjustment if necessary; when motor runs properly at all speeds, report to the officer to whom the aeroplane is assigned that it is ready for service.

Seaplanes being used by Belgians in the campaign in East Africa on the shore of Lake Tanganyika.

General Rules

(1) Attention of all mechanicians is called to the fact that their duties in connection with aeroplanes and power plants is fully as important as regards efficiency and safety in flight as the duties of the pilot. Inspections before and after flight shall always be made in accordance with the prescribed Inspection Routine for the types of power plant and aeroplane concerned.

(2) Chief mechanicians shall see that the rule is strictly enforced that no members of the crew shall engage in any occupation or conversation not directly connected with the work on the aëroplane during the inspections before and after flight and during morning routine.

(3) No mechanician shall smoke while engaged in work on an aeroplane.

(4) Gasoline shall be stowed only in the authorized receptacles and in specified localities; these receptacles shall be kept closed tight when not in use.

(5) Smoking inside hangars is prohibited.

(6) No blow torch shall be used during the filling of the gasoline tanks or when gasoline receptacles or vents are open or where there is any possibility of flame from the torch causing a gasoline or oil fire.

(7) None other than the specified adjustments of any part of the aeroplane or the power plant shall be made unless authorized by proper authority; all changes of adjustments will be recorded in the work report (N. O. A. No. 4).

(8) Mechanicians shall not make any private collection of tools or spare parts for use in effecting repairs. Only such wire, bolts, nuts, cotter pins and other material that is issued from stock shall be used for replacements in any part of the aeroplane or power plant. Cotter pins or safety wires shall never be used more than once.

(9) Mechanicians will not touch any part of an aeroplane or its equipment to which they are not assigned without the consent of the chief mechanician in charge of it.

(10) Visitors to hangars will not be permitted to handle any part of an aeroplane or any of the apparatus in the hangar.

(11) All mechanicians shall aid in enforcing the Station Regulations relative to admission of visitors in hangars and shops, and shall require all strangers or unauthorized persons to show the standard pass signed by the commandant or the Captain of the Yard before allowing admission to hangars, shops, Flying School beach and offices.

Regulations for Flying School and Reports to Be Made

Name	*Information Contained*	*Disposition*
Record of flights of Individuals.	Date, No. of flight, aeroplane, length, height, glide, wind, temperature, gusts, nature.	1—Officers personal record. 1—Station files, Flying School. 1—Central Office.
Flight Card.	Same as above, also orders, water, fuel, weights carried.	1—Flying School.
Hangar Aeroplane Log.	Record of each flight and total hours; record of work on aeroplane. Hangar Motor Log.	Kept in hangar in rough.
Hangar Motor Log.	Record of each motor run, oil, gas, total time and overhaul work.	Kept in hangar in rough.
Aeroplane Log.	Record of each flight, aeroplane No., motor No., flight No., date, duration of flight, weather and air conditions, initial of pilot, nature of flight.	1—Central Office. 1—Aeroplane Log, Flying School. 1—Motor Log, Flying School. 1—Operations.
Motor Log.	Same as Aeroplane Log.	1—Motor Log, Flying School, as above.
Aeroplane Work Report.	Page, nature of work, weekly, aeroplane No., item No. dates, name of part, description of work, reason, authority, inspector's initials.	1—Aeroplane Work Report Book, Flying School. 1—Central Office.
Motor Work Report.	Same as Aëroplane Work Report, except Motor Number.	1—Motor Work Report Book, Flying School.
Trouble Report.	File No., flight No., aeroplane or motor No., part, name and location, record of trouble, Trouble Board and Recommendations.	1—Central Office. 1—Aeroplane Trouble Report Book, or Motor Trouble Report Book, Flying School. 1—Bureau concerned. 1—Central Office. 1—Operations. 1—C. O. North Carolina.
Semi-Monthly Report of Flying.	Names of officers and men, number of aeroplanes, time in air since last report, total time to date.	1—Bureau of Nav. 1—Central Office. 1—Flying School files.
Weekly Report of Flying School Operations.	Letter, telling aeroplanes available, flights made, hours of flight, remarks.	1—Commandant. 1—Flying School files.
Daily Report of Aeroplanes.	Aeroplanes available for use; aeroplanes available for use naval aviators.	1—Commandant. 1—Flying School files.
Gas and Oil Report.	Gas and oil served out by gas house each day—Installed by officer in charge gas and oil.	1—Planning Division. 1—Flying School files.
Weekly Requisitions.	Supplies required by aeroplanes for upkeep and repair.	1—Planning Division.
Monthly Inventory of Tools.	Tools on hand and tools lost or worn out during month for each aeroplane.	1—Planning Division. 1—Flying School files.

Reports are to be made as follows:

Aeroplane Work Reports, by officer in charge of aeroplane, and turned in on Monday A. M. They will be typewritten and put on file in Flying School for signature.

Aeroplane Log Sheets, by officer in charge of aëroplane, and turned in as soon as page is finished—9 flights.

Individual Flight Reports will be filled in after each flight. When page is full—17 flights—turn page over to Flying School yeoman who will typewrite copies for signature and filing.

Trouble Reports, by officer in charge of aeroplane, as soon as aeroplane reaches beach and brought to senior officer of Trouble Board on beach for filling in "Cause and recommendations."

Student naval aviators on reporting to Flying School will read over all Station Orders, which are to be found on file in office, paying special attention to orders regarding flying, Rules of Air and Beach, safety orders and inspections. Also students will obtain from office an individual flight record book and start same.

When assigned to duty in charge of an aeroplane, an officer will assure himself that the aeroplane's records are up to date, and that tool inventory is correct and show a full allowance. He will familiarize himself with the type of aeroplane, its routine operation and maintenance, all orders relative to daily inspections and inspections before and after flight. He will see that hangar records of work on aeroplane and motor, of flights and of gasoline and oil used are carefully kept by the chief mechanician.

Memoranda:

CHAPTER XII

RULES FOR FLYING ISSUED BY BRITISH ROYAL FLYING CORPS

1. *Aircraft Meeting Each Other.*—Two aircraft meeting each other end on, and thereby running the risk of a collision, must always steer to the right. They must, in addition to this, pass at a distance of at least 100 yards.

2. *Aircraft Overtaking Each Other.*—Any aircraft overtaking another aircraft is responsible for keeping clear and must not approach within 100 yards on the right or 350 yards on the left of the overtaken aircraft, and must not pass directly underneath or over, save when the vertical distance is in excess of 800 feet. No aircraft shall remain persistently below or above another. In no case must the overtaking aircraft turn in across the bows of the other aircraft after passing it or move so as to foul it in any way.

3. *Aircraft Approaching Each Other in a Cross Direction.*—When any aircraft are approaching one another in cross directions, then the aircraft that sees another aircraft on its right-hand forward quadrant—from 0 degrees (i.e., straight ahead) to 90 degrees on the right-hand constitutes the right-hand forward quadrant—must give way, and the other aircraft must keep on its course at the same level till both are well clear.

4. *Distance to be Maintained from Airships.*—When one of the aircraft is an airship, the distance of 100 yards prescribed above shall be increased to 600 yards.

5. *Long Glides and Quick Rises.*—Except when prearranged for instructional purposes or in cases of emergency, long glides and quick rises will be practised only to and from the usual landing area.

6. *Position of Other Aircraft to be Noted Before Starting.*—Aeroplane pilots will, when starting, carefully note the position of other aircraft and will be responsible for keeping clear of them.

7. *Danger Flag to be Hoisted Before Aeroplane Flying Commences.*—No aeroplane flying will take place without a red flag being hoisted at the appointed place as a warning to all concerned. In cases where the flag is likely to be mistaken for other danger flags, the flag of the Royal Flying Corps will be hoisted immediately below the red flag.

When the flag is flying, no unauthorized persons are to be allowed in the prohibited area.

8. *Officer Responsible for Regulation of Aeroplane Flying.*—The senior officer belonging to an aeroplane squadron of the Royal Flying Corps (Military Wing) present on duty in the landing area will be responsible for the control of the flying of all aeroplanes using an aerodrome or landing area reserved for War Department use (except aeroplanes flying at Farnborough under the control of the Superintendent of the Royal Aircraft Factory), and persons on duty in connection with such flying.

9. *"Stop" Signal.*—The "stop" flag (International Code flag "S," i.e., a square white flag with a blue square in the center) will be the signal for all aeroplanes in the air to return to the landing area; it will be hoisted when necessary by the order of the officer referred to in paragraph 8. This officer may also suspend any one from flying pending enquiry.

10. *Rolling Practice.*—Rolling practice will not take place on the landing area whilst aeroplanes are flying.

11. *Beginners Practice Area.*—Beginners will be restricted to such area as may be prescribed by the officer referred to in paragraph 8.

12. *Landing Marks.*—Permanent marks will be made on the ground at the usual landing place to indicate the nearest points at which it is safe for aeroplanes to land in directions facing the sheds, etc. An aeroplane landing in such a direction must be on the ground before it reaches the point in question.

13. *Flying over towns.*—Flying unnecessarily over towns and villages is to be avoided.

14. *Dogs.*—No dog not on a leash is allowed in the starting and landing area while flying is in progress.

CHAPTER XIII

TRAINING OF AVIATORS

The training of aviators is fast becoming an exact science. When early in the war aviators were needed in large numbers and were employed mainly for scouting, aerial coast patrol and spotting of shots even the European navies were willing to forego practically all the qualifications apart from flying. Later, as the duties of aviators increased rapidly, and the shortage of trained men made it necessary to "break in" civilians, their training had to be carried out on scientific lines.

Whenever the personnel available is untrained in naval matters, it is necessary to teach the students the rudiments of naval discipline and naval regulations as well as aeronautics. Great Britain has been obliged to do so to obtain military aviators, and the British system, which has been adopted by her Allies and Canada, is undoubtedly the best system to follow to-day.

General W. S. Brancker, R.A., Director of British Air Organization, has given a clear idea of the extent to which civilians are considered suitable for service at the front. He says in part:

"The civilian who wishes to join the Army or Navy Air Service in Great Britain or Canada at present has first to join the Service as a cadet and go through a course in the cadets' school, at which military subjects as well as aeronautic subjects are taught. He gets a grounding in drill and discipline, care of arms, interior economy, military law, and the use of the machine gun; this course lasts about two months. From this the cadet is sent to a Flying Corps Training School, School of Military Aeronautics, where he begins his technical training on the ground. In Canada, and in some cases in England, he gets the first mentioned military training at the same time as he gets the rudimentary training in flying or operation of dirigibles and observation balloons. He goes through a course in the care of engines and rigging, is given some ideas of the theory of flight, and is taught wireless signaling and receiving.

"He gets instruction in the care of machine guns, in the use of the camera, in map reading, in the observation of artillery wire with models, and in his spare moments he gets a certain amount of drill. This course lasts another two months, and if he gets through this successfully, he is given a commission on the General List. He then joins a preliminary training squadron as a pupil and starts his instruction usually on the slow Maurice Farman aeroplane, his training both in military and technical subjects going on concurrently. After reaching a certain standard of efficiency and having completed a certain number of hours in the air, he is sent on to an advanced training squadron or service squadron, where he learns to fly Service types of machines for military purposes, and eventually qualifies for his wings. He is then gazetted as a flying officer of the Royal Flying Corps and posted to a service squadron. If he shows exceptional promise as a pilot after his qualification, he is sent to the Central Flying School, where he is given extra higher instruction on flying scouts. During the period of advanced training, he goes through a course of aerial gunnery away from his squadron. The total time in the air usually required to reach the qualification stage is about thirty hours' solo in present circumstances, but, of course, the length of time that it takes to reach this standard depends entirely on the weather and the number of aeroplanes available. During the winter it works out to about four months, but in the summer it is considerably shorter."

All this may seem a long process, but it is doubtless the best and will prove the shortest in the end in producing well-trained aviators.

CHAPTER XIV

COURSES OF INSTRUCTION AND REQUIRED QUALIFICATIONS OF PERSONNEL FOR THE AIR SERVICE OF THE UNITED STATES NAVY

1. In accordance with the department's order of April 10, 1913, the following instructions for the training of officers and enlisted men for the Air Service are issued:

(*a*) Classes of officers and men to be trained for the Air Service will be detailed every three months beginning January 1, 1916.

(*b*) The course of instruction will not exceed 2 years for officers and 18 months for enlisted men.

(*c*) Only officers and men who hold certificates of qualification as herein prescribed or have heretofore qualified will be eligible to detail for duty in aircraft in actual service.

(*d*) The classes of officers will be composed of eight line officers. The officers must have served at least two years in seagoing ships.

(*e*) An officer desiring instruction in aeronautics must make official application and pass the physical examination prescribed by the Bureau of Medicine and Surgery. The senior officer present will at once have applicants examined physically and forward the report of such examination with application.

(*f*) The classes of enlisted men will be composed of:

- Eight chief petty officers, seaman branch.
- Two chief petty officers, preferably machinist's mates.
- Two petty officers, first class, preferably carpenter's mates.
- Two petty officers, second class, preferably electrician's or gunner's mates.
- Two seamen.

(*g*) Enlisted men, to be eligible for this duty, must have had at least two years' service in a seagoing ship, must be under forty years of age, and must be recommended by their commanding officers on account of their very good record.

(*h*) The eight chief petty officers of the seaman branch will be trained to steer aircraft and will be required to pass the same physical examination required of officers detailed to aeronautic duty. The remainder of each class will be trained in handling aircraft machinery.

(*i*) The commander in chief, Atlantic Fleet, will select these classes of enlisted men, and all requests for this detail should be made to him.

(*j*) A senior officer present may act immediately upon requests for instruction in aeronautics if away from an Atlantic home port and facilities exist for carrying on the instructions under his command.

(*k*) Each officer or man regularly ordered to duty involving actual flying in aircraft will be given orders by the commanding officer of his ship or station when he takes up that duty, appointing him as a student or qualified aviator or as a student or qualified airman involving actual flying in aircraft in accordance with act of Congress passed March 3, 1915. These orders will be forwarded to the department for approval before extra compensation is paid.

(l) Officers detailed for aeronautic duty will be classed as, viz.:

- Student naval aviators.
- Naval aviators.
- Navy air pilots, aeroplane.
- Navy air pilots, dirigible.
- Military aviator.

(*m*) Enlisted men detailed for aeronautic duty will be classed as, viz:

- Student airmen.
- Airmen.
- Quartermasters, aeroplane.
- Quartermasters, dirigible.
- Machinists, aeronautic.

These classifications designate the duty that enlisted men are qualified to perform in aircraft and do not affect their regular ratings in the service.

(*n*) This circular will be revised each six months. For this purpose the commandant of the United States Navy Aeronautic Station, Pensacola, will convene a board of naval aviators on the 1st of June and 1st of December to recommend the necessary corrections and revision for publication of this circular on the 1st of July and 1st of January, respectively. Thus the requirements should be established every six months to keep up with progress, and also anticipate the most probable progress of the near future, so that the requirements will be possible of accomplishment by those officers and men detailed for training on the date of issue of the circular.

2. Course of Instruction of Student Naval Aviators

(1) Upon reporting at the aeronautic station student naval aviators will supply themselves with such

textbooks as are prescribed by the commandant of the station.

(2) The course of instruction begins in aeroplanes and will be grouped under the following heads:

(*a*) Shopwork.
(*b*) Lectures.
(*c*) Flying lessons.
(*d*) Elementary flying.
(*e*) Advanced flying.
(*f*) Aircraft station administration.
(*g*) Examinations.

(3) Aeroplane shopwork will be divided into three parts: (*a*) Machinery, (*b*) Structural, (*c*) Instrument. They will be required to do the actual work as far as possible with necessary advice from the instructors.

(*a*) Machinery work: Disassembly, reassembly, installation, and adjustment of all parts of each type of machinery plant at the station.

(*b*) Structural work: Disassembly, reassembly, installation, and adjustment of each type of aerodynamic instrument at the station.

(4) *Lectures.*—The officer in charge of Flying School will lecture or arrange for lectures once each week. These lectures will be prepared so as to assist in the progress of the course of instruction, also to stimulate original thought and development. Copies of lectures will be furnished the students.

(5) *Flying Lessons.*—During their first week at the station students will be given occasional flights as passengers for the purpose of giving them an idea of the "air feel" and for general observation of the handling of an aeroplane. The actual flying lessons will then begin and progress through the following:

(*a*) Adjustment of safety jacket and straps.

(*b*) Inspection of machine required in safety orders.

(*c*) Handling controls while machine is on the ground.

(*d*) Handling controls in straight horizontal flights.

(*e*) Handling controls and throttle in straight horizontal flights.

(*f*) Turning, right and left.

(*g*) Figure eights.

(*h*) Straight glides.

(*i*) Get-aways.

(*j*) Landings, with and without power.

(*k*) Spirals.

(*l*) Rough-weather flying.

(*m*) Taking care of all casualties or unusual conditions that will be encountered in flight that can be safely demonstrated by the instructor.

At least once each week during the first month the student is flying, and thereafter once each month, he shall be examined physically immediately after a flight by the medical officer of the station, who will keep a

1 Six Cylinder Vertical Water Cooled Motor (Hall Scott).
2 Motor Exhaust Pipes to protect the pilot from gases.
3 Tractor Propellor directly connected to motor
4 Vertical Radiators, one on either side of fuselage, for cooling the water, circulating around the cylinders.
5 Passengers Cock-Pit (located at the center of gravity).
6 Pilots Cock-Pit (behind the passengers seat).
7 Vertical Finn or Stabilizer which keeps the aircraft on a straight course.
8 Rudder Hinged to the Vertical Finn (for steering to right or left).
9 Tail Float (keeps the tail of the fuselage out of the water).
10 Main Float or Pontoon, Martin type (supports the aeroplane on the water).
11 Pontoon Supports (keep the float in position).
12 Step (to assist in getting out of the water).
13 Left Wing Float.
14 Right Wing Float (keeps the wings from striking the water on landing).
15 Wing Float Wire Bracing (to keep floats in position).
16 Small Central Wing Panel or "Engine Panel."
17 Upper Right Wing Panel.
18 Upper Left Wing Panel.
19 Lower Left Wing Panel.
20 Upper Left Wing Flap (to restore equilibrium).
21 Upper Right Wing Flap.
22 Lower Right Wing Flap.
23 Lower Right Wing Flap Lever.
24 Upper Right Wing Flap Lever (for operating the ailerons).
25 Wire Connections between Upper and Lower Ailerons (to cause them to work in unison).
26 Horizontal Stabilizer or Tail Plane (to keep aeroplane on straight course).
27 Right Outer Forward Interplane Strut.
28 Left Outer Rear Interplane Strut (for spacing planes).
29 Interplane Bracing Cables (lift wires) (for carrying lift).
30 Interplane Bracing Cables.
31 Interplane Bracing Cables (incident wires) (for bracing between forward and rear struts).
32 Pontoon Guys or Bracing Wires.
33 Metal Cowling of the Fuselage (to guard against fire).
34 Fuselage or Body (covered with fabric simiiliar to wings).

1 Center Wing Panel (between motor-carrying struts).
2 Left Upper Wing Section (between motor-carrying struts and outer struts).
3 Left Overhanging Upper Wing Section (extending beyond the outer struts).
4 Left Wing-Flap or Aileron (for lateral or rolling control).
5 Right Upper Wing Section.
6 Right Overhanging Wing Section (the above mentioned surfaces comprise the lifting area of the upper plane).
7 Wing-Flap Control Arms (for operating wing flaps).
8 Wing-Flap Control Cables (connecting ailerons with pilot's control).
9 Overhang Bracing Wires.
10 "Non-Skid" or "Side-Slip" Planes (to check skidding or side-slipping).
11 Masts to Support Side-Slip Planes and Carrying Overhang Bracing Wires.
12 Lower "Sidewalk" Section, extending from the hull of the motor-carrying struts.
13 Lower Left Wing Section.
14 Right Lower Wing Section (the surfaces 12, 13 and 14, comprise the lifting area of the lower plane).
15 Central Interplane Struts.
16 Motor-Carrying Interplane Struts.
17 Front Intermediate Strut.
18 Rear Intermediate Strut.
19 Front End or Outer Strut.
20 Rear End or Outer Strut (the struts mentioned above are part of the system of bracing between the planes).
21 Fore-and-Aft Wing Bracing Cables ("Incidence Wires").
22 "Lift Wires" (to transmit the lifting effect of the wings to the hull).
23 Landing Wires (to brace the wings in landing).
24 Wing Pontoon or Float (to keep the wing tips clear of the water).
25 Left Propeller — revolves in clockwise direction when viewed from the rear. Propellers being forward of the planes are in "tractor" position.
26 Right Propeller — rotates in anti-clock direction.
27 Curtiss Motor 100 h. p. ("V" Type, 8 cylinders).
28 Copper Tips on Propeller to prevent splitting.
29 Propeller Hub connecting propeller directly to the crankshaft of the motor.
30 Radiator for cooling water which circulates around the cylinders of the motor.
31 Gasoline "Gravity Tank." (Supplied from the main fuel tank by means of a windmill pump).
32 Outrigger. This supports the tail group.
33 Tail Plane or Horizontal Stabilizer to keep a machine on a direct course.
34 Tail Flap or Elevator for pitching control or ascending and descending.
35 Vertical Fin to assist in maintaining a straight course.
36 Rudder for steering to the right and left.
37 Tail Braces which run from the hull to the tail plane to support the tail group.
38 Boat Hull. Frame and keel of ash, planking of mahogany and cedar, covered with canvas.
39 Planing Fins to assist the machine in rising from the water.
40 Metal Tube Bracing from the hull to support the motor-carrying struts.
41 Supports for the Planing Fins.

careful record of his physical condition, and report to the commandant if there is any change that is detrimental to flying.

When an instructor is satisfied that one of his students is far enough advanced in the above items of training to fly alone, he will notify the officer in charge of Flying School, who will take one or more flights with the student; if he is then satisfied with the student's ability and the medical officer reports him physically fit, he will be permitted to fly alone and go ahead with elementary flying.

(6) *Elementary Flying.*—During this period the student will make only such manœuvers and stay in the air for such periods as are directed by his "flight orders." The following will be covered in order:

(*a*) Get-aways, straight courses, turns and landings into the wind; good weather.

(*b*) Figure eights and landings with the wind; good weather.

(*c*) Spirals; good weather.

(*d*) Higher altitude flying; good weather.

(*e*) Rough weather flying.

(*f*) Course by compass.

(*g*) Endurance flights.

During this period, when sufficiently advanced in the shop course, the student will have an aeroplane assigned to him for care, preservation, and keeping its logs and records. When his instructor and the officer in charge of the Flying School are satisfied as to his progress in elementary flying and as to his practical and theoretical knowledge as shown by the examinations he will be permitted to go ahead with the advanced flying under the supervision of the officer in charge of the Flying School.

(7) *Advanced Flying.*—

(*a*) Starts from catapult.

(*b*) Landings in deep-sea waves.

(*c*) Bomb-dropping practice.

(*d*) Flying in formation.

(*e*) Sending and receiving radio messages in the air.

(8) *Station Administration.*—During the period the student is carrying on advanced flying work he will be relieved from charge of an aeroplane and will go through the following details:

(*a*) Subinspector of machinery work.

(*b*) Subinspector of aeroplane work.

(*c*) Assistant planning superintendent.

While on these details he will familiarize himself with the following subjects:

(*a*) Inspection of repairs, alterations, assembly, and test of machinery.

(*b*) Inspection of repairs, alterations, assembly, and test of aeroplanes.

(*c*) Routing and filing correspondence.

(*d*) Planning division's methods of originating work.

(*e*) Preparations of specifications for requisitions.

(*f*) Preparation of requisitions and purchase of material.

(*g*) Accounting methods.

(9) *Examinations.*—Written examinations in theoretical and practical aviation will be given once each month, the questions being prepared by the officer in charge of the Flying School and approved by the commandant. In theory the questions will be limited to subjects discussed in the prescribed textbooks and lectures. The practical questions will be limited to subjects involving shopwork and practical flying.

Students will be marked monthly as follows:

Subject.	*Weight.*	*Marked by—*
(1) Adaptability for Air Service	3	Officer in charge of Flying School and instructor assigned to.
(2) Bearing and conduct.	2	Commandant.
(3) Flying	2	Officer in charge of Flying School and instructor assigned to.
(4) Practical knowledge.	2	Instructor assigned to.
(5) Written examination.	1	Commandant and officer in charge of Flying School.

Marks will be given on a scale of 4, anything below 2.5 being unsatisfactory. Any student whose average is unsatisfactory in any subject at the end of any month will be reported to the department and may be recommended for detachment. Monthly marks will be averaged, and the result will be the official mark for the student naval aviator's course. After passing the final written and practical examinations, and after having had at least 50 hours' flying, students, upon recommendation of the officer in charge of Flying School and approval of commandant will take the flying tests prescribed for qualification as naval aviators.

(10) *Flying Tests for Qualification as Naval Aviator.*—The tests will be conducted by a board of not less than two naval aviators, designated by the commandant, and the following will be done, in the order named:

(*a*) Climb to an altitude of 6,000 feet, as shown by a recording barograph, and glide with motor idling to a normal landing within 200 feet of a mark previously designated by the board; horizontal flights to be re-

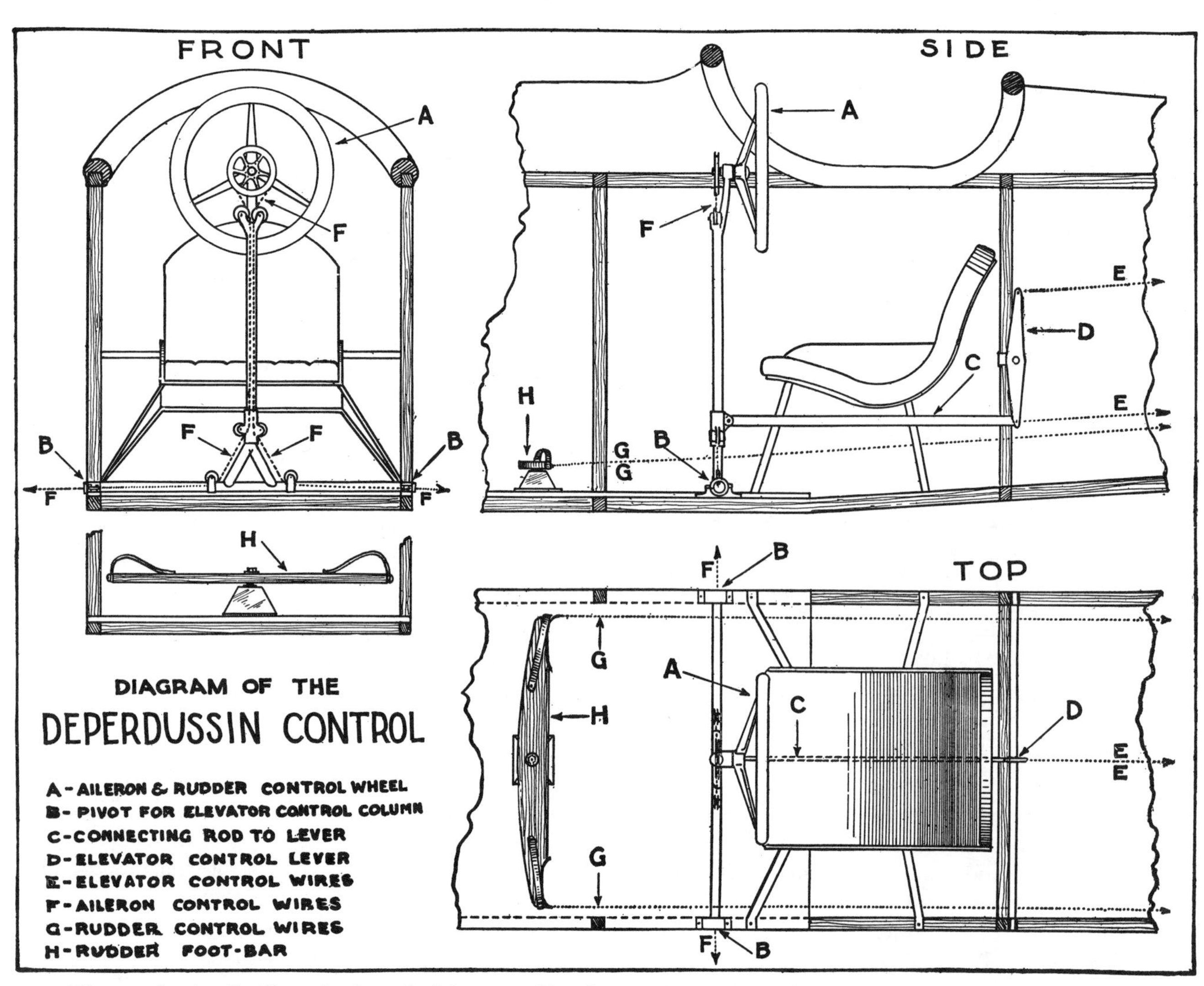

Diagram showing the Deperdussin method for controlling the aeroplane, adopted by the United States Army and Navy.

sumed twice during the descent, but not within the last 1,000 feet.

(*b*) Make a spiral glide with motor cut off (stopped) from an altitude of 3,000 feet, as shown by a recording barograph, and make normal landing within 200 feet of a given mark previously designated.

(*c*) Make a landing in a seaway where height of wave is at least 3 feet, without damage to any part of aeroplane.

(*d*) Make a straight course and return between two objects not less than 5 miles apart in a wind of not less than 20 miles per hour and not more than four points forward or abaft the beam, in order to demonstrate ability to maintain a given course.

(*e*) Demonstrate to the satisfaction of the board ability to fly in very bad weather.

(*f*) Start a flight from the catapult after personally making all adjustments.

Upon the completion of these tests the officer will be designated a naval aviator, and be eligible for further training in aeronautics to qualify as a Navy air pilot or as a military aviator by taking a course at the United States Army Aviation School.

3. Course of Instruction of Naval Aviators for Navy Air Pilots (Aeroplane)

(1) Only commissioned line officers who have qualified as aviators will be detailed to this course.

(2) The commandant of the aeronautic station will recommend for this course of instruction aviators who have qualified, the recommendation to be based upon his opinion of their aptness for this course in the proportion of three-fourths of each class qualified as aviators.

(3) The course of instruction shall consist of:
(*a*) Taking sights in aeroplanes.
(*b*) Working out sights while flying.
(*c*) Compensating aeroplane compasses.
(*d*) Installing aeroplane compasses.
(*e*) Open-sea scouting flights.
(*f*) Solution of scouting problems.
(*g*) Controlling the fire of the guns of an aeroplane.

(4) Upon the completion of this course the naval aviator for Navy air pilot (aeroplanes) shall be given

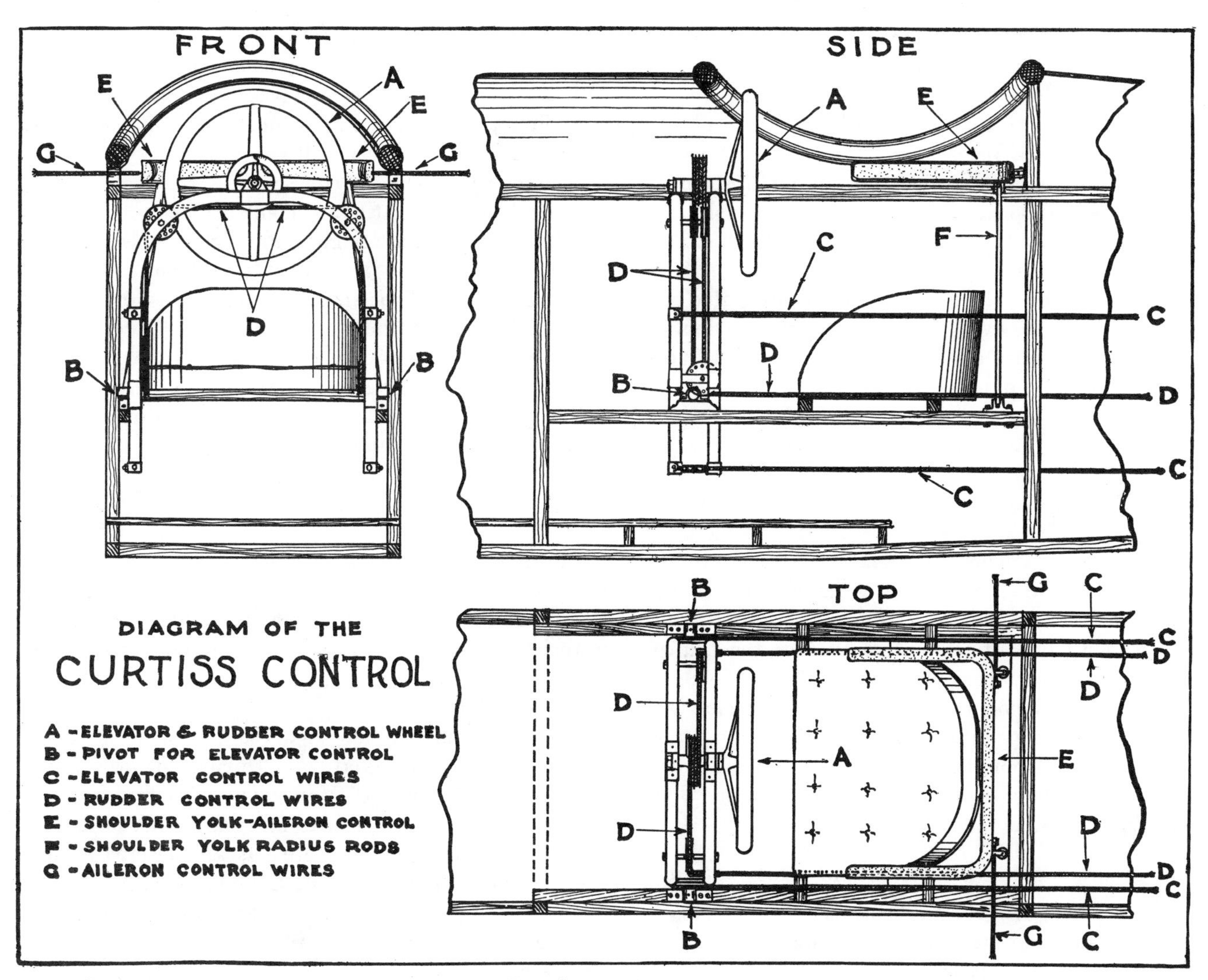

Diagram showing the Curtiss method for controlling the aeroplane. Also used by the United States Army and Navy.

an examination, theoretical and practical, covering all points of his instruction. This mark added to 4 times his final mark for naval aviator's course, divided by 5, will give his grand final mark for determination of air pilot certificate number.

Upon successful completion of the examination the naval aviator will be designated Navy air pilot (aeroplane) and issued a certificate, numbered according to his standing in the class with which he qualified as Navy air pilot. Only those line officers who have qualified as Navy air pilots (aeroplane) are eligible to command Navy aeroplanes.

The course of instruction for Navy air pilots is at present simple, because the aeroplanes now available have not greater capacity. These requirements will be more comprehensive in the next issue of this circular as a result of better air-craft equipment and more experience in the art of aerial navigation.

4. Couse of Instruction of Naval Aviators for Navy Air Pilots (Dirigible)

(1) Only commissioned line officers who have qualified as naval aviators are eligible for detail to this course.

(2) The commandant of the aeronautic station will recommend for this course of instruction aviators who have qualified, based upon his opinion of their aptness for the course in the proportion of one-fourth of each class qualified as aviators.

(3) This course will include instruction in the operation of free balloons, kite balloons, and dirigibles. A more detailed course of instruction will be drawn up when the capabilities of free balloons, kite balloons, and dirigibles have been more fully investigated and these air craft are available for use at the station.

(4) Officers must qualify as Navy air pilots (dirigible) in order to be eligible for duty in these air craft in actual service.

5. Course of Instruction for Student Airmen

(1) Upon reporting at the aeronautic station enlisted men will be assigned for instruction in the crews of aeroplanes in commission, and will become familiar with the general characteristics of the aeroplanes, their care, preservation, and repair. After one month of duty as members of crews of aeroplanes, the men will be assigned as members of crews of free balloons or kite balloons. Upon the completion of one month in such crews the men will be reassigned among the crews of dirigibles. Upon the completion of one month's service as members of dirigible crews the men will be given shop work, as follows:

AERONAUTIC MACHINERY-ERECTING SHOP COURSE

(*a*) Work in dismounting motors.
(*b*) Cleaning up various parts of motors.
(*c*) Rebabbitting bearings.
(*d*) Grinding in valves.
(*e*) Assembling motors.
(*f*) Setting valves and timing motors.
(*g*) Instruction in principle and operation of magnetos.
(*h*) Dismounting and assembling of carburetors.
(*i*) Testing motor on stand.

In all the above student airmen will act as assistants to airmen, air machinists, or yard machinists in their regular work.

Upon the completion of the machinery-erecting shop course (one month), student airmen will be assigned to the aeroplane-erecting shop for instruction. Their instruction will be grouped under the following general heads:

(*a*) Assembly and disassembly of the various types of aeroplanes.
(*b*) Renewing control, brace, and plane wires.
(*c*) Lining up planes.
(*d*) Lining up tail.
(*e*) Setting balancing flaps.
(*f*) Application of dope to fabric.
(*g*) Patching planes.
(*h*) Installing power plant.

In all the above student airmen will act as assistants to airmen, air quartermasters, or yard shopmen in their regular work. After one month in the erecting shop, student airmen will be given an examination covering all points in the course of instruction. Upon passing this examination, student airmen will be designated airmen and assigned as regular members of crews of aeroplanes, free and kite balloons, and dirigibles, and selected for special course of instruction as machinists (aeronautic).

Throughout the course of instruction for student airmen they will be given as many flights as practicable in aëroplanes, free and kite balloons, and dirigibles to accustom them to actual air work.

6. Course of Instruction for Quartermasters (Aeroplanes)

(1) From among those airmen having had at least three months' service as such, and whose Navy ratings are C. P. O. and P. O., first class, of the seaman branch, shall be selected the requisite number to take the course of instruction for training as quartermasters (aeroplane).

The flying instruction shall be the same as that given student naval aviators. During the period of this instruction airmen for quartermasters (aeroplane) shall be given lectures and shall be required to study such books, papers, etc., as are designated by the officer in charge of Flying School. When the airman has completed this instruction he shall be given an examination embracing all points covered by the course of instruction.

Upon the completion of the course and passing the required mental examination, and when the airman has had 50 hours flying, if recommended by the officer in charge of Flying School and approved by the commandant, he shall take the flying test prescribed for student naval aviators. If successful the airman is now given a quartermaster's (aeroplanes) certificate.

7. COURSE OF INSTRUCTION FOR QUARTERMASTERS (DIRIGIBLES)

(1) From among those airmen having had at least three months' service as such, and whose Navy ratings are C. P. O. or P. O., first class, of the seaman branch, shall be selected the required number to take the course of instruction for training as quartermasters (dirigibles).

The course of instruction in operation of free balloons, kite balloons, and dirigibles shall be the same as that required of naval aviators undergoing instruction for designation as Navy air pilots (dirigibles), except that they shall not be required to navigate a dirigible. The training will be in handling these aircraft and in steering dirigibles. Upon the completion of this course the airman shall be given an examination embracing all points of his instruction.

Upon passing the examination, and if recommended by the officer in charge of the Flying School and approved by the commandant, the airman shall take the prescribed test for qualification as quartermaster (dirigibles).

This test will be decided upon later when the capabilities of free balloons, kite balloons, and dirigibles have been fully investigated.

Upon successful completion of test the airman shall be given a certificate of qualification as quartermaster (dirigibles).

8. COURSE OF INSTRUCTION FOR MACHINISTS (AERONAUTIC)

(1) From among those airmen of the engineering and artificer branches who have had at least three months' service as airmen shall be chosen as many men as are required to take the course of instruction for machinists (aeronautic). These men shall be assigned to the machinery erecting shop for a period of one month. The instruction in this shop shall be under the following general heads:

(*a*) Explanation of theory of valve and magneto settings.

(*b*) Practice in the above.

(*c*) Overhauling all types magnetos in use at station.

(*d*) Explanation of theory of all types carburetors in use at station.

(*e*) Adjustment of all types carburetors on motors on testing stands.

(*f*) Take charge of group in the disassembly, overhaul, reassembly, and adjustment of motors (routine shopwork).

(2) The airman under instruction for machinist is now assigned to the machine shop, where he is first instructed in the use of, then required to operate for routine work, the following tools:

(*a*) Screw-cutting lathe.
(*b*) Lathe for general work.
(*c*) Radial drill.
(*d*) Sensitive drill.
(*e*) Shaper.
(*f*) Grinding machine.
(*g*) Boring machine.

(3) After the machine-shop course the airman is next assigned to the copper shop for a period of two weeks, where he is first instructed, then required to assist in the making of gasoline tanks, repair of gas tanks and radiators, putting copper sheathing on propeller blades, balancing propellers, copper sheathing aeroplane spar and rib joints, engine bed joints, etc.

(4) Following the copper-shop course the airman is given two weeks in the blacksmith shop, where he is first instructed, then required to turn out for actual use aeroplane fittings, steel-tube braces, engine-bed joint fittings, steel-tube braces, engine-bed joint fittings, etc.

(5) The airman is then passed on to the gas plant for balloons and dirigibles, where he is first instructed, then required to take charge of operation of the plant for the inflation and deflation of balloons and dirigibles.

(6) The airman is next given an examination covering all points embodied in the course of instruction, and if he passes is given a machinist's (aeronautic) certificate.

NOTE.—If the equipment of the station precludes the possibility of any part of the foregoing instruction, such part as is impossible to give shall be omitted from the course and the airman advanced to machinist regardless of the omission.

VICTOR BLUE,
Chief of Bureau.

Course of Instruction for United States Student Naval Airmen

Herewith is given an outline of the course of instruction for student airman at the Pensacola, Florida, station.

Upon reporting to the flying school students are divided into sections, each section spending about twelve weeks on shop instruction and about a month on the beach with aeroplanes. When the marks in all shops and on the beach are satisfactory, the student is qualified as an airman.

The terms of the course are as follows:

The time spent in the various departments is as follows:

Erecting Shop.	3 weeks	Multiple 3
Motor Erecting Shop	3 "	" 2
Carburetors & Magnetos	1 week	" 1
Joiners, Cooper & Fabric Shops.	2 weeks	" 1
Balloons	1 week	" 1
Hangars	4 weeks	" 2

2. Upon becoming qualified airmen the required number of Student Quartermasters will be selected, the rest of the qualified airmen will undergo a course of instruction for the rating of machinist (aeronautic).

3. Student Quartermaster will receive instructions in flying and will undergo a theoretical course of instruction in aeronautics.

4. The course of instruction for machinist's (aeronautic) will be published later.

5. Marks on all shops will be on a scale of 5. A mark of three in each and every subject is necessary to pass.

Aeroplane Erection Shop

Shop Instructor (Lieut., j.g.). Asst. Shop Instructor (Lieut., j.g.). Assistant Instructors: (petty officers or civilians).

I

(a) Names of all parts of aeroplanes.

(b) Kinds of material used in wings, pontoons, propellers, struts, fuselage, etc.

(c) Instruments carried in the different aeroplanes and the purpose of each.

(d) Approximate overall dimension of the various types of aeroplanes.

II

(a) Method of assembling.

(b) Method of wiring and location of struts.

(c) Method of securing wings to fuselage in different types of aeroplanes.

(d) Method of securing pontoons to planes in different types of aëroplanes.

(e) Method of securing horizontal rudder, vertical rudder, ailerons or wing flaps, vertical fin, and stabilizer of various aëroplanes.

III

(a) Factor of safety usually required in aeroplane construction for the use of the Navy.

(b) Kind and approximate size and strength of control wires and other wires.

(c) Describe three ways of making a wire terminal and tell which way is most efficient.

(d) The general design of various fittings used, such as wing fittings, strut fittings, pontoon fittings, fuselage fittings, control surface fittings, etc.

(e) Which aeroplane used turnbuckles in their wiring and which do not. How may an aeroplane be assembled if no turnbuckles are used?

(f) At what angle are pontoons set relative to wings, and reason for same.

IV

(a) How are propellers balanced?

(b) How are propellers secured to the various motors.

(c) How are propellers marked and what record is kept of them?

(d) How are aëroplanes marked?

V

(a) Trace the gas loads from tank to motor in the different aeroplanes.

(b) What method is employed to fight fire in Navy tractors?

(c) Names and general description of the different methods of controlling an aeroplane in flight.

(d) What arrangement is made to control certain aeroplanes on the water?

VI

In addition to the above general question on construction and assembly of aircraft, the knowledge of the following with reference to experiments and tests is required:—

(a) Method of testing material with Tinis Olsen Testing Machine and general description of same.

(b) Method of testing material for compression, and description of machine for same.

(c) Method of stretching wire, including description of machine and reason for stretching.

(d) Method of bending wire; description of wire bending machine.

(e) Method of finding center of gravity of an aeroplane.

Motor Erecting Shop

Shop Instructor (Lieut. j.g.). Asst. Shop Instructor (Lieut. j.g.). Assistant Instructors: (petty officers or civilians).

I

(a) Lecture on the principles of the internal combustion engine or gasoline motor.

1. Definition of a motor.
2. Explanation of the term cycle.
3. Two and four cycle motors.

(b) Practical explanation of motor.

1. Function of different parts.
2. General nomenclature.

II

(a) Disassembling of a motor.

1. Steps in disassembling.
2. Detailed nomenclature.
3. Material used in different parts and why?
4. Forms and size of bolts, parts, etc., and reason for same; locking devices and reasons for same.
5. Notes on any worn, cracked, bent, twisted or found weakened part and the reason for this condition and how to correct.

III

(a) Overhaul of motors.

1. Cleaning of parts.
2. Tests for alignment, as far as cam-shaft, main shaft, face plates, and gear balance.
3. Trace oiling and water systems.
4. Thorough examinations of all parts for flaws or worn places as scorings in cylinders, etc.

IV

(a) Assembling of motor.

1. Spotting in main-shaft.
2. Fitting Cam-shaft.
3. Reaming out wrist-pin bearings.
4. Test main-shaft for balance.
5. Grinding in of valves and testing same.
6. Testing valve springs, exhaust and intake.
7. Fitting thrust block.
8. General assemblage.

(b) Timing of motor and valve setting.

1. Method of adjusting valve clearance and reason for clearances.
2. Method of timing magneto and reasons for advance.

(c) Final adjustments and connections before going to test stand.

V

(a) Test stand.

1. Connecting up motor.
2. Securing propeller.
3. Data required for motor tests and the requirements for an aeronautic motor.

(b) Motor troubles.

1. Causes.
2. Results.
3. Connections.

(c) Care and upkeep.

Carburetors and Magnetos

Shop Instructor (Lieut. j.g.). Assist. Shop Instructor (Lieut. j.g.). Assistant Instructors (petty officers or civilians).

(a) Carburetors.

1. Theory of the carburetor.
2. Theory of the Zenith carburetor.
3. Jets, chokes and their uses.
4. Carburetion troubles and their remedies.

II

(a) Magnetos.

1. Electrical principles of the magneto.
2. High and low tension magnetos.
3. Dixie principle.
4. Bosch principle.
5. The construction of both magnetos, and the difference between them.
6. Distributor, its construction and use.
7. Magneto circuit.
8. Spark plugs.
9. Coils.
10. Timing and firing order.
11. Magneto troubles and remedies.

Joiner, Copper and Fabric Shops

Shop Instructor (Lieut. j.g.). Assistant Instructor.

I

Joiner Shop.

(a) Nomenclature of parts.
(b) Wings, fins, struts and stabilizers.

1. Construction of ribs.
2. Longitudinals.
3. Spars.

(c) Floats.

1. Outside: keel, steps.
2. Inside: bulkheads, frames, supports.
3. Drains.
4. Methods of fastening to aëroplane.
5. Various types.

II.

Copper shop.

(a) Gasoline tanks.
(b) Fittings.
(c) Leads, overflow pipes, exhaust pipes, and drains.
(d) Patterns and shapes.
(e) Brazing.

III.

Fabric Shop.

(a) Material used in covering.
(b) Method of securing to surfaces.
(c) Doping and varnishing.

Balloons

Shop Instructor (Captain), Asst. Shop Instructor (Lieut. j.g.), Assistant Instructors (petty officers and civilians).

Balloons.

(a) Lecture on free and kite balloons.
(b) Prepare balloon for flight.

1. Lay out to inflate.
2. Inflate when possible.
3. Handling balloon during ascension.
4. Make up as after flight.
5. Repairs to fabric.
6. Cementing on rip panel.
7. Nomenclature of all parts, lines and gear.
8. Two hours or more in hydrogen plant.

Memoranda:

Lawrence B. Sperry and Captain L. A. Dewey, Acting Judge Advocate, Eastern Department, U. S. Army during their flight from Amityville, Long Island, to Boston, Mass., on September 30, 1916, made a flight of over one hour and one quarter in pitch darkness before they landed at Block Island. The above photograph shows the ruggedness of the place where they landed at night. The flight was arranged as a military experiment and Major Carl F. Hartmann, Signal Corps, was to be the passenger and to have charge of the experiment, but circumstances prevented his carrying out the plan.

CHAPTER XV

COURSE OF INSTRUCTION FOR THE TRAINING OF AVIATORS

BY LAWRENCE B. SPERRY

Much has been said and written in the last two years on the training of aviators, and we of the United States are vitally interested in the subject, since the recent appropriations have assured the formation of a sizeable air service. An investigation into the present civilian aviation schools reveals the fact that the men are turned out knowing only the rudiments of flying and without being instructed in the finer points that are essential in the mak-

The General Aeroplane Company flying boat equipped with 100 horse-power Curtiss motor.

To clear small boats the seaplane pilot must have a trained sense of distances.

ing of a capable aviator. Unnecessary risks are taken by pupils in many of the schools, which could easily be avoided if the proper instructions were given.

The following outline of instruction is one that the writer, who has made a study of the matter in the principal aerodromes of Europe since the beginning of the war, considers would render training in flying not only safe to the pupil, but calculated to make him an experienced pilot ready for military work.

First Stage.—The preliminary part of the course may be divided into three parts:

1. The student pilot should take a few trips as a passenger to accustom himself to air travel and height, as well as to the ordinary manœuvers of the machine.

Cornell men have been learning to fly at the Thomas Aviation School at Ithaca. This photograph shows a delightful spot on Lake Ithaca where students learn to fly. The machine is a Thomas seaplane.

2. He should fly a dual controlled machine under the guidance of the instructor, in order to learn the control movements. This machine should be equipped with some means of intercommunication so that the pilot may instantly direct the pupil. He should be taught how to get off the ground, by running along the ground until full speed is attained before elevating. This point is important for two reasons:

(a) The aeroplane will get off the ground at a lower speed than that at which it will retain buoyancy, because of the "ground bank." This "ground bank" consists of a blanket of air, close to the surface of the earth, of greater density when the machine is passing over it than that encountered at a hundred feet elevation. A machine traveling at a speed capable of sustaining flight a few feet from the ground will go into a "stall" when it rises out of this "bank."

(b) If the machine is elevated from the ground before it obtains full flying speed, the energy of the motor is expended in lifting the machine, instead of going into speed as it does when the machine is allowed to run on the ground a sufficient length of time. It must be remembered that the quickest climb for getting over trees and the like is obtained by allowing the machine full acceleration on the ground.

3. The pupil should then be taught to maintain a straight course, to make proper turns, and to land properly. This instruction in landing is most important, and the following points should be brought out:

(a) The wheels of the landing chassis and the tail skid should strike the ground simultaneously.

A Martin seaplane equipped with 125 horse-power Hall-Scott motor.

(b) There is a speed at which the maximum lift angle is able only to sustain the plane; this is the critical speed at which the controls should be brought quickly in toward the chest of the pupil in order to increase the angle beyond the maximum lift angle, thereby increasing the resistance of the wings and throwing the greater part of this resistance into drift, which slows the machine down to a minimum.

(c) Such landing is desirable not only because of the reduced speed, which is advantageous when one is landing on rough ground, but also because the tail, on account of being low, throws the center of gravity far back in relation to the wheels, thereby reducing the possibility of the machine nosing over. This is also very advantageous as a means of protecting the propeller, especially on tractors with small propeller clearance.

(d) The pupil should have at least six hours' training in a dual controlled machine; this time being divided into lessons of not more than fifteen minutes' duration.

Second Stage.—The pupil should fly alone at a large field in a "Penguin," i.e., a machine the wing spread of which is so reduced that it is capable of flying not more than five feet from the ground. This teaches the pupil not only the art of getting the most results from small horse-power, which is of great advantage in case of motor disability in a large machine, but it also teaches him principally what it means to get into a "stall" near the ground.

Third Stage.—The pupil should be allowed to fly alone under ideal air conditions, in a higher powered machine in a large field at least one-half mile wide by one mile long, having been previously instructed simply to make "straightaways" up and down the field, turn-

The Burgess-Dunne seaplane equipped with 140 horse-power Sturtevant motor.

The 125 horse-power Aeromarine hydroaeroplane, twin pontoons, two passengers.

ing the machine around at each end on the ground.

Fourth Stage.—The pupil should be instructed to climb to an altitude of at least 600 feet, and then to make a very slow turn and land in the opposite direction. This is watched by an instructor through a pair of binoculars to enable him to observe the pupil carefully and to advise him of his mistakes after his return.

At the same time that the air instructor is directing the student's field work, the pupil is learning practical theory, with the aid of a blackboard, about subjects such as tail spins and their recovery, spirals, the prevention of stalling, the making of a landing in a small field against the wind, without the necessity of spiraling, by zigzagging back and forth against the wind, etc. This instruction is incorporated in a complete course of aerodynamics, aeroplane design, and construction.

The Sperry instructograph to facilitate the instruction of pupils.

The pupil is now ready for spirals, which should be made at an altitude of not less than 3000 feet and finished at an altitude of not less than 1000 feet. The pupil should not make continuous spirals, but should turn first 90 degrees to the right and 90 degrees to the left with the motor shut off; then 180 degrees to the right and 180 degrees to the left; after that 360 degrees each way, gradually working up to three spirals to the right and three to the left, always finishing with an altitude of 1000 feet.

Fifth Stage.—Part of the field, if the field is large, is marked off one-quarter mile by one-quarter mile. The pupil is instructed to fly off the field with the wind in different directions, and on the sounding of a gun to shut his motor instantly and to make the field safely without turning more than 90 degrees. In case he finds that he is not within gliding distance of the field, he should be instructed to plug in his motor again and to continue his flight, rather than to take any chances in landing.

Sixth Stage.—Landing on a mark is practised.

Seventh Stage.—Each of the requirements of the Junior Military Licenses is practised, except the cross-country flights, which are withheld until the course of training is completed.

Eighth Stage.—The pupil flies at 5000 feet and deliberately "stalls" in the air four times, each time at a steeper angle than the one before until a tail slide is made. These stalls are to be made with the use of an angle of incidence indicator, the pupil carefully watching the indicator and slowly placing the machine at the stalling angle.

Ninth Stage.—The pupil is instructed to "loop" at an altitude of not less than 5000 feet, after which he should stall the machine, at the same time throwing his rudder hard over one way, thus putting his machine into a tail spin, if the machine is capable of executing this manœuver.

To recover from the tail spin is the salient point that the pupil is now to learn. This is effected by throwing his rudder over in the opposite direction, which requires a tremendous pressure on the rudder bar. The writer has found it necessary to exert the force of both feet. The pupil is now ready to take his military license.

It might be well to remember at this point that "looping," tail sliding, and such alleged "stunting" is only carried on with definite objects in view, namely:

1. To give the pilot confidence.
2. To teach him quick recoveries from unsafe positions, especially near the ground, should he find himself in such positions in the course of carrying out his work as a military aviator.

Tenth Stage.—After the pupil has secured his military license, instruction in night flying should be given. This is both one of the most difficult feats of military aviation and one of the most important; a great deal of time and attention being given to its development by all of the warring European nations. Night flying should be taught in four stages, as follows:

1. The pupil should fly as a passenger with a competent pilot, thoroughly familiar with night flying.
2. The pupil should fly alone on a moonlight night, or one in which the stars are bright enough to give considerable light.
3. The pupil should begin flying before sundown on a cloudy day, and should continue flying until the sun sets, not landing until total darkness has fallen. No night flying should be undertaken in a machine not equipped with a night flying outfit.
4. The pupil should practise night flying, using his night flying equipment to signal an aerodrome when he intends landing. In case of war, the aerodromes are unlighted, except for a single beam of light, which is thrown verti-

The Sturtevant seaplane equipped with 140 horse-power Sturtevant motor.

Burgess training hydroaeroplane.

cally; machines also are not illuminated, but, on desiring to land, their pilots flash on their night flying equipments, signaling by prearranged code to the aerodrome, whereupon the grounds are illuminated for landing. The pupil should be cautioned against inadvertently getting into fog. There are foggy days when, although it is not possible to see more than six hundred feet straight ahead, it is frequently practicable to fly. These days can be gaged by looking straight upward at the sky. If the azure blue can be seen, the fog is limited to a blanket that does not extend over five or six hundred feet above the ground. While the pupil is climbing in this fog, he can see nothing but fog surrounding him. When he has climbed just above the fog, he begins to be able to see around him. This is due to the fact that it is only possible for him to see through the blanket of fog at an obtuse angle, when a large area of the ground becomes visible from an elevation of some 2000 feet.

Flying in fog is very similar to night flying, as it requires a finely developed sense of equilibrium. In both night flying and flying in fog it is extremely advantageous to use a machine equipped with an automatic pilot.

During the latter part of the tenth stage, the pupil should be allowed to carry an observer as passenger, so as to familiarize himself with the various manœuvers that are necessary in order that he may obtain such information as he required. This needs a system of cooperation and some means of communication between the pilot and observer.

Coming into a small field against the wind the pilot pupil should either zigzag or reduce his altitude to two or three hundred feet, by going abeam or at right angles to the wind. This prevents him from getting into the predicament of going short of his mark.

It should be remembered that in learning to turn, if the pupil attempts to make his turns sharp or short at first, no matter how much he practises he will never learn how to accomplish a perfect turn. The only way he can expect to learn to turn satisfactorily is first to make the turns very big, watching all the time his revolutions, and his air drift indicator, and also his angle of incidence indicator, if he has one, being sure all the time to keep his relations at all

The Lawrence-Lewis flying boat *A-1* in flight.

parts of the turn exactly right. Otherwise if if he tries to do short turns he will think he is doing correctly when he is really on the brink of dangerous positions.

The pupil should be watched at all times through glasses and made to repeat any faulty manœuver until he does it perfectly.

Any tendency toward over-confidence and carelessness in landing should be followed by a suspension of three days to one week.

Memoranda:

Photo of a schooner sunk by a Cunard boat off Governor's Island, taken by Charles Reed at a height of 500 feet.

CHAPTER XVI

AERIAL NAVIGATION OVER WATER

By Elmer A. Sperry

Abstract

The author calls attention to the unreliability of the magnetic compass when used for aerial navigation and to the possible development of the gyroscopic compass for this purpose. He then explains how the drift of an aëroplane in flight makes it difficult to follow with accuracy a given course devoid of landmarks, unless an accurate drift indicator using the principle of the stroboscope is available.

The development of such an instrument is then described, as are also means for synchronizing it with the compass. The use of the automatic synchronized instrument in flight over land is outlined, and its application to flight over water is described in considerable detail. Rules for aerial navigation over water, observation as to movement of wave crest and determination of wind velocity and direction are considered in their relation to the use of the instrument.

The author, because of the development and practical application of his gyro-compass, has been brought to consider, more or less broadly, the whole science of navigation. For many decades past, this science has been one of high exactitude, limited only by the accuracy of the instruments used in obtaining its ground work or in giving it its base lines.

It has long been known that the magnetic compass is unreliable; as ships represent greater and greater masses of steel, this inaccuracy has become more and more aggravated. Methods of checking its accuracy have been diligently sought for and made as nearly perfect as possible. These, however, depend upon observations and fail usually just at the time they are most needed; namely, when observations of the heavenly bodies are impossible. All this emphasizes strongly the desirability of an instrument of precision that will function as a compass. The adaptation of the gyroscope is found to fulfil this satisfactorily.

Limitations of Magnetic Compass

Among the difficulties met in using the magnetic compass in the air is that known as the heeling error. When a magnetic compass is used on aeroplanes, and the machine is even mildly "banked," and persists in such a position for an appreciable period, the heeling error is found to be of such magnitude as to render the compass useless. In some instances it will amount to 360 degrees, or around the entire circle, giving no clue whatever to the aviator as to the true azimuth, or where to stop on the turn and straighten out into the tangent. Thus the compass fails him utterly at just the critical time when it is most needed. After the tangent is persisted in for a sufficiently long time for the compass to settle, then and then only does it again become useful.

Moreover, the lag or tardiness of action of the magnetic compass is a serious drawback. The magnetic compass, to be reliable, we find, must be of the so-called liquid type; upon spiralling and making two or three turns the liquid is found to take up the swivelling motion, carrying the card with it round and round, and becoming a serious disturbing factor for some time after the aeroplane straightens out on the tangent. This is not meant to imply that the magnetic compass is not an extremely useful instrument upon aircraft, but for best results it should be understood by the aviator and not relied upon when conditions are such that it is impossible for it to function.

This condition has become so aggravating that the United States Navy is ordering a gyroscopic compass to be employed on aeroplanes. Much interest centers in the results that it will be possible to achieve by substituting this instrument for the magnetic compass under practical service conditions in the air. By extreme refinement in execution and design it is expected by the Navy Department that the weight of this instrument will be reduced to 20 or 30 pounds.

When one is navigating the air and holding an absolutely true course, that is, with the lubber line of the compass precisely upon the desired heading in azimuth, the direction of flight coincides with this heading, under the condition of absence of movement of the medium, namely the atmosphere, through which the flight takes place. At first thought, one would think that at the moment the medium itself was moving, and especially when this movement was normal or at a small angle to the direction of flight, the compass would instantly indicate the resulting deviation from the true course. This conception, however, is not correct.

An aviator can hold his course true to the compass and still be following a course having a wide angle of deviation from the course in which he thinks he is flying. For instance, Carlstrom, in his noteworthy flight from Chicago to New York (equipped with instruments described later), found he was drifting 17½ degrees, when flying over Cleveland; his apparent course had to be changed to this extent to neutralize drift and to maintain the true direction along the south shore of Lake Erie. This angle was given him by his drift set, which, although not indispensable, he used throughout his flight. Carlstrom, however, had the shore line and general landmarks to aid in his guidance. The case would have been different had he been flying at sea out of sight of land, where no landmarks could possibly be seen. Then certain aids to navigation are indispensable, and it is the province of this paper to discuss briefly some general aspects of these instruments and their uses, no effort being made to present a mathematical or an exhaustive treatise on the subject.

Determination of Drift

Let us assume an aeroplane maintaining a compass course due north with the lubber line of the compass maintained at zero on the card, and for the moment the compass properly functions without variation or deviation. In still air the course of this machine over the surface of the earth will be north. But suppose, wholly without the knowledge of the aviator, the medium in which he is flying is itself in motion toward the east with a velocity equal to that of the aircraft. It is quite evident that, although he holds his course with exactitude, the craft itself is passing over the surface of the earth on a diagonal, that is, his actual course is northeast; the easterly component of his course depends upon the eastward velocity of the medium. Again, if this velocity be half of that of the aircraft, then his real course is 22½ degrees, or "north-northeast." The question arises: how is the pilot to obtain this knowledge, or knowledge that he is drifting at all, so as to make the necessary correction of course in the absence of landmarks or other indications to guide him?

A device, worked out by the author some time ago to give pilots this knowledge, has now come into general use, and is found to perform its function satisfactorily. Fig. 1 shows an earlier form of this device, in which a series of moving observing tubes acted as a stroboscope. By the backward movement of these tubes a point upon the earth's surface could be made to appear as though it stood still; the backwardly moving point of persistence of vision being exactly equal to the forward advance of the machine. Knowing the angular velocity of these vision tubes, or simple telescopes, and the altitude, the actual speed over the earth's surface was at once obtained. When compared with the anemometer speed, that is, the real speed of flight through the atmosphere, a clew was at once available as to the actual movements of the atmosphere itself.

This instrument was then carried a step further. The telescopes were furnished with cross-hairs and mounted upon a swiveling base with a pointer *B* moving over an azimuth scale *C*. When the slow motion of the telescopes was arrested and one telescope furnished with one or more fore-and-aft cross-hairs pointed directly downward, it was found easy to make a

New Type of Drift Indicator.

Synchronized Drift Set.

peculiar reading known as the "stream-line observation."

When one is looking downward at the surface of the earth, through a tube or telescope, if, instead of looking for specific objects, he simply observes the passage of *all* objects across the field of the tube or telescope, he at once becomes conscious of the passage of all of these objects taking place in certain clearly-defined parallel lines, which I have denominated "stream-lines." When one gets somewhat familiar with this kind of observation, he can see almost nothing but these stream-lines, and the nearer the surface or greater the speed, the more tense and clearly-defined these lines become. Now, if a good, heavy cross-hair be stretched across the tube or telescope, and the tube be made so that it can be rotated upon its major axis, then it is found easily possible so to rotate the tube as to bring the cross-hair exactly coincident or parallel with the stream-lines.

The rotating tube or telescope is furnished with a stationary scale, the zero of which is coincident with the longitudinal axis of the aircraft. By taking readings on this scale with a pointer on the tube opposite the cross-hair, it becomes easy to determine the angle between the stream-lines and the major axis of the aircraft, since the latter always lies in the apparent direction of flight, the angle being between the stream-lines, or actual direction of flight, and the aircraft. The determination of such

Drift Compass with Adjustable Lubber.

an angle as this is extremely useful, as it at once gives the aviator a clew as to what change to make in his course so that his direction of flight is such as to neutralize the drift of the medium through which he is flying, his actual course being thus brought into exact harmony with the direction required to reach his destination.

In securing this parallelism the pointer ***B*** is swung upon the scale *C* and the angle of drift in degrees can be immediately read on the scale from the longitudinal axis of the aircraft. Many surprises are in store for the pilot or observer when he first makes this observation, as he often is certain that it cannot be correct. He cannot believe it possible from his compass heading, to which he is holding with great accuracy, that he is actually making headway at so large an angle from its readings. And if he has had experience at sea with the compass, this impression is all the more startling, because it is always true that a ship is traveling practically on the exact course indicated by its compass. But here the pilot is holding his course absolutely true to the compass, and yet the drift indicator shows a quite different condition of affairs; namely, that he is actually traveling at a considerable angle to his supposed course.

Fig. 2 shows the instrument described mounted upon an aeroplane and being used by the observer. The observer in this instance is Captain Creagh-Osborne, R. N., head of the Hydrographic Office of the British Admiralty. The pilot is Lieutenant John H. Towers, U. S. N.

With the earlier instruments the pilot would change his course and by trial in error finally reach a flight heading on the compass card, resulting in actual headway of the aircraft along the true course; that is, along the line that he was originally instructed to pursue to reach the desired destination.

Thus it will be seen that the accurate determination of drift is an important factor in aerial navigation. It is true that observations are somewhat more difficult at great heights and over rough water, but one soon becomes accustomed to obtaining a mean reading, which is found to be very accurate. It has also been definitely ascertained that when flying with the automatic pilot, with which side disturbances are practically eliminated, the stream-line observation becomes very much simplified and accurate at practically all altitudes.

The Synchronized Drift Set

A later form of the apparatus is shown in Fig. 3, in which a single stationary telescope, provided with the cross-hairs, is employed for the moving series of telescopes, inasmuch as the drift factor is found to be of far more importance than the actual speed of advance with reference to the earth's surface.

In Fig. 4 is shown a drift-compass with an adjustable lubber line and with a little tiller wheel extending from the side of the case, by means of which the lubber line is set from the indications upon the scale *C* of the drift indicator. The scale upon the bezel of the compass is used for reading the deflection of the lubber line. Care must be taken in moving the lubber

Synchronized Drift Set with Two Compasses.

Wave crests viewed from aeroplane.

line to be sure that its setting is correct as to direction, and that it is not set on the right side of the zero when it should be on the left.

A pilot, especially when without the automatic pilot and working alone, has enough cares without the added one of worrying whether the compass has been accurately synchronized with the drift indicator, and whether the direction of the adjustment is also correct. So my son Lawrence conceived the idea of coupling them mechanically, in order that they might be at all times automatically synchronized. The combined instrument, namely, the synchronized drift-set has now become the most useful form of the apparatus and the one most widely adopted. In this instrument the most minute azimuth movement introduced by the observer at once causes a corresponding alteration of the position of the lubber line of the compass, thus eliminating the possibility of error either in the direction of this movement or in its exact amount. This is found very practical, inasmuch as it vastly simplifies the pilot's operation, he needing only to continue without change to hold the lubber line upon the originally selected point on the compass card. The fact that the lubber line is displaced, especially if under the control of the observer, is something with which the pilot has nothing to do and is not concerned—he simply continues on his original compass course. This arrangement automatically introduces all of the deviations in course to correct fully for drift, and is found to save much valuable time and fuel, and to allow the pilot to reach his destination by a true meridional course. With the actual drift known and corrected for, the correction that should be given the anemometer speed, which is always known to the operator, can be determined easily. Thus he has all the knowledge of the actual forward advance that is roughly necessary in short flights.

Fig. 5 shows the synchronized drift set with the observing telescope to the left, the compass to the right, and the compass lubber line thrown around 30 degrees from the longitudinal axis of the aircraft indicated by the zero on the scale shown on the bezel. When the observer and the pilot sit in tandem relation in the aircraft, then it becomes desirable that each have a compass. We have made a number of sets, Fig. 6, in which two compasses are synchronized by means of a single drift indicator.

Precautions in Flight Over Water

We will now turn our attention to some refinements of the use of this apparatus when one is flying under actual service conditions over water. It is always wise to note the direction of the wind before starting, and also something as to the length of the wave, that is, the distance from wave crest to wave crest. A pilot of seaplanes or an observer, or both, should become accustomed to estimating this wave length or distance from crest to crest, say for this country, in feet, and should also invariably note the direction of the wind in terms of compass azimuth.

In leaving the surface of the water, it should also be the duty of the personnel to see that the aneroid is adjusted exactly on zero. This observation should also be made whenever the plane is brought close to the surface, thus eliminating effects of changes in barometer in determing actual heights. The uses of these observations will be presently apparent. Let us now divide the problem of speed control into two classes.

In the first class stroboscope methods are used. One form of the stroboscope is shown above in Fig. 1. Another instrument working upon this principle is illustrated in Fig. 7. Still another simpler form is in course of being developed for the United States service. The speed with reference to the surface of the earth can be ascertained with a good degree of accuracy over water. It is, of course, necessary to know the movement of the sea, or, rather, its apparent movement, as for instance, the movement of the wave crest. In every instance the actual angular velocity of the stroboscope, or of the apparent passage of the earth's surface, and the careful reading of the aneroid are used as prime factors.

The second class is where a close approximation of the actual speed can be ascertained quickly. Using the anemometer speed as a base, we determine whether the actual speed is the same as, or greater or less than the anemometer speed, and also obtain a close approximation of how much the variation is. It is assumed that the anemometer speed is always available to the navigator. A good anemometer is known to possess a high degree of accuracy.

When the wave crests are small and cannot be seen directly, the same telescope that is used to ascertain the stream-line directions can be employed to observe the direction of the wave crest. I have had some discussion as to the visibility of the wave crests with some who are unfamiliar with it, but, as a matter of fact, the wave crests are clearly visible. Fig. 8 is a reproduction of a photograph taken from a considerable altitude, and shows how clearly visible these wave crests are. They are an excellent indicator of two valuable factors: First, from their length from crest to crest we can ascertain their speed; and second, their direction always lies directly normal to that of the wind. In all probability there also exists a reasonable relation between the velocity of the wind and the velocity of the waves, but we will not concern ourselves with this at the present time. The waves shown in Fig. 8 are about 10 feet from crest to crest, giving us a velocity of 7.2 feet per second. At first glance it might be difficult to know in which direction these waves are moving, but with the conditions such as in Fig. 8, there is the simplest possible clew consisting of the vapors flowing from the funnels of the two small craft shown. The line of vapors is seen to be normal to the line of the wave crests.

Bearing these points in mind, the following rules can be applied to aerial navigation:

1. Note on which side of the keel line or longitudinal axis of the aircraft lies the actual or true course, or on which side of the zero on the bezel of the drift compass lies the adjustable lubber line. This is the "drift side," and can, of course, be to the right or to the left. (The zero on the bezel indicates the keel line.)

2. Note closely the apparent alignment of the wave crests while the aircraft is being maintained on its course.

3. Note the relative angle between the wave crests and the keel line, and also between the wave crests and the "drift line," or the alignment of the adjustable lubber line.

Speed Indicator Using Stroboscope Method.

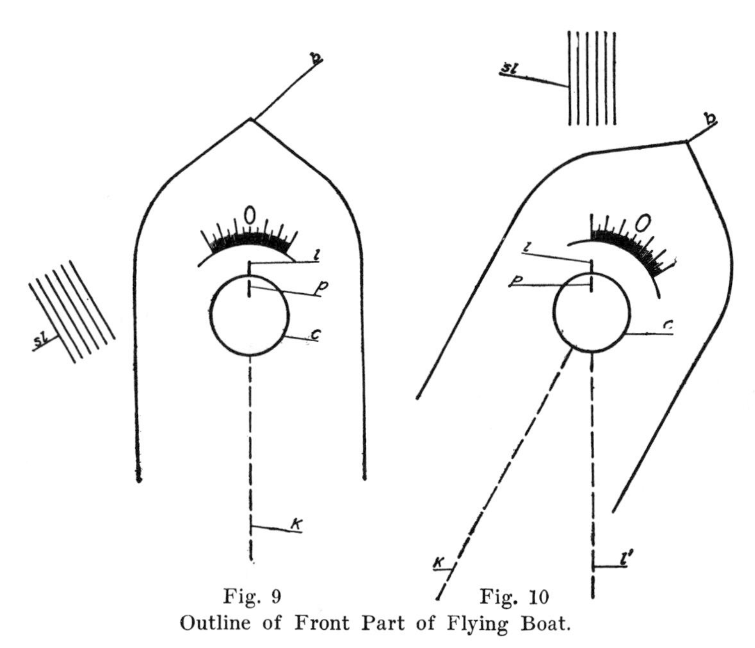

Fig. 9 Fig. 10
Outline of Front Part of Flying Boat.

With these observations well in hand, the following deductions can be made:

(a) If the crests lie within the angle between the keel line and the drift line, then the real speed is approximately identical with the anemometer speed. If these crests exactly bisect this angle, then the real speed on the actual course is exactly the anemometer speed.

(b) If the aft ends of the crests be toward the drift side, then the actual speed over the earth's surface is always greater than the anemometer speed, because of an abaft wind.

(c) If the aft ends of the crests are away from the drift side to a greater extent than is the true course angle or drift line, then the actual speed is always smaller or less than the anemometer speed, because of head wind.

(d) If the crests lie exactly at right angles to the keel line, of course no drift angle exists, and the stream-lines lie parallel to the keel. However, it is always known which way the wind is blowing, from observations made before starting (which have been referred to), and in this manner we at once know whether we are passing over the surface of the earth at a faster or a slower speed than that indicated by the anemometer. Suppose, however, the navigator should have mislaid his data or forgotten the direction of the wind or its true azimuth. A clew to the direction can be obtained as follows: Change the course sufficiently to make the stream-lines veer away from the keel line. If the leading ends of the stream-lines are seen to move to the left for a right turn, the wind is a *following* wind, and if they move to the right for a right turn it is a *head* wind. Of course, for a left turn just the reverse is true. As to whether the velocity is great or little can easily be judged from the sea conditions.

In Figs. 9 and 10, *b* is the forward part of the flying boat in outline; *k* is the keel line; *l* is the lubber line; *c* is the compass card. The zero on the compass bezel is on the keel line; *sl* indicates the stream lines and their direction; *p* is the point on the compass card indicating the desired direction of flight. In Fig. 10, *l* and *l'* indicate the "drift line."

Movements of the Wave Crests

As a still further refinement in ascertaining speed and direction, account should be taken of the movement of the wave crests themselves. This can ordinarily be found by determining the length between the crests, inasmuch as the velocity varies about as the square root of the wave length or distance between crests; the velocity in feet per second equals the wave length in feet at a point where each is expressed by the figure $5\frac{1}{8}$—to be accurate, 5.123—that is, in measuring the distance from crest to crest, their velocity is also 5.123 feet per second. The 10-feet wave shown in Fig. 8 would therefore have a velocity of about 7.2 feet per second.

We know the approximate length of this wave, having observed it before leaving the water; but if this has not been done, we can ascertain the length by dropping down near enough to the surface of the water to get a fairly accurate determination of the mean lengths between wave crests, taking this opportunity also to note that the aneroid is at zero. Having thus determined the velocity of the wave crests, we proceed to determine the corrections.

Corrections for Movement

With reference to the speed, a correction should be made in proportion to the sine of the angle between the wave crest and the stream-line. To obtain the actual speed over the earth's surface, we should add the speed of the wave crests to the speeds obtained by stroboscopic methods in all cases of following wind, and subtract in case of head wind. The amounts that are subtracted are equal to the velocity of the wave crests only when the wave crests are at right angles to the stream-lines. At other angles the velocities subtracted or added are as the sine of the angle between the wave crests and the stream-line.

As to the influence of the moving wave crests on the apparent direction of the stream-lines, it should be noted that (a) the angle between the stream-lines and the wave crests should be determined; and (b) starting with the position where the stream-lines and wave crests lie parallel to each other, as being the condition of maximum correction, we derive this correction from the known velocity of the wave crests. The correction, it should be remembered, is an angular correction to be applied to the apparent direction of the stream-line, resulting in a refinement of heading, which gives the true meridional course to the point of destination.

This correction is always proportional to the ratio of the velocities of wave crests and aircraft determined, say by stroboscopic methods or by the anemometer, after the proper corrections have been applied. When the wave crests and stream-lines are parallel, this ratio is applied directly. When other angles obtain, the correction should be multiplied by the cosine of the angle between the wave crests and the stream-lines.

One factor now remains, that is, as to whether the correction is to be applied clockwise or anticlockwise. Assuming the observer is facing the direction of the stream-line or actual flight, then if the wind is from the right the correction should be counter-clockwise, and if from his left the correction should be clockwise. The navigator knows the direction of the wind from his original observation of its true compass azimuth before rising from the sea. If, however, this information is lacking, the direction of the wind can be found as follows:

Notice the direction of the stream-lines with respect to the keel line. If this is to the left, the wind is from the right; and if to the right, the wind is from the left. If the stream-lines happen to be parallel to the keel, then the aeroplane should be veered to the right or to the left until the keel line and stream-line lie at an angle to each other, whereupon the above observation can be made.

There are a number of other interesting aids to navigation of aircraft, such as clinometers, gyroscopic base lines, and artificial horizons, banking indicators, angle of incidence indicators, and the like, but their uses are apparent and need little explanation.

Memoranda:

CHAPTER XVII

AEROPLANE GUNS AND AERIAL GUNNERY

Now that aeroplanes are increasing in size and number to tremendous proportions, the aeroplane gun becomes a most important factor. As has been aptly pointed out by Rear-Admiral Bradley A. Fiske and other authorities, the equipping of the large present-day seaplanes with a three-inch gun introduces a new revolutionary factor in naval warfare. This is especially true because the aircraft is the most effective weapon against the submarine, and armed air cruisers equipped with guns ranging from the one-pounder size to the three-inch size can protect ships against submarine attacks as far as the air cruisers can go—which is a few hundred miles at present but will soon extend beyond one thousand miles.

A British seaplane of the 1914 type equipped with Vick's marine aeroplane gun.

Two classes of guns are needed, those which will represent the artillery of the air and those which will represent the machine-gun service of the air. At present we have only the latter, but experiments are being conducted in the developing of guns of from one-inch to three-inch caliber, and the prospects are that there will soon be such guns available.

A British authority stated recently that there are 40,000 Lewis aeroplane guns in use in Europe. We know, from the size of the orders placed by the United States Army and Navy that there soon will be 10,000 Lewis guns in use in the United States Army and Navy, part of which will be for the Air Service. This gives an idea of the swift development of the light machine gun.

Besides the Lewis gun, there are being used in the present war as aeroplane guns, the Vickers-Maxim, the Colt, the Benet Mercier and the Davis. Owing to the necessity of going to press with the "Textbook of Naval Aeronautics," the discussion of the different types of aeroplane guns, available and prospective, and aerial gunnery in general will be included in the "Textbook of Military Aeronautics" which is to follow, also being published by the Century Company, New York, price $6.00.

Memoranda:

CHAPTER XVIII

SPOTTING THE FALL OF SHOTS

Spotting the fall of shots was one of the first recognized uses for naval aeroplanes. The employment of aeroplanes for this purpose greatly extended the range of vision of ships and became invaluable in long range, indirect, high-angle firing.

At first this work was hampered by the lack of efficient wireless sets to be carried on seaplanes, to make it possible to the aviators to communicate with the man behind the gun. The weight of wireless sets up to 1914 was between three and five pounds per mile of transmission, which was almost prohibitive, as the seaplanes at the time had a small margin of carrying capacity. By 1916 the weight of sets was cut down to one pound per mile of transmission, and the margin of carrying capacity was increased through general improvements in the construction of seaplanes.

The first actual tests of seaplanes to spot the fall of shots took place in July, 1915, when British seaplanes were used to direct the guns of monitors to attack the German cruiser *Königsberg* which was hidden up the Rufigi River, in East Africa. The writer is fortunate in being able to present herewith the first complete report and illustrations of this historic event.

How the Aeroplanes Made It Possible to Wreck the Königsberg

The wrecking of the German cruiser by two British monitors, one of the most remarkable events of the war, was made possible by seaplanes. Following is a letter written home by an English naval officer, which describes the aid rendered by the two aeroplanes, and shows how closely the gunners of the sea, as well as the gunners of the land, have been working as a team with the air scouts. The action described was the attack by monitors upon the German cruiser *Königsberg*. It may be remembered she took refuge up the river on the east coast of Central Africa and was a menace to British interests. It was found after many months up the river where she was hidden from the monitors by palm trees. Aeroplanes were procured after many weeks and action started. The officer of the monitor *Savern* writes:

"We went on higher up the river, and finally anchored. Two shells fell within eight feet of the side and drenched the quarterdeck. It was a very critical time. If she hit us we were probably finished.

"We had no sooner anchored than the aeroplane signaled she was ready to spot. Our first four salvos, at about one minute intervals, were all signaled as, 'Did not observe fall of shot.' We came down 400, then another 400 and more to the left. The next was spotted as 200 yards over and about 200 to the right. The next 150 short and 100 to the left. At the seventh salvo we hit with one and were just over with the other. We hit eight times in the next twelve shots. It was frightfully exciting. The *Königsberg* was now firing salvos of three only. The aeroplanes signaled all hits were forward,

Remarkable snapshot of the *Königsberg* hidden up the tortuous Rufigi River taken from one of the British seaplanes. The *Königsberg* is shown by the arrow, on the right; nearer is a supply ship.

so we came a little left to get her amidships. The aeroplane suddenly signaled, 'Am hit; coming down; send a boat.' As they fell they continued to signal our shots, we, of course, kept on firing. The aeroplane fell in the water about 150 yards from the *Mersey;* one man was thrown clear, but the other had a struggle to get free. Finally both got away and were swimming for ten minutes before the *Mersey's* motorboat reached them—beating ours by a short head. They were uninjured and as merry as crickets."

Following is the official report of the British Admiralty describing the work of the two aeroplanes:

"The position of the *Königsberg* was accurately located by aircraft, and as soon as the monitors were ready the operations were begun. On July 4, as the *Königsberg* was surrounded by jungle the aeroplanes experienced very great difficulty in spotting the fall of the shot. She was hit five times early in the action, but after the monitors had fired for six hours the aeroplanes reported that the *Königsberg's* masts were still standing.

"In order to complete the destruction of the *Königsberg* the commander-in-chief ordered a further attack on July 11, and a telegram has now been received from him stating that the ship is a total wreck."

The following are the extracts relating to the work of aerial scouting on the occasion of the destruction of the *Königsberg* from Vice-Admiral King-Hall's official despatch published on December 9:

"At 5:25 A.M. (on July 6) an aeroplane, with Flight Commander Harold E. M. Watkins as pilot, and carrying six bombs, left the aerodrome on Mafia Island. The bombs were dropped at the *Königsberg* with the intention of hampering any interference she might attempt with the monitors while they were getting into position.

"At 5:40 A.M. another aeroplane, with Flight-Commander John T. Cull as pilot and Flight Sub-Lieutenant Harwood J. Arnold as observer, left the aerodrome for the purpose of spotting for the monitors.

"Returning to the operations of the monitors, fire was opened as before stated at 6:30 A.M., but as the *Königsberg* was out of sight it was very difficult to obtain satisfactory results, and the difficulties of the observers in the aeroplanes in marking the fall of the shots which fell amongst the trees were very great, and made systematic shooting most difficult.

"There being only two aeroplanes available, considerable intervals elapsed between the departure of one and the arrival of its relief from the aerodrome, thirty miles distant, and this resulted in a loss of shooting efficiency.

"At 12:35 one of the aeroplanes broke down, and at 3:50 the second one also. I signaled to Captain Fullerton to move further up the river,

Two Sopwith seaplanes, which "spotted" for the gunners on the British monitors *Savern* and *Mersey* and made it possible to destroy the German cruiser *Königsberg* which was hidden up the Rufigi River in German East Africa.

The *Königsberg* after the bombardment of the British monitors *Mersey* and *Savern,* which was made possible by the aeroplanes. As a ship the *Königsberg* ceased to exist when she was riddled and set on fire by the monitors' guns.

which he did, until about 12:50 the tops of the *Königsberg's* masts were visible.

"As it was necessary to make a fresh attack on the *Königsberg* to complete her destruction, further operations were carried out on July 11, by which date the aeroplanes were again ready for service, and the monitors had made good certain defects and completed with coal.

"The observers in the aeroplanes, by their excellent spotting, soon got the guns on the target, and hit after hit was rapidly signaled. At 12:50 it was reported that the *Königsberg* was on fire.

"I have much pleasure in bringing to the notice of their lordships the names of the following officers and men: Squadron Commander Robert Gordon, in command of the air squadron; Flight Commander John T. Cull, Flight Lieutenant Vivian G. Blackburn, Flight Sub-Lieutenant Harwood J. Arnold, Flight Lieutenant Harold E. M. Watkins.

"Assistant Paymaster Harold G. Badger, H. M. S. *Hyacinth.* This officer volunteered to observe during the first attack on the *Königsberg* though he had had no previous experience of flying.

"Acting Lieutenant Alan G. Bishop, R. M. L. I. of H. M. S. *Hyacinth.* This officer volunteered to observe during the second attack on the *Königsberg* though he had had no previous experience of flying.

"Air Mechanic Ebenezer Henry Alexander Boggis, Chatham 14841, who went up on April 25 with Flight Commander Cull, and photographed the *Königsberg* at a height of 700 feet. They were heavily fired on, and the engine of the machine was badly damaged.

"Most serious risks have been run by the officers and men who have flown in this climate, where the effect of the atmosphere and the extreme heat of the sun are quite unknown to those whose flying experience is limited to moderate climates. "Bumps" of 250 feet have been experienced several times and the temperature varies from extreme cold when flying at a height to a great heat, with burning, tropical sun when on land.

"In the operations against the *Königsberg*

One of the destroyed *Königsberg's* guns mounted in a position inland. It is a 4.1 inch, and was transported to one of the German ports in the Kilimanjaro district in East Africa, used for the campaign and later captured by the British under General Smuts.

The Anti-Aircraft gunner ready to snipe the enemy's aircraft. He is warned of the effect of hostile aircraft by wireless from stations on land or sea.

on July 6 both the personnel and material of the Royal Naval Air Service were worked to the extreme limit of endurance. The total distance covered by the two available aeroplanes on that date was no less than 950 miles, and the time in the air, working watch and watch, was thirteen hours. I will sum up by saying that the flying officers, one and all, have earned my highest commendation."

"Spotting" from a Dirigible

Dirigibles are also very extensively employed in spotting and observing (see Chapter on Naval Dirigibles).

"Spotting" from a Kite Balloon

The kite balloon is used extensively for spotting the fall of shots in naval operations, although there has hardly been any opportunity for the employment of kite balloons to spot in an actual naval engagement. The spotting has been done mainly in connection with the protecting of naval bases and ships from submarine attacks, with an occasional case of warning battleships and directing their fire on some hostile raider which was stealing toward some naval base unseen by the patroling ships. This was particularly true in the Dardanelles and Salonika campaign, where a good number of kite balloons were used, most of them sent up from barges stationed out at sea and kite-balloon ships, from place to place, as explained in the Chapter on Kite-Balloon Ships.

The kite balloon is sent up to whatever height is necessary, 2000 feet or more, and from there the observer officer in the basket telephones to the officer in charge of the kite-balloon ship what he sees, and the officer in charge signals usually by wireless to the battleship's fire control.

Owing to the extreme altitude to which a kite balloon can go, it can see much more extensively than can be seen from a battleship's fire-control station. A kite balloon on any point on Long Island could, for instance, see practically every movement of vessels on both Long Island Sound and the Atlantic, whereas the fire-control station of a ship could only see a few miles. Again, the kite balloon placed on Block Island could detect the movement of ships within a radius of fifty miles, thus warning the battleships and coast defenses of the approach of hostile ships. In the Dardanelles and at Salonika the observations were especially valuable in watching the ports and many bays to prevent Turkish ships from landing munitions and troops from Asia.

A naval kite balloon tethered to a ship guarding the entrance of a French harbor and directing the guns of the ships.

When a hostile ship is detected or the target is on land, the observer in the basket of the kite balloon notifies the fire control of the ship of the location of the target and then of the result of the fire, telling quickly whether the target has been hit or whether the shot was short or over and whether the line is right. This is repeated at every shot and the location of the target and its movements, in the case of a ship moving away, are given very rapidly, so that the gunners of the ship can adjust their range accordingly.

While the information regarding the method of locating batteries and directing gunfire from kite balloons over land are given in the text book on "Military Aeronautics," there may be mentioned here the record of a kite balloon observer, Sub-Lieutenant Maurice Arondel, who is mentioned in the despatches for his long service record and exceptional merit. He carried out his duties from March, 1915, to the end of 1916, when he was mentioned in the despatches.

CHAPTER XIX

BOMB DROPPING FROM AIRCRAFT

It will be remembered that the German official excuse for declaring war on France was an unsubstantiated claim that French aviators had dropped bombs on German soil. The war was only a few days old when the dropping of bombs begun. The first case of bomb dropping reported was on August 13, when a German aviator threw a bomb upon the railroad station at Vesoul, the capital of the Department of Haute-Saone and two bombs in the town of Lure, fifteen miles northeast of Vesoul. This was followed by the dropping of bombs by German aviators on Namur on August 15, 1914. On that day two French aviators flew from Verdun to Metz and dropped two bombs on the Zeppelin sheds there.

On August 25, 1914, took place the first Zeppelin raid, a Zeppelin dropping bombs on Antwerp. On August 30 a German aviator dropped bombs on Paris—which was followed for a time by an almost daily succession of dropping of bombs on the French capitol. On November 21, 1914, three British naval aviators —Squadron-Commander E. F. Briggs, Flight-Commander J. T. Babbington, and Flight-Lieutenant V. S. Sippe—flew from French territory to Friedrichshafen and dropped bombs on that chief German Zeppelin center. In this flight the naval aviators penetrated 120 miles into German territory, crossing mountains under difficult weather conditions. Commander Briggs was shot down and made a prisoner.

A French aeroplane bomb holder.

On Christmas day, 1914, took place the raid on Cuxhaven, reported in the chapter on "Aerial Operations Independent of the Fleet."

On January 11, 1915, fourteen German biplanes raided Dunkirk, being the largest number of aeroplanes employed in one raid up to that date.

On February 12, 1915, thirty-four British aeroplanes and seaplanes, under the command of Wing-Commander Samson, assisted by Wing-Commander Longmore and Squadron-Commanders Porte, Courtney, and Rathborne, raided Bruges, Zeebrugge, Blankenberghe, and Ostend districts. On the same day the French aviators dropped 240 bombs on the German aerodrome at Ghistelle. On March 21–22 took place the first Zeppelin attacks on Paris.

The first air raid on British shores took place on Christmas eve, 1914, when a German aeroplane dropped bombs on Dover, which did no damage. This was followed by a dropping of bombs by a German aeroplane on the Thames Estuary district on December 25. The first Zeppelin raid on British soil took place on January 19, 1915, when Zeppelins visited Norfolk towns, attacking the seaport of Yarmouth, then the Royal Summer Palace at Sandringham and Kings Lynn and Sheringham. In all, nine towns received bombs, Yarmouth being the greatest sufferer. The next attack was made on the night of February 21 and 22, when a Zeppelin dropped bombs on Braintree, Colchester, and Marks Tey.

The air raids on all fronts since 1915 have been so numerous that it would take a large book to report them. The following account of a day's raiding on Zeebrugge, gives an idea of how easily aeroplanes can attack an enemy when no ship can approach the enemy's stronghold. It is the report of the raids of February

An unexploded French aeroplane bomb.

11 and 13, 1915, given in the "London Daily Mail":

On Thursday morning (February 11th) at about 8:30 ten aeroplanes passed high over Dunkirk, coming from the west, and proceeded to Belgium via the coast. These were British machines which had flown direct from England, and they were soon lost in the clouds. An hour later, however, all the aeroplanes were back again, as they had met heavy snow-clouds and only three or four had been able to carry out their mission. One of the machines fell into the sea off Dunkirk. The airman was picked up and the waterplane was towed in. The raid was therefore postponed until night, and at 10 P. M. a second start was made. A methodical bombardment of Zeebrugge was then begun. *Each of the waterplanes in turn rose from the sea, dashed into Zeebrugge, dropped its bombs, and returned to the base at sea.* As soon as one machine returned another left, and thus seventeen consecutive visits were paid to Zeebrugge. While this was going on from the sea British and French aeroplanes left the aerodrome on land and completed the work of their waterplane comrades.

On Friday a further raid was carried out. The entire fleet of waterplanes and the full fleet of British biplanes and French monoplanes took the air together and started all over the German positions in Flanders. Some went as far as Zeebrugge again, while others visited Ostend and Blankenberg. One hundred and forty bombs, of which thirty were very large, were dropped on various ammunition and food depots. The extent of the damage done is not known, but there were German submarines at Zeebrugge, while the Ostend railway station, which was set on fire, was still burning this afternoon, when some French airmen made a reconnaissance as far as Ostend.

The naval operators who participated in raiding expeditions during 1914–15 used, like all aviators, steel arrows. These arrows were about six inches in length, rounded at one end and brought to a needle point. The other end for about four inches was deeply grooved.

Three unexploded bombs found after the bringing down of Zeppelin "L 85" at Salonika, in 1916. A native is using one of the bombs as a pillow.

Effects of Zeppelin bombs on Antwerp, August 25, 1914.

They weighed about six ounces each and were carried in boxes large enough to hold between 500 and 1000. They were dropped on the enemy, by a simple device which opened the bottom of the box.

The sizes and shape and nature of bombs dropped from aeroplanes and dirigibles have undergone a continuous change. At first the bombs weighed mostly from 20 to 25 pounds; later the weight was increased to 250 pounds. The Zeppelins also dropped incendiary bombs intended to set places on fire.

An extended discussion of this subject and a chapter discussing the instruments used in bomb dropping will be given in the forthcoming "Textbook of Military Aeronautics," also published by the Century Company, New York.

CHAPTER XX

AERIAL PHOTOGRAPHY

Aircraft have made photography of tremendous value in warfare. Whereas at the beginning of the present war commanders thought that it was wonderful to be able to get clear written or sketched reports of conditions from air scouts who had flown over the enemy lines, to-day they expect photographic evidence.

It is not sufficient for an aviator to return from a raid and report that he dropped bombs on a given place and did certain damage. He must bring back photographic proofs, and he usually does. Likewise, the observer sent out on an aerial reconnaissance brings back a series of photographs, which are promptly put together and enlarged by experts, and photographic maps are constructed therefrom, showing the exact topography of the country, existing conditions, location, composition, and disposition of fleets or steamers and transports, and of land defenses protecting the approaches to harbors of places of strategic importance. The accuracy obtained by aerial photography is so revolutionary from a military standpoint that it would have been inconceivable before the present war to be able to realize it. As a matter of fact, the commander can have a moving picture taken in a few hours which will show the exact conditions between any two given points, and enable him to plan his operations accordingly.

Of course, the commander of the opposing forces has the same privilege—unless one side has the mastery of the air and is in a position to prevent the aviators of the other side from fly-

This most remarkable photograph of a squadron of Russisan battleships on the Baltic Sea taken at a height of over 4000 feet, is most significant. It shows how clearly the camera reports the composition and the disposition of the enemy's fleet.

A railroad bridge destroyed by an Allied naval aviator in the Balkan theater of war. Aviators must now bring back photographic evidence of their accomplishments whenever possible.

ing over their own lines and taking photographs. To prevent this is of as much importance as preventing the enemy aviators from dropping bombs. Therefore, every effort is made to command the air. The accompanying photographs show the photographic proofs of the aviator's accomplishments.

In a report of a bombing raid on the Turkish lines by a seaplane squadron, which started from a seaplane carrier at Salonika, Lieut. François

Photograph of Gallipoli taken by one of the Allied aviators from a height of 2600 feet. Every inlet is clearly shown, also the ships in the bay.

A British naval aviator starting out to film a military observation.

Bernou, who was on the seaplane carrier *Ben-Ma-Chree,* relates how, after the aviators had returned from the raid they were ordered back to get photographic evidence of the damage done. Returning to the spot to get photographic evidence, instead of waiting to take it after the smoke of the bombs has cleared away, is often wise, as the aircraft guns and the enemy's fighting aeroplanes are usually in action soon after the aviators begin to drop their bombs. This is to be decided after considering the conditions obtaining in each case.

The speed and accuracy in locating things permitted by aerial photography is positively revolutionary. Within a few minutes after the photograph is received, the experts, with a knowledge of the height at which the photograph was taken, and by means of special devices, promptly find the compass direction, location, size of objects, and distance between objects shown on the photograph.

Memoranda:

CHAPTER XXI

RADIO TELEGRAPHY

To the United States belongs the distinction of having made the first experiments to enable a seaplane to communicate with a ship by radio.

The first wireless message ever sent from a hydroaeroplane was received at Annapolis on July 26, 1912, on the United States torpedo boat *Stringham*. The message sent by Ensign Charles Hamilton Maddox from a height of 300 feet was as follows: "We are off the water, going ahead full speed on course for Naval Academy." The pilot of the hydroaeroplane was Lieutenant John Rodgers, U. S. N., an officer of the Navy Aviation School. The apparatus was designed by Ensign Maddox and had several new features, including type of aerial and a receiving device to overcome the noise of the engine. That achievement was especially remarkable because the hydroaeroplanes of that time had a limited lifting capacity and the weight required for dry batteries or storage cells to furnish the current since there were no small generators available to be driven by the engine of the aeroplane, was prohibitive.

Of course Zeppelins and large dirigibles had been carrying radio sets for a number of years, capable of receiving as well as transmitting. It will be remembered that the passenger-carrying Zeppelins often carried such sets, and it was one of the marvels of the passengers of the airships to learn from the crew that the airship was in constant wireless communication with different stations. But the dirigible sets weighed complete between two and three hundred pounds, which would have been prohibitive to the small seaplanes used before 1914.

Until the beginning of the war we marveled

First test of wireless made from a hydroaeroplane in the United States Navy, July, 1912. Ensign Maddox is shown sitting in the navy's first Wright machine, ready for a flight to test its wireless outfit. The complete receiving apparatus is suspended in front of the operator by a strap passing over his shoulders, which protects the delicate device from harmful vibration. The double head telephone receivers are worn under a specially constructed cap, which assists in keeping out external noises. The sending key is mounted on a T-shaped baseboard, the vertical part of which is gripped by the operator's knees. A hot-wire ammeter is mounted beside the key. A switch within reach of the operator throws from sending to receiving. The aerial in this case is permanently fixed to the planes. The insert shows an aerial permanently attached to a navy aeroplane, which required no trailing wire. This type of aerial proved successful for moderate distances of communication.

The cock-pit with arrangement of the 140-mile sending set and the experimental receiving wireless apparatus, invention of Captain C. C. Culver, United States Army.

at a small aeroplane radio set transmitting one mile per pound weight, but the progress made in the past two years has been extensive and at present we get three miles per pound weight. In different chapters are given further details regarding the use of seaplanes and dirigibles equipped with radio for different purposes. In the Chapter on "Spotting the Fall of Shots" are given details about the application of radio from aircraft for spotting the fall of shots.

Lack of time prevents the author from giving in the "Textbook of Naval Aeronautics" a more extensive chapter on Radio Telegraphy. This Chapter will be given in the forthcoming "Textbook of Military Aeronautics," which is being published by the Century Company, New York, price $6.00.

A British authority recently defined briefly four general rules to be adopted by the wireless operators from aircraft as follows:

(1) See before starting that the wireless instrument is properly adjusted to send strong signals.

(2) Don't send when turning; always send when the nose of the machine is towards the receiving station.

(3) Don't send too near the receiving station; a minimum distance of from 2,000 to 3,000 yards gives better results.

(4) Don't send jerkily; send evenly and remember that slow, bad sending is quite as undesirable as quick bad sending. In sending slowly, don't stop in the middle of a word, a set of code letters, figures or coordinates.

A seaplane of the Allies returning to the seaplane carrier at night, in the Mediterranean.

CHAPTER XXII

NIGHT FLYING

The night affords the best opportunities for effective work, and night flying is, therefore, common on the war fronts. Before the war only dirigibles navigated the air at night, but now all types of aircraft go up in the dark. Even kite balloons are sent up at night to scan the face of the waters for ships.

In leaving a seaplane station or seaplane carrier at night, successfully navigating the air, and returning to the base, three distinct things are accomplished. On leaving the base all lights must be subdued as much as possible, to avoid attracting the attention of enemy aircraft, but, otherwise, the task is fairly easy.

To navigate the air in the dark successfully with a seaplane requires experience and knowledge of compass and navigating instruments. The most difficult task is to find the base and return to it. Lights on land or water can be seen from a height of 5000 feet or more, but they may be the lights of hostile ships or enemy bases. To identify them the aviator may drop a flare parachute, which lights the objects below but does not disclose the position of the airman. It is seldom convenient for an enemy's vessel to admit its presence, but when it is, the sky is flooded with the beams of searchlights, and the anti-aircraft guns fire at the aircraft.

The aviator reports his findings to the ship or station by wireless, and when ready to land flashes a Veri light according to the predetermined signals. Then the searchlights of the ship or station are turned on the spot where the aviator is to land, and the landing is made.

First Night Flight Over Water in the United States

The first night flight over water in the United States was made by Lawrence B. Sperry on the evening of September 1, 1916. He flew from Moriches to Amityville, fifty miles away, in pitch dark, lighting his way over the dark waters of the bay with specially arranged lights attached to his aeroplane, and guiding his course by compass.

Mr. Sperry, accompanied by his mechanic, started from Moriches at 8:22 on the evening of September 1, to fly to his hangar at Amityville. His flying boat was equipped with a new night-flying outfit, constructed by Mr.

A parachute flare dropped by an airman to find the nature of the ship below him.

Sperry. After the lights were switched on and the aeroplane started, the machine sped through the black sky with weird effect. The machine, entirely operated by the Sperry automatic pilot, which controls its course and maintains its even keel, and directed by compass, flew without trouble to and landed at Amityville.

The Sperry night-flying outfit consists of a bank of three stream-lined searchlights of 50 candle-power each. Through the use of parabolic reflectors each lamp throws a light beam of approximately 40,000 candle-power. These lights are mounted on a cleverly designed fitting which secures them to the leading edge of either the upper or lower plane. This mounting is so constructed that the lights can be tilted in a vertical plane, making it possible to use them for signaling purposes and at the same time rendering them most efficient for landing. The tilting of the lights is secured by turning a small knob fastened within easy reach of the pilot so that the lights can be operated without interfering with the control of the machine.

The lights themselves are controlled by a specially designed push switch, normally held open by a spring, which is operated like a telegraph key for signaling and, by giving the top a quarter turn, locks in a closed position when desired.

The current supply is secured from a very efficiently designed generator of 150-watt capacity, mounted on a convenient part of the machine, where it will not be in the slip stream, and is driven by means of a wind turbine at 4000 revolutions per minute. By means of an automatic cutout, one of the three lamps remain

Curtiss F boat equipped with Sperry night flying equipment, consisting of the bank of three lights which can be moved in vertical plane, current being supplied by a wind turbine driven generator, shown to left of radiator on top plane. Both the lights and the generator can be seen mounted on the leading edge of the upper plane.

lighted should anything happen to cut off the main current supply. A compact storage battery is automatically thrown into circuit which is otherwise floating on the line.

The subject of night-flying will be discussed thoroughly in all its phases in the "Textbook of Military Aeronautics" issued by the Century Company.

Memoranda:

The instrument board of an aeroplane.

1, Watch; 2, Altimeter (registering height); 3, Compass; 4, Pressure Gages or Two Gasoline Tanks; 5, Dial Registering Engine Revolutions; 6, Inclinometer Registering Level Fore and Aft; 7, Oil Pulsator; 8, Control Stick with Thumb Switch; 9, Switches, Two Magnetos; 10, Air Speed Indicator; 11, Gasoline Supply Pipe.

CHAPTER XXIII

INSTRUMENTS FOR AERIAL NAVIGATION

In the official report of the investigations of the British Royal Flying Corps, made at the close of 1916, there occurs the following detailed description of the aeronautical instruments and maps available to-day for naval air pilots:

"*Maps.*—The question of maps has been unfavorably commented on and some witnesses consider that, if maps had been clearer, pilots would have found it easier to recognize places, and an accident, such as occurred on May 31 last, when Lieutenant Littlewood lost his way and was captured with his machine in Lille, would never have happened.

"The directorate have had great trouble in getting suitable maps, owing to the fact that air operations extend over so many different countries, the maps of which differ in style and scale. For instance, in the course of a flight, machines frequently pass over parts of *England, France, and Belgium.* On the whole, though the construction of maps was necessarily rather slow, we do not consider that any fault can be found with the Royal Flying Corps, nor can we attribute the loss of the machine in Lille to the map, which was the same which all pilots flying across to France use, and seems to us reasonably sufficient.

"Some further adverse comment has been made by witnesses because the maps for the theater of war have not been constructed on the process by which Lord Montagu has produced maps for air pilots over this country. There are several reasons given why these undoubtedly excellent air pilot maps have not been constructed, and with these we are in sympathy.

"(a) The process was a private invention and

quite unknown to the military authorities until December, 1915.

"(b) Even if the invention had been known, it could not have been adopted for military pilots, as it is essential, whilst learning to fly, that the latter should use the same type of map as they will find in vogue in the theater of war.

"(c) To produce maps of the theater of war on Lord Montagu's plan would take time, though it could probably be done.

"(d) The maps, if produced, would be useless, owing to its being essential for the purpose of orders, reports, descriptions, etc., that pilots should use the same maps as the army to which they belong unless Lord Montagu's maps should prove to be suitable for the ordinary work of any army. This we think possible although they have not yet been adopted.

"*Compasses.*—The provision of a suitable compass has presented very real difficulties, and only quite lately has it been possible to invent a really satisfactory one. It appears that the twisting and turning of an aeroplane are so sharp and sudden that no existing compass was trustworthy in an aeroplane; and it has been urged that an indifferent compass was useless, and that the number of high-class compasses was very limited. There are certainly instances in the earlier days of the war of machines flying without compasses owing to there being none to give them; and, later on, isolated instances due to the negligence or rashness of local officers who were responsible for seeing that a machine was properly equipped before leaving the ground.

"*Altimeters.*—The complaint that altimeters were limited to registering a height of 10,000 feet, and that they burst if an aircraft rose above that height, appears to be borne out by fact, and, here again, was a surprise of the war. When hostilities commenced 10,000 feet was considered an ample maximum height, but the range and accuracy of anti-aircraft guns increased, until machines were hit at over 20,000 feet, and the range of the altimeter had to be increased. Consequently, there was a period when there were no suitable altimeters for flights above 10,000 feet, but the committee have no reason to suppose that the period was unduly protracted."

The foregoing official report demonstrates the importance of aeronautical instruments and the development that has taken place in the past few years.

The number and variety of instruments available for navigating aircraft is extraordinary, as is shown by the following specifications prepared by the National Advisory Committee on Aeronautics:

For the information of those concerned with the use or production of instruments used in the navigation and operation of aircraft, the following general list and specifications have been prepared with a view to indicating the lines on which development is required, and the restrictions and difficulties to be overcome in the design and construction of aeronautical instruments:

Barometer or altimeter.
Compass.
Air-speed meter.
Inclinometer.
Drift meter.
Tachometer.
Oil gage.
Oil pressure gage.
Gasoline gage.
Gasoline flow indicator.
Distance indicator.
Barograph.
Angle of attack indicator.
Radiator temperature indicator.
Gasoline feed system pressure indicator.
Sextant.
Aeroplane director.
Stallometer.

General Requirements

All indicating instruments required in the navigation of aircraft should be as compact, rugged, and light as is consistent with accuracy, reliability, and durability, and with ease of reading. Such instruments must be free from the influence of the following disturbing effects, excepting, of course, those effects on which they depend for their operation, viz., vibration, change of altitude, and change of temperature.

Barometer or Altimeter

Barometers or altimeters must be sensitive and of open scale, and the lag in their operation should be the absolute minimum obtainable. When operating in a fog it is essential that the distance above the surface should be known

Aviation barometer or altitude meter.

within very close limits. Such instruments, of course, are dependent on barometric pressure and on variations of barometric pressure from the time of the start of a flight until the completion of a flight, which cannot be provided for, but aside from this error their indication should be substantially accurate once they are adjusted at the point of departure. It is, therefore, necessary that the scale should be of equal divisions, as otherwise a change of zero to meet change of barometric height will introduce an error. Their location on the aeroplane must be carefully chosen so that their indications will not be influenced by the velocity pressures in flight.

Compass

Compasses should have as high a directive force as is consistent with restricted dimensions. Provision should also be made in the compass mounting for compensation for the presence of magnetic material in the construction of the aeroplane, particularly compensation for heeling and dipping errors. In order that the directive force shall not be abnormally reduced by such compensation, it is, of course, desirable that the structure should avoid the use of magnetic materials in moving parts near the compass location, such as the control columns, shafts, and leads.

Air-Speed Meter

An air-speed meter should indicate reliably the speed through the air, and should be free from the effects of accelerations, as when the machine is banking strongly in a turn the effect of gravitation is augmented by the presence of the centrifugal force. As the sustaining power of an aeroplane is dependent upon the density of the atmosphere, it is considered that air-speed meters which are dependent on the pressure due to velocity will be a safer form of indicator than a true anemometer type.

It is essential that the indicators shall be particularly sensitive and have an open scale reading at velocities approaching a stalling speed, which is the lower limit of safe flying speed. It is also necessary that they should indicate high speeds accurately, in order that excessive speed may be avoided when gliding. Excessive speed in gliding involves danger when a machine is brought up too sharply, as the combination of high speed and the maximum lift factor may readily stress the machine beyond safe limits. Also, when flying at high speed the angles of attack are small, and there is danger of the aeroplane entering a critical condition in which the flow of air may develop radical changes of state,

New type of aeroplane compass.

and consequently great changes in the lifting power available. Air-speed meters should be capable of calibration immediately prior to a flight. Air-speed meters of the Pilot type dependent on a fluid are subject to gravitational errors when banking. They are also subject to error due to heeling or diving. Unless the leads from the Pilot tube to the indicating instruments are sufficiently large, there is also danger of a serious lag in indications.

Inclinometer

Inclinometers of the pendulum or spirit-level type are inaccurate in the presence of accelerations and are only useful as a general check as to the attitude of the machine when flying in a fog.

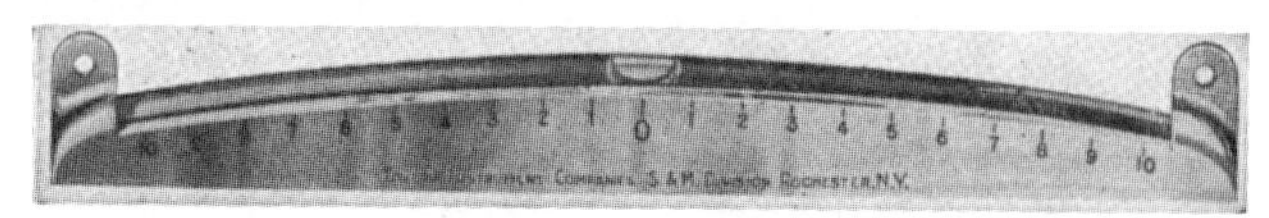

Aeroplane Inclinometer.

It is very desirable that an indicator free from these defects should be developed. A gyroscopic base line is considered desirable not only for purposes of indicating inclination but as affording a base line for sighting and for the use of instruments of navigation.

Drift Meter

Drift meters are of two types—one designed for the purpose of indicating leeway over the surface for use in connection with navigation, and the other, more properly termed "side-slip indicator," for the purpose of indicating whether or not the machine is flying square to the wind. The latter designation is considered preferable for indicating the attitude of the machine. For navigating over the ground the course is readily determined by ascertaining the apparent motion of objects on the surface, and the same method is available for navigating over the water, provided there is a definite object on which to sight. One type of drift meter indicates by the streaking of waves across the objective glass of the instrument as apparent drift, but as the particles of waves themselves which indicate this streaking have a velocity of their own, such indications are subject to error. If the surface wind direction or velocity were known, correction might be made, but when flying at an altitude of several thousand feet it is very likely that the aeroplane itself may be in an entirely different current of air than that present at the surface. In addition to this, tidal currents may also affect the velocity of the water particles. Two forms of side-slip indicators exist, the simplest form being that of the well-known string or pennant, but the latter cannot be used satisfactorily in the wake of a tractor propeller. The other type consists of a very sensitive pendulum which indicates whether or not lateral accelerations are present, as will be the case for a machine which is not properly balanced laterally, but such an instrument is subject to the defect that if the machine is side slipping laterally at a constant speed, lateral acceleration is no longer present. It can only be depended on to indicate initial disturbances.

Tachometer

Tachometers should be absolute in their indications, and if electrical should not be subject to disturbances in the conductivity of circuits from any cause, or to deterioration of magnetism of a permanent magnet.

Oil Gage

Oil gages must definitely indicate the amount of oil present in the crank case

Oil gage used on aeroplanes.

Oil-Pressure Gage

Oil-pressure gages must acurately indicate the pressure in the oil system and should also indicate that the flow of oil is undisturbed.

Gasoline Gage

Gasoline gages should indicate the amount of gasoline available in the main tanks, and should not depend on the visibility of gasoline in a glass tube, as, due to the transparency of gasoline, a full tank and an empty tank would give the same indications. Mechanical indicators are considered preferable.

Gasoline-flow Indicator

Gasoline-flow indicators should depend on mechanical means of indicating that the gasoline is being supplied from the main tanks to the service tanks.

Distance Indicator

For navigation at sea or over unknown country, it is desirable that a record of distance flown through the air should be available. If it were not for the fact that the slip of the propeller depends largely on the load of the machine, and whether or not the machine is climbing or gliding, an engine counter would serve this purpose, but it is considered preferable to have a counter or recorder actuated by an anemometer for this purpose. In either case, actual distance over the surface will require correction for the wind velocity and direction.

Barograph

Barographs are subject to the same general specifications as altimeters.

Angle of Attack Indicator

An angle of attack indicator should be dead beat, free from the effects of gravitation, and accurately respond to and indicate any change of the directions and flow of air to the supporting surfaces. It should be light, rugged, and its indications should be clearly legible to the pilot. It should be designed for attachment in advance of the wings on a tractor biplane and clear of the influence of the propeller or the fuselage.

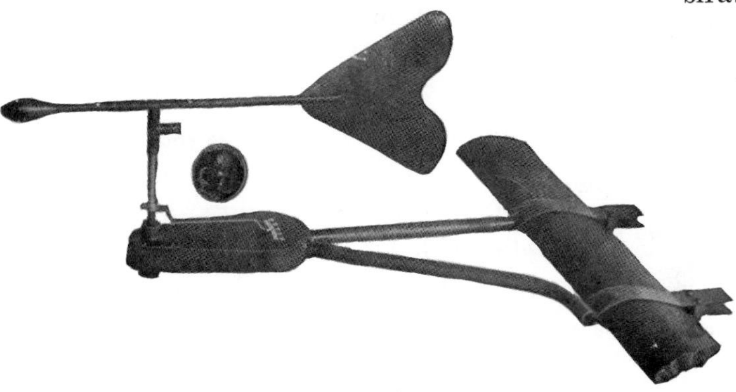

Incidence indicator.

Radiator Temperature Indicator

A radiator temperature indicator should be readily inserted in the top of the radiator and should clearly indicate the best operating temperature. The thermometer should conform to best practice, and the entire instrument be sufficiently rugged to withstand reasonable vibration and shock.

Gasoline Feed System Pressure Indicator

Where the gasoline feed is not gravitational, the indications of the pressure available must be accurate. The gasoline feed system pressure indicator must not be affected by vibration or change of temperature. It must have a good scale and a deadbeat action.

Sextant

Sextants should be as light and small as possible commensurate with proper accuracy. A sextant for measuring the altitude of a heavenly body above a horizontal plane without the use of the sea horizon or an artificial horizon would be most desirable.

Aeroplane Director

An aeroplane director for the mechanical solution of the course and distance made good, based on the course and speed of the aeroplane and the force and direction of the wind, is a desirable development.

Warns the Aviator Against Stalling

The stallometer is an instrument that warns the aviator when his machine is approaching a stalling condition by indicating that the minimum air speed has been reached. It is mounted in any convenient position where the air flow is unobstructed.

The Sperry Stallometer.

The Sperry Clinometer.

The Sperry stallometer is adjustable for any desired air speed, depending on the aeroplane on which it is installed. When the predetermined speed is reached an electric contact is made in the stallemometer, closing the circuit to an indicating lamp which is mounted on the instrument board.

Signals Aviator to Keep the Aeroplane Level

Aviators wishing to know at any time the correct fore-and-aft position of the machine, with reference to the horizontal, can read it on the scale of the Sperry deadbeat clinometer.

Cockpit of Curtiss aeroplane flown by Carlstrom in Chicago-New York flight.

The operation of this instrument is simple. Whenever the clinometer is tipped forward or backward by the motion of the aeroplane this movement is registered on a scale mounted on a wheel which is damped by floating in a liquid.

If the aeroplane tips forward, the scale moves upward indicating in degrees below the zero line the exact angle. If the machine tips backward, the scale moves downward, the exact amount which is likewise shown in degrees. The scale is painted in radium so that it is visible at night by its own light.

The case, measuring four and three-eighths inches in over-all diameter, is made of a bronze spinning and painted black. The clinometer is usually mounted on the instrument board in the pilot's cockpit. But it may be placed elsewhere in the fuselage, providing that location is such that the instrument can be seen at all times. Though comparatively new, the Sperry clinometer has already rendered valuable and efficient service in its particular field.

The United States Navy aero camp and hangar tents at Guantanamo Bay.

CHAPTER XXIV

UNITED STATES NAVY AERONAUTICS

To the United States Navy belongs the distinction of having been the first to take steps to organize an aviation section. This was early in 1911, when Congress made the first appropriation for naval aeronautics, the amount of which was $25,000.

Three officers—Lieutenants T. G. Ellyson, John Rodgers, and John H. Towers, U. S. N.—were ordered to aeroplane factories for instruction, and three machines—two Curtiss and one Wright—were purchased. A land aerodrome was established on Greenbury Point, Annapolis, Maryland, with three hangars for the machines, which did not yet have pontoons.

These officers made extensive experiments in launching aeroplanes by means of a cable-launching device at Hammondsport, New York, which was the forerunner of the catapult developed by Captain W. I. Chambers, U. S. N., who was in charge of naval aviation at that time.

Subsequently, in the spring of 1912, the personnel of the Aviation Section was increased by the addition of Ensign V. D. Herbster, U. S. N.; Assistant Naval Constructor H. C. Richardson, U. S. N.; Lieut. I. F. Dortsch, U. S. N.; Lieut. L. N. McNair, and First-Lieut. A. A. Cunningham, U. S. Marine Corps.

Between April 12, 1911, and the latter part of August, 1912, a total of 593 flights were made by the instruction officers—Lieutenants Ellyson and Towers in the Curtiss machines and Lieutenants Rodgers and Herbster in the Wright machine. In December, 1911, the three machines and the aviators were transferred to San Diego, and a camp was established there for the winter season. Then it was transferred to Annapolis, near the Engineering Experiment Station, on the north shore of the Severn River.

During this period the United States Navy led the navies of the world in naval aviation. Its experimentations, while not extensive, were of fundamental importance, and its aviators made history in many ways.

On May 9, 1912, Rear-Admiral Bradley A. Fiske, commander of Second Squadron, Atlantic Fleet, made the first aerial reconnaissance ever made by a naval officer of rank. W. Starling Burgess and P. W. Page flew to and landed alongside the U. S. S. *Georgia,* the flagship, anchored at Salem Harbor, and subsequently Rear-Admiral Bradley A. Fiske made a flight on the Burgess-Wright hydroaeroplane piloted by P. W. Page.

On June 28, 1912, Lieut. John Rodgers flew from the Aero Station near Annapolis to the

The Naval Aviation Camp at Annapolis in 1912—lower left hand corner; the old Aerodrome on Greenbury Point marked by arrow.

battleship *Louisiana.* Ascending at 11 o'clock, the aviator shot his plane directly ahead at an altitude of about 400 feet, and alighted on the starboard side of the battleship about thirty-five minutes later. The aviator subsequently performed a number of evolutions and returned to the experiment station opposite Annapolis at four o'clock.

On October 6, 1912, Lieut. John H. Towers made a world's duration record for hydroaeroplanes, and an American endurance record for aeroplanes of any kind by flying a Curtiss hydroaeroplane six hours and ten minutes, flying from Annapolis over the Chesapeake Bay. On December 17 of that year, the anniversary of the first aeroplane flight, there was tried at the Washington Navy Yard Captain W. Irving Chamber's catapult device for launching aeroplanes from battleships. The new Curtiss flying boat, carrying Lieutenant Ellyson, was launched successfully. (See chapter on Launching Aeroplanes From Ships.) There were also conducted valuable experiments in radio telegraphy communication between hydroaeroplanes and ships.

On October 9, 1913, the Navy Board convened a Board of Aeronautics consisting of Capt. W. Irving Chambers, U. S. N., Aeronautic Expert; Commander C. B. Brittain, U. S. N., Asst. Chief of Bureau of Navigation; Commander S. S. Robison, U. S. N., Assistant Chief of Bureau of Steam Engineering; Lieut. M. H. Simons, U. S. N., of the Bureau of Ordnance; Naval Constructor H. C. Richardson, U. S. N., Bureau of C. and R.; Lieut. J. H. Towers, U. S. N., Aviator; First-Lieut. A. A. Cunningham, U. S. M. C. This board rendered a report, making many important recommendations, including the assigning of a reserve ship to aeronautic duty, the establishing of an aeronautic station at Pensacola Navy Yard, and the acquisition of dirigibles, observation balloons, and other equipment for a substantial air service. There was also recommended the establishing of an office of navy aeronautics, "to be under the charge of a director of naval aeronautics with the rank of captain, if practicable, who shall coordinate the work of the office for the Secretary of the Navy in conformity with the departmental organization and in cooperation with the necessary assistants representing the bureaus."

The U. S. S. *Mississippi* was detached from the reserve fleet and assigned as aeronautic station ship at Pensacola, Florida. It remained on aeronautic duty until the summer of 1914, when it was sold to Greece, and the *North Carolina* was assigned to take its place.

United States Navy Aviation Section Holds Distinction of Being First to Operate Under Conditions Approximating Warfare.

The Aviation Section of the United States Navy distinguished itself in the summer of 1914 at Vera Cruz. What the naval aviators accomplished at Vera Cruz, as well as an interesting discussion of the status of seaplanes at that time, and an expression of the United States Navy's needs were given in the following letter from Secretary of the Navy Daniels to the writer, under date of May 19, 1914:

NAVY DEPARTMENT

Washington

May 19, 1914.

To Mr. Henry Woodhouse,
297 Madison Ave.,
New York City.

Dear Mr. Woodhouse:

(1) Your letter of the 8th instant has been received.

(2) The Navy had just about established an aeronautic center at Pensacola, Florida, when the mobilization of the Fleet in Mexican waters became necessary. Aeroplanes are now considered one of the arms of the Fleet the same as battleships, destroyers, submarines, and cruisers. That the Navy's aeronautical service was well organized was shown by the prompt way in which the Aeroplane Division got away from Pensacola to take part in the mobilization in Mexican waters. At noon, Sunday, the 19th of April, Patrol orders were received at Pensacola for the First Aeroplane Section to embark on the *Birmingham.* In six hours two aeroplanes, with all spare parts for active service, two hangar tents, and tents and camp equipment for three officers and ten men of the Section, were on dock ready to go on board the *Birmingham.* The Second Aeroplane Section was just as quick when the orders were issued. The Aeronautical training ship *Mississippi* under the command of Lieut-Commander H. C. Mustin was also ordered from Pensacola to join Admiral Badger's Fleet off the east coast of Mexico. The Second Aeroplane Section was embarked on board the *Mississippi.* The First Aeroplane Section is under the command of Lieut. J. H. Towers, U. S. N., and the Second Section under the command of Lieut. P. N. L. Bellinger, and these two sections forming a division are under the command of Lieut-Commander H. C. Mustin.

(3) The First Aeroplane Section on board the *Birmingham* has been stationed off Tampico, and there has been no necessity for any work by this section.

(4) The *Mississippi* with the Second Aeroplane Section on board arrived off Vera Cruz on the 25th of April and within five minutes after the anchor was dropped one of the aeroplanes was in the air. Every day since then and often more than once a day the navy aeroplanes have scouted along the outposts and far beyond our lines, mapping the country and observing the motions of the Mexican forces.

(5) The latter part of last year, Mr. A. B. Lambert of St. Louis, with much public spirit, organized a United States Aviation Reserve Corps. Recently, when it seemed likely that our armed forces might be called out, Mr. Lambert volunteered to assist the Navy Department to ascertain the aviators, mechanicians, and aeroplanes that would be available if the Navy required them. Mr. Lambert at his own expense has traveled widely over the country, gathering valuable information for the Navy. In addition, there has been a large number of aviators who have personally offered their services to the Government. All such applications are on file ready for reference if the necessity arises to call for volunteers. It is gratifying to find that the Navy could be most ably reinforced in its Air Service. It does not seem probable that volunteers will be called for, but this experience has been invaluable in collecting such useful information, and will also lead to better plans for a reserve or volunteer force for future eventualities. If volunteers ever are needed the greater the knowledge they have of flying in the open sea the better they will

The hangars of the Annapolis aerodrome in 1911. The navy Wright machine before hydroplanes were placed on it.

be able to acquit themselves in assisting the Navy.

(6) The greatest amount of flying for pleasure, sport, or commercial purposes is over land or inland waters. The aeronautic service for the Navy must operate over the open sea. Therefore, the greatest developments in aeroplanes have been along lines that have not given the best aid to the Navy. Hydroaeroplanes and flying boats well fitted for use on inland waters fail in rough water in the open sea. At Monaco last April, the most experienced aviators like Prevost, Jansor, Moinau, and Berlin did not start a race because of a rough sea, and Garros only succeeded after a second trial, while Brindejonc des Moulinais gave it up after trying three times. Hirth, the German aviator that made a world's record flight from Gotha to Marseilles, capsized when landing on the water in Samaris Bay. Moineau made a fine landing in the same bay, but was capsized by a large wave while trying to get away. Those aviators who made the successful flights from Marseilles to Monaco in the "Aerial Rally" were fortunate in not finding rough water when they had to make the landing on the water required by the rules.

(7) It is probable that many of the problems of aeronautics in the Navy will have to be solved without much aid from outside sources. Anyway, this will be so until such time as cross-ocean flights become a commercial success, or at least readily accomplished. The Navy now feels that a flying radius of 800 miles is required for its aeroplane service. It is not too much to expect, but to accomplish this there must be much improvement in motors and refinement of design and of construction of the aeroplanes. I am sorry we have no photographs of the operations in Mexican waters to send you.

Very truly yours,

(SIGNED) JOSEPHUS DANIELS,
Secretary of the Navy.

Besides the officers mentioned in Secretary Daniels's letter the following officers participated and distinguished themselves in the operations in Mexican waters: Lt. R. C. Saufley, Ensigns M. L. Stolz and W. D. Lamont.

Captain W. Irving Chambers having retired, Captain Mark L. Bristol was appointed in charge of the Office of Naval Aeronautics, as director of Naval Aeronautics; Lieut. John H. Towers was sent to London, England as assistant naval attaché, and the number of aviators was again increased, but on account of the shortage of personnel in the Navy, it was impossible to assign to aeronautic duty the number of officers and men which the Department felt should be assigned.

Owing to lack of appropriations, the aeronautic station at Pensacola was operated on a very limited basis until 1916, the work being further delayed by a succession of storms, which wrecked the tent hangars, buildings, aeroplanes, and equipment. Another storm on July 5, 1916, found the station better, but not entirely, prepared for it. Buildings had taken the place of tent hangars, but the storm demolished buildings, aeroplanes, and run-ways. Now the hangar doors are east to west, therefore least exposed to the storms.

Navy Department Decides Against Government Construction of Aircraft

At the close of 1914, there was suggested in the House of Representatives the establishment

The first hydroaeroplane of the United States Navy Aviation Section, 1911.

Ensign V. D. Herbster, U. S. N., making a landing in a Wright machine, after a trial flight in 1912. The Naval Academy and station-ship *Hartford* on the right background.

of a government factory for aircraft. Secretary Daniels had the subject considered and transmitted to the House of Representatives under date of December 14, 1914, the following report, which advised against the establishing of a Government aeroplane factory:

December 14, 1914.

FROM: Bureau of Construction and Repair and Bureau of Steam Engineering.
TO: Navy Department (material).
SUBJECT: Aeroplanes.
REFERENCE: (a) Department's memorandum, December 12, 1914.

(1) While the initial successes in air-craft work were attained in this country, the design and construction here on a successful scale are still in the development stage. Foreign countries are far in advance of our builders. The marked progress of this class of work abroad is due mainly, if not solely, to the encouragement given to private manufacturers by foreign governments. While there are only a few companies in this country that can at present be considered as competent designers and builders, their number is sufficient to stimulate competition and bring about great improvement in design, provided there is a reasonable amount of Government business in sight. Furthermore, there are other companies that are only awaiting the existence of sufficient business to develop their ideas along the same line.

(2) While the Government has resources, including a few officers specially trained in aeronautical-design work, this force can at present be considered only a nucleus and is capable of carrying on only a very limited volume of work. It would be a tremendous loss to the advancement of aeronautical work to lose the ideas and results of private invention and experiment.

(3) In view of the above and in view of the extremely hazardous nature of aircraft work, involving the loss of life and property, if not designed and manufactured with extreme care and along what experience has taught to be the safest lines, the bureaus believe that it would be a great mistake for the Department to undertake at the present time a manufacture of aircraft except on an experimental scale.

(4) Preparations have already been under way for about two years looking to the design and construction of an experimental machine, with a view to developing ultimately the necessary plans, specifications and detail instructions for the manufacture of aeroplanes, but hulls and power plants, by private manufacturers, including shipyards, and by navy yards in an emergency. This experimental work includes a continued series of laboratory experiments on a large scale at the navy yard, Washington. The preliminary work toward the experimental construction above mentioned is already in hand, and it has been the bureau's intention to take up the manufacture of such an experimental aeroplane at the Washington Navy Yard in the near future.

(5) The establishment of a Government plant for the general manufacture of aircraft would require a complement of officers that can ill be spared at the present time, and not only because the Navy has a very limited number of specially trained designers in this class of work, but because such a plant would call for the diversion from actual flying work of many of the most competent operators. As stated above, the establishment of such a plant would tend greatly to discourage the valuable initiative and resources of private manufacturers, who should be encouraged and stimulated as a most valuable asset not only in the development of aircraft but also for turning out such craft in quantities in time of an emergency. Any government plant which could be established in the near future would be entirely inadequate in war time, as aircraft would be required in large quantities in such an emergency.

(6) It is therefore recommended that the utilization of existing plants for aeroplane work be confined to the construction of an aeroplane engine at one of the navy yards, with a view to the preparation of de-

The earliest experiments in launching a hydroaeroplane, made in 1911, when the United States Navy led the world's navies in aeronautics. Lieut. T. G. Ellyson, U. S. N., who shared with Lieut. John H. Towers, now Lieut.-Commander Towers, the distinction of being the earliest naval aviators, is shown at the wheel. The machine is about to be launched on the cable. Lieutenant Towers is shown on the left, holding one of the ropes. Augustus Post is shown below the machine, in the center. This device did not prove practical but afforded valuable experience.

partmental plans, specifications, and manufacturing instructions in sufficient detail for use in an emergency.

(7) However, if the Department directs the establishment of a plant for the manufacture of aircraft, it is recommended that the work be done either at the navy yard, Philadelphia, or the navy yard, Norfolk, these yards having a moderate amount of space for testing work. A considerable portion of the necessary plant is already available at these yards, but certain special tools would be required, some delay would be experienced in training a special force of mechanics, who would have to be instilled with the supreme importance of perfect workmanship. The approximate estimated cost of putting the shops at one of these yards in order and establishing an aircraft factory with a capacity of two or three machines per month is placed at $30,000. The estimated cost of turning out such machines under the present navy yard cost system is about $6000. This does not include the cost of the commissioned personnel, classified employees, leave, holiday, and disability, and certain other overhead charges not at present included in the cost of work, and does not include the question of patent rights; all of these would probably run the actual cost much above the above figures.

(SIGNED) SCHAEFER, *Acting*,
R. S. GRIFFIN.

The project was again suggested in the early part of 1915. The writer asked Secretary Daniels for a statement, and received the following:

THE SECRETARY OF THE NAVY,
WASHINGTON,
April 27, 1915.

Dear Mr. Woodhouse:

I thank you for the newspaper clippings which you enclosed and for your letter of the 17th instant. The Advisory Committee on Aeronautics had not organized at the time this statement was made in the clippings you were good enough to send. This Advisory Committee, provided for as you will undoubtedly recall in the Naval Appropriations Bill upon my recommendation, has its duties and powers defined in that bill. It is the belief of the Department that its advice will be of very great value, within the limitations of its functions.

The question of whether aircraft should be manufactured by the Navy Department is not a question that would come before this Advisory Committee in any way. The position of the Department was clearly expressed in my approval of the reports of the Bureau of Construction and Repair and the Bureau of Steam Engineering, transmitted to Congress last December. I have seen no reason to change the position then taken. The sole desire of the Department is to use the money appropriated by Congress in a way that will, as rapidly as possible, make this arm of the service as effective as possible? The Director of Naval Aeronautics and the experts at the aeronautic station at Pensacola and our observers abroad are giving everything regarding the improvement of aircraft in the Navy their earnest attention and consideration.

We are much gratified in the assurance we have received from designers and constructors that they will give us a suitable type of aircraft for the Navy, and that it will be possible to make a rapid increase in our aeroplane fleet without much delay.

I thank you sincerely for the deep interest you show in our aeronautic service, and will always appreciate your interest and suggestions.

With sentiments of esteem and high regard, believe me,

Cordially yours,

(SIGNED) JOSEPHUS DANIELS.

The First United States Navy Dirigible

In 1914–15 attention was given to getting a dirigible. Specifications were issued by the office of Naval Aeronautics on March 20, 1915. Bids for one or two dirigibles were asked and opened at the Navy Department in Washington on April 20, 1915. The general specifications required that the dirigible should be of the nonrigid type, about 175 feet long and 50 feet high, 35 feet in diameter, with a useful load of about 2000 pounds. It was specified that the dirigibles should have a speed of 25 miles an hour or more, and to be capable of rising 3000 feet without disposing of ballast. The bids submitted were as follows:

American Dirigible Balloon Syndicate, Inc., New York, N. Y.
One machine—$41,000.00.
One machine (larger)—$45,000.00.

The Connecticut Aircraft Company, New Haven, Conn.
One machine—$45,636.25.
Two machines—$82,215.12.

The Goodyear Tire & Rubber Company, Akron, Ohio.
One machine—$200,000.00.

(This bid was subject to a reduction which will make the total cost to the Government equal to the cost of the machine to the Goodyear Tire & Rubber Company plus 50 per cent. The amount entered as the bid is the maximum to be charged under any condition.)

Contract for one dirigible was awarded to the Connecticut Aircraft Company.

Specifications for Hydroaeroplanes, 1915

Specifications for hydroaeroplanes were also issued and bids for supplying three or six machines were opened on February 27, 1915. The following bids were received:

Firms	Item 1	Item 2	Item 3	Item 4	Item 1a	Item 2a	Item 3a	Item 4a	Remarks
Aircraft Company, Inc.....	$6,962.00	$5,142.00	$716.00	$2,760.00	$7,962.00	$5,000.00	$725.00	$3,000.00	Should automatic stabilizer be accepted with each aeroplane the cost of Item 2 power plant in each case will be reduced $190.00.
	......	......	.....	......	6,780.00	4,837.00	716.00	2,760.00	
Burgess Company	6,400.00	4,325.00	280.00	......	5,350.00	4,325.00	280.00	......	Has inherent stability. Wireless outfit and lighting outfit not included as no definite approved type is specified.
Curtiss Aeroplane Co......	10,500.00	7,000.00	425.00	3,000.00	10,500.00	7,000.00	425.00	3,000.00	Informal—No Guarantee.
Gallaudet Company, Inc....	18,000.00	......	.....	......	......	......	.....	......	For one machine.
Grinnell Aeroplane Company	6,500.00	8,000.00	500.00	......	......	......	.....	......	
William C. Hurst..........	7,500.00	3,500.00	.....	......	......	......	.....	......	Informal—No Guarantee.
Peoli Aeroplane Corporation	3,100.00	3,700.00	500.00	......	3,100.00	3,700.00	500.00	......	
Shaw Aeroplane Company..	4,499.00	3,415.00	586.00	......	......	......	.....	......	Informal—No Guarantee.
B. F. Sturtevant Co........	......	4,325.00	.....	......	......	4,325.00	.....	......	Price does not include wireless and lighting outfits, but includes fitting of such outfits if furnished by Government.
Thomas Bros. Aeroplane Co.	4,600.00	3,550.00	750.00	......	......	......	.....	......	Type H. S.
	5,850.00	6,380.00	750.00	......	......	......	.....	......	Type S. Prices do not include wireless or lighting outfits.
The Tygard Engine........	......	14,000.00	.....	......	......	14,000.00	.....	......	
The Wright Company......	9,740.00	5,200.00	60.00	......	7,500.00	4,940.00	60.00	......	Price does not include wireless and lighting outfits. Price does not include compass, chart holder and sextant.
B. Stephens & Son.........	3,000.00	3,400.00	200.00	......	......	......	.....	......	If Sturtevant motor is furnished Item 2 will be $4,200.00.
G. H. Armitage............	3,800.00	4,300.00	250.00	......	......	......	.....	......	Informal—No Guarantee.

On September 1, 1915, an order went into effect providing that officers attached to the aeronautic station at Pensacola should perform administrative and executive duty in the upkeep and proper maintenance of the Pensacola station, these duties to be performed as secondary to the aeronautical work at such time as flying was not possible, due to unfavorable weather conditions. This was done to give the officers experience in navy yard administration as a preparation for the officers who might in the future be placed in command of shore stations. The complete aeronautic commissioned personnel in September, 1915, was as follows: Lieutenant Commander H. C. Mustin, U. S. N., Naval Aviator and Commandant of the U. S. Navy Aeronautic Station and U. S. Naval Reservation; Lieutenant K. Whiting, U. S. N., Naval Aviator and Captain of the Yard; Lieutenant J. H. Towers, U. S. N., on duty, London, England, Assistant Naval Attaché; Lieutenant A. C. Read, Student Naval Aviator; Lieutenant E. F. Johnson, Student Naval Aviator; Lieutenant L. H. Maxfield, U. S. N., on duty Akron, Ohio, Goodyear Tire & Rubber Company; Lieutenant (j. g.) P. N. L. Bellinger, Naval Aviator in Charge Erecting and Test Division; Lieutenant (j. g.) R. C. Saufley, Naval Aviator, in Charge Flying School; Lieutenant (j. g.) V. D. Herbster, Naval Aviator, duty Berlin, Germany, Assistant Naval Attaché; Lieutenant (j. g.) P. R. Paunack, Student Naval Aviator; Lieutenant (j. g.) F. G. Haas, Student Naval Aviator; Lieutenant (j. g.) C. K. Bronson, Naval Aviator, Assistant to Officer in Charge Erecting and Test Division; Lieutenant (j. g.) W. Capehart, Naval Aviator, Planning Superintendent; Lieutenant (j. g.) W. M. Corry, Student Naval Aviator; Lieutenant (j. g.) J. E. Norfleet, U. S. N., Student Naval Aviator; Lieutenant (j. g.) Lieutenant (j. g.) G. de C. Chevalier, U. S. N., Naval Aviator; Inspection Duty, Marblehead, Massachusetts; Lieutenant (j. g.) W. A. Edwards, U. S. N., Student Naval Aviator and Assistant to Captain of Yard; Lieutenant (j. g.) E. W. Spencer, Jr., U. S. N., Student Naval Aviator, and Assistant to Officer in Charge Erecting and Test Division; Lieutenant (j. g.) G. D. Murray, U. S. N., Student Naval Aviator and Asst. to Officer in Charge, Erecting and Test Division; Lieutenant (j. g.) H. T. Bartlett, U. S. N., Student Naval Aviator and Assistant to Officer in Charge Motor Erecting Shop; Lieut. (j. g.) E. O. McDonnell, U. S. N., Student Naval Aviator, and Asst. to Planning Superintendent; Lieut. (j. g.) H. W. Scofield, Student Naval Aviator; First-Lieut. A. A. Cunningham, U. S. M. C., Student Naval Aviator and in Charge Motor Erecting Shop; First-Lieut. B. L. Smith, U. S. M. C., duty, Paris, France, Asst. Naval Attaché; First Lieut. F. T. Evans, U. S. M. C., in Charge Barracks, Building 45; Second-Lieut. W. M. McIlvain, U. S. M. C., Naval Aviator on Inspection Duty, Hammondsport, New York.

Lieut. Alfred A. Cunningham, U. S. M. C. flying over the battleship *Connecticut* in a Burgess Hydroaeroplane in 1913.

United States Navy Experimental Wind Tunnel

Early in 1916 there was established at the Experimental Model Basin, where warship models are tested, a large experimental wind tunnel. The tunnel has a section eight feet square at the point where the models are placed for testing, and is equipped with a 500 horse-power motor

One of the U. S. N. Air Scouts at Vera Cruz during the 1914 expedition. U. S. N. ships shown in the distance. U. S. S. *Utah's* special corps underneath in the Municipal Building.

driven fan, giving wind speeds up to seventy-five miles an hour.

Specifications for Seaplanes, 1916

In August, 1916, the Navy Department issued specifications and asked bids for supplying three, six, nine, and twelve aeroplanes and power plants. The bids were opened on September 5, 1916. The performance required was as follows:

Maximum Speed—Not less than 52.1 knots (60 miles per hour, nor more than 60.7 knots (70 miles) per hour.

Minimum Speed—Not more than 34.7 knots (40 miles) per hour.

Climb—2500 feet in first ten minutes from surface.

Landing—Not over 34.7 knots (40 miles) per hour.

Radius—4 hours with full power.

Fly in wind of 30.4 knots (35 miles) per hour.

Drift in wind of 21.7 knots (25 miles) per hour.

Get-away and land in wind of 21.7 knots (25 miles) per hour.

United States Navy Experimental Seaplane

During the summer of 1915 naval construction, H. C. Richardson designed a large twin-motored seaplane. On October 27, 1915, Secretary Daniels signed an order for the construction of this seaplane and it was built at the Washington navy yard.

Act Increasing Pay of Naval Aviators

The act granting special pay and allowances to officers of the Navy and Marine Corps detailed to aviation duty which became a law in 1915 is as follows:

Hereafter officers of the Navy and Marine Corps appointed student and naval aviators, while lawfully detailed for duty involving actual flying in aircraft, including balloons, dirigibles, and aeroplanes, shall receive the pay and allowances of their rank and service plus 35 per centum increase thereof; and those officers who have heretofore qualified or may hereafter qualify, as naval aviators, under such rules and regulations as have been or may be prescribed by the Secretary of the Navy, shall, while lawfully detailed for duty involving actual flying in air craft, receive the pay and allowances of their rank and service plus fifty per centum increase thereof. Hereafter enlisted men of the Navy or Marine Corps, while detailed for duty involving actual flying in air craft, shall receive the pay, and the permanent additions thereto, including allowances, of their rating and service, or rank and service as the case may be, plus fifty per centum increase thereof: Provided, That not more than a yearly average of 48 officers and 96 enlisted men of the Navy, and 12 officers and 24 enlisted men of the Marine Corps, detailed for duty involving actual flying in air craft, shall receive any increase in pay while on duty involving actual flying in air craft, nor shall any officer in the Navy senior in rank to lieutenant commander, nor any officer in the Marine Corps senior in rank to major, receive any increase in pay or allowances by reason of such detail or duty.

In the event of the death of an officer or enlisted man of the Navy or Marine Corps from wounds or disease, the result of an aviation accident, not the result of his own misconduct, received while engaged in actual flying in or in handling air craft, the gratuity to be paid under the provisions of the Act approved Aug. 22, 1912, entitled "An Act making appropriations for the naval service for the fiscal year ending June 30, 1913, and for other purposes" shall be an amount equal to one year's pay at the rate received by such officer or enlisted man at the time of the accident resulting in his death. In all cases where an officer or enlisted man of the Navy or Marine Corps dies, or where an enlisted man of the Navy or Marine Corps is disabled by reason of an injury received or disease contracted in line of duty, the result of an aviation accident, received while employed in actual flying in or in handling air craft, the amount of pension allowed shall be double that authorized to be paid should death or the disability have occurred by reason of an injury received or disease contracted in line of duty, not the result of an aviation accident.

All Acts or parts of Acts in so far as they are inconsistent with the provisions of this Act are hereby repealed.

During 1916, the Navy Department placed contracts for the following seaplanes and kite balloons:

Names of Companies	Number	Type
Aeromarine Plane and Motor Co.	3	100 h.p. Tractors.
The Burgess Co.	6	125 h.p. Tractors.
Curtiss Aeroplane Co.	9	100 h.p. Pushers.
	1	2-100 h.p. Twin Tractor.
	30	100 h.p. Tractors.
Gallaudet Aircraft Corporation	1	300 h.p. Twin Tractor.
Goodyear Tire and Rubber Co.	3	Kite Balloons.
Standard Aeroplane Corpn.	4	125 h.p. Tractors.
	1	2-135 h.p. Twin Tractor.
The Sturtevant Aeroplane Co.	6	140 h.p. Tractors.
	6	145 h.p. Tractors.
Thomas Bros. Aeroplane Co.	3	135 h.p. Tractors.
	2	2-135 h.p. Twin Tractors.

On December 1, 1916, Captain J. S. McKean, in charge of aviation in the Navy, stated to the Committee on Naval Aeronautics of the House of Representatives, that on that date the aeronautic equipment of the United States Navy was as follows:

There are in use at the Pensacola Aeronautic Station the following: Aeroplanes, 12; lighter-than-air craft, 2; on board the *North Carolina*, 12; ready for service on board the *Washington*—she went to sea without getting them and they are at Portsmouth, New York, awaiting her return, 5; on board the *Nevada*, 1 kite balloon; on board the *Oklahoma*, 1 kite balloon.

The report to Congress dated December 1, 1916, of Secretary Daniels made the following statement giving the number of naval officers, officers of the Marine Corps, officers of the Coast Guard and officers of the Naval Militia under training, as follows:

The first session of the Sixty-fourth Congress authorized the admission of civilians into the Naval Flying Corps. Some of the most expert aviators in Europe are young men who have the requisite quality of skill and daring needed as scouts of the air. They were neither Navy or Army officers. This country is now to utilize men of this gift in its Naval Flying Corps, in addition to the classes of educated and trained officers who go into this branch of the service. This training is now being extended to officers and men of the Naval Militia and the Coast Guard. There are at present at the Pensacola Aviation Station nine officers qualified as naval aviators and sixteen officers under instruction. A new class consisting of twelve naval officers, eighty men, four marine officers and sixteen men, four Naval Militia officers and sixteen men and two Coast Guard officers and eight men is now about to be sent to the station. There are now also two marine officers receiving instruction in land machine flying at the Army School at San Diego, California. Qualified officers for aeronautic work at sea from ships and for inspection duty on shore are continually required, and this demand is continually increasing. The supply of officers to be trained must be maintained and the rate of supply must be increased to meet the increasing demand.

Appropriations and Expenditures for Naval Aeronautics

The expenditures for naval aeronautics during the fiscal year ending June 30, 1912, amounted to $24,532.79; for 1913, $56,032.90; for 1914, $194,492.46; for 1915, $219,429.20; for 1916—out of the million dollar appropriation allowed—there was expended $684,679.28; for 1917 there was appropriated $3,500,000 for aeronautics and $420,000 for the aeronautic station at Pensacola. The provision for aviation in the Naval Bill for the fiscal year ending 1917 was as follows:

"For aviation, to be expended under the direction of the Secretary of the Navy for procuring, producing, constructing, operating, preserving, storing, and handling aircraft, *including dirigibles,* and appurtenances, maintenance of aircraft stations and experimental work in development of aviation for naval purposes, $3,500,000: *Provided,* That the sum to be paid out of this appropriation under the direction of the Secretary of the Navy for drafting, clerical, inspection, and messenger service for aircraft stations shall not exceed $25,000.

The part of the Naval Bill which became a law in 1916 providing for the extension of the Naval Flying Corps to include civilians reads as follows:

The Naval Flying Corps shall be composed of one hundred and fifty officers and three hundred and fifty enlisted men, detailed, appointed, commissioned, enlisted, and distributed in the various grades, ranks, and ratings of the Navy and Marine Corps as hereafter provided. The said number of officers, student flyers, and enlisted men shall be in addition to the total number of officers and enlisted men which is now or may hereafter be provided by law for the other branches of the naval service.

The number of officers detailed to duty in aircraft involving actual flying in any one year shall be in accordance with the requirements of the Air Service as determined by the Secretary of the Navy: *Provided,* That the officers so detailed from the line of the Navy and from the Marine Corps shall not exceed the total number herein prescribed for the Naval Flying Corps: *Provided further,* That the proportion of line officers of the Navy and of the Marine Corps thus detailed shall be the same as the proportion established for the regular service: *And provided further,* That the student flyers hereinafter provided for shall be in addition to the officers and enlisted men comprising the Naval Flying Corps.

The officers detailed and the enlisted men of the Naval Flying Corps shall receive the same pay and allowances that are now provided by law for officers and enlisted men of the same grade or rank and rating in the Navy and Marine Corps detailed to duty with aircraft involving actual flying.

The Secretary of the Navy is hereby authorized to appoint annually in the line of the Navy and the Marine Corps for a period of two years following the passage of this Act, in order to merit as determined by such competitive examinations as he may prescribe, fifteen acting ensigns or acting second lieutenants for the performance of aeronautic duties only. Persons so appointed must be citizens of the United States, and may be appointed from warrant officers or enlisted men of the naval service or from civil life, and must, at the time of appointment, be not less than eighteen or more than twenty-four years of age: *Provided,* That no person shall be so appointed until he has been found physically qualified by a board of medical officers of the Navy for the performance of the duties required: *Provided further,* That the number of such appointments to the line of the Navy and of the Marine Corps shall be in the proportion decided for the regular services. Such appointments shall be for a probationary period of three years and may be revoked at any time by the Secretary of the Navy.

Such acting ensigns and acting second lieutenants shall be detailed to duty in the Naval Flying Corps in aircraft involving actual flying.

Such acting ensigns of the Navy and acting second lieutenants of the Marine Corps shall, upon completion of the probationary period of three years, be appointed acting lieutenants of the junior grade, or acting first lieutenants, respectively, by the Secretary of the Navy for the performance of aeronautic duties only, after satisfactorily passing such examinations as he may prescribe, and after having been recommended for promotion by the examining board and found physically qualified by a board of medical officers of the Navy. Such appointments shall be for a probationary period of four years and may be revoked at any time by the Secretary of the Navy.

Such acting lieutenants (junior grade) and acting first lieutenants may elect to qualify for aeronautic duty only or to qualify for all the duties of officers of the same grade in the Navy and in the Marine Corps, respectively. Those officers who elect to qual-

Officers in the U. S. Navy Aeronautical Establishment, Fall of 1914. Standing, left to right: Lieutenant Commander H. C. Mustin, in Charge, U. S. Navy Aeronautic Station, Pensacola, Florida; Lieutenant N. P. L. Bellinger, Lieutenant R. C. Saufley, Captain Mark L. Bristol, in Charge of Aeronautics, Navy Department, Washington; Lieutenant W. M. McIlvain, U. S. M. C.; Lieutenant B. L. Smith, U. S. M. C. Sitting, left to right: Lieutenant V. D. Herbster, Ensign G. de Chevalier, Ensign M. L. Stolz. Lieutenant John H. Towers was absent on duty when this photo was taken.

ify for aeronautic duty only shall be detailed to duty in the Naval Flying Corps involving actual flying in aircraft. Those officers who elect to qualify for the regular duties of their grade shall be detailed to duty in the regular service for at least two years to allow them to prepare for such qualification.

Such acting lieutenants (junior grade) and acting first lieutenants who have elected to qualify for aeronautic duty only shall, upon completion of the probationary period of four years, be commissioned in the grade of lieutenant of the line of the Navy or captain of the Marine Corps for aeronautic duties only, after satisfactorily passing such competitive examination as may be prescribed by the Secretary of the Navy to determine their moral, physical, and professional qualifications for such commissions and the order of rank in which they shall be commissioned. Such lieutenants for aeronautic duty only shall be borne on the list as extra numbers, taking rank with and next after officers of the same date of commission.

Such acting lieutenants (junior grade) and acting first lieutenants who have elected to qualify for the regular duties of the line of the Navy and of the Marine Corps, respectively, shall, upon completion of the probationary period of four years, two years of which shall have been on such regular duties, be commissioned in the grade of lieutenant of the line of the Navy and captain of the Marine Corps, after passing satisfactorily such competitive examinations as may be prescribed by the Secretary of the Navy to determine their moral, physical, and professional qualifications for such commissions and to determine the order of rank in which they shall be commissioned. Such lieutenants of the line of the Navy and captains of the Marine Corps will be borne upon the lists of their respective corps as extra numbers, taking rank with and next after officers of the regular services of the same date of commissions.

Acting lieutenants (junior grade) of the line of the Navy for aeronautic duties only and acting first lieutenants of the Marine Corps for aeronautic duty only who have completed the probationary period of four years may, upon examination for commissions to the next higher grade, if recommended by the board of examination, be transferred to the Naval Reserve Flying Corps and commissioned in the same grade or the next higher grade as may be recommended in accordance with their qualifications as determined by the examination: *Provided*, That at any time during such probationary period any such officer can, upon his own request, if his record warrants it, be transferred to the Naval Reserve Flying Corps and commissioned in the acting grade he then holds. Any officer of the Naval Flying Corps holding an appointment of student flyer or acting ensign, second lieutenant, lieutenant (junior grade), or first lieutenant, who, upon examination for promotion, is found not qualified shall, if not recommended by the examining board for transfer to the Naval Reserve Flying Corps, be honorably discharged from the naval service.

Officers commissioned for aeronautic duty only shall be eligible for advancement to the higher grades, not above captain in the Navy or colonel in the Marine Corps, in the same manner as other officers whose employment is not so restricted, except that they shall be eligible to promotion without restriction as to sea duty, and their professional examinations shall be restricted to the duty to which personally assigned: *Provided,* That any such officer must serve at least three years in any grade before being eligible to promotion to the next higher grade.

Nothing in this Act shall be so construed as to prevent the detail of officers and enlisted men of other branches of the Navy as student aviators or student airmen in such numbers as the needs of the service may require.

Such officers and enlisted men, while detailed as student aviators, and student airmen involving actually flying in aircraft, shall receive the same pay and allowances that are now provided by law for officers and enlisted men of the same grade or rank and rating in the Navy detailed for duty with aircraft.

The Secretary of the Navy is hereby authorized to appoint annually for a period of four years, from enlisted men of the naval service, or from citizens of the United States in civil life, not to exceed thirty student flyers for instruction and training in aeronautics who shall receive the same pay and allowances as midshipmen at the United States Naval Academy: *Provided,* That persons so appointed must, at the time of appointment, be not less than seventeen or more than twenty-one years of age: *Provided further,* That no person shall be appointed a student flyer until he shall have qualified therefor by such examination as may be prescribed by the Secretary of the Navy.

The appointment of student flyers shall continue in force for two years, unless sooner revoked by the Secretary of the Navy, in his discretion, and at the end of such period student flyers shall be examined for qualification as qualified aviators: *Provided,* That if such student flyers are not qualified, their appointment will be revoked, or, if recommended by the examining board, they shall be transferred to the Naval Reserve Flying Corps and commissioned as ensigns therein.

Student flyers shall, after receiving a certificate of qualification as an aviator for actual flying in aircraft, rank with midshipmen and shall receive the same pay and allowances as midshipmen, plus fifty per centum thereof: *Provided,* That student flyers who have qualified as aviators under the provisions of this Act shall be commissioned acting ensigns for aeronautic duties only, after three years' service: *Provided further,* That they shall have been examined by a board of officers of the Naval Flying Corps to determine by a competitive examination prescribed by

Pensacola, Florida, the Navy's only aeronautical station, photographed from a naval aeroplane in 1915. The aeroplane tent hangars are shown in the center.

the Secretary of the Navy their moral, physical, and professional fitness and the order of rank in which they shall be commissioned: *And provided further,* That any student flyer qualified as an aviator may at any time, in the discretion of the Secretary of the Navy, if his record warrants it, at his own request, be transferred to the Naval Reserve Flying Corps and be commissioned as ensign therein: *And provided further,* That student flyers not considered qualified for commissions as acting ensigns for aeronautic duties only may, upon recommendation of the examining board, be transferred to the Naval Reserve Flying Corps and be commissioned as ensigns therein.

The Secretary of the Navy is hereby authorized to established aeronautic schools for the instruction and training of student flyers and prescribe the course of instruction and qualifications for certificate of graduation as a qualified aviator.

Nothing in this or any other Act shall be so construed as to prevent the temporary detail of officers and enlisted men of any branch of the Navy for duty with aircraft.

In the event of the death of an officer or enlisted man or student flyer of the Naval Flying Corps from wounds or disease, the result of an aviation accident, not the result of his own misconduct, received while engaged in actual flying in or in handling aircraft, the gratuity to be paid under the provisions of the Act approved August twenty-second, nineteen hundred and twelve, entitled "An Act making appropriations for the naval service for the fiscal year ending June thirtieth, nineteen hundred and thirteen, and for other purposes," shall be an amount equal to one year's pay at the rate received by such officer or enlisted man or student flyer at the time of the accident resulting in his death. In all cases where an officer or enlisted man or student flyer of the Navy or Marine Corps dies, or where a student flyer or an enlisted man of the Navy or Marine Corps is disabled by reason of any injury received or disease contracted in line of duty, the result of an aviation accident, received while employed in actual flying in or in handling aircraft, the amount of pension allowed shall be double that authorized to be paid should death or the disability have occurred by reason of an injury received or disease contracted in line of duty not the result of an aviation accident.

Student flyers and the acting ensigns and acting lieutenants (junior grade) and acting second and first lieutenants for aeronautic duties only provided for herein shall be subject to the laws and regulations and orders for the government of the Navy, but shall not be entitled to retirement or retired pay.

The enlisted personnel of the Naval Flying Corps shall be distributed by the Secretary of the Navy in the various ratings as now obtain in the Navy in so far as such ratings are applicable to duties connected with aircraft.

Within the first two years after the passage of this Act enlisted men may be transferred from other branches of the Naval Service to the Naval Flying Corps, under regulations established by the Secretary of the Navy governing such transfer and the qualifications for this corps: *Provided,* That the number so transferred shall not exceed one-half the total number of enlisted men allowed by this Act.

The Secretary of the Navy shall establish regulations governing the term of enlistment, the qualifications, and advancement of the enlisted men of the Flying Corps.

Any enlisted man who passes satisfactorily the prescribed examination and is recommended by a board of officers may be appointed a student flyer as herein provided.

Navy Orders Sixteen Coast Patrol Dirigibles

On February 24, 1917, the Department issued specifications for dirigibles and asked for bids, which were opened on March 6th. Sixteen dirigibles were ordered under these bids as follows:

The Curtiss Aeroplane Company of Buffalo

Flying officers and staff officers at Pensacola in 1915.

Enlisted personnel United States Navy Aero Station at Pensacola in 1915.

was awarded three for a total price of $122,250; the Connecticut Aircraft Company, New Haven, two for a total price of $84,000; the Goodyear Tire and Rubber Company, Akron, Ohio, nine for a total price of $360,000; and the B. F. Goodrich Company, of Akron, two, at a price for both of $83,000.

The specifications for the Coast Patrol dirigibles was given in the chapter on "Naval Dirigibles."

In the early part of 1917, the War and Navy Departments decided to combine efforts to get large dirigibles, the work to be started immediately on the first airship of that type.

Officers in Charge of Naval Aeronautics

On January 13, 1917, Lieutenant-Commander Henry C. Mustin, who had been in charge of the Pensacola Aeronautic Station for over two years was succeeded by Captain Joseph L. Jayne.

The administration of the Office of Aeronautics in the Bureau of Operation on May 15, 1917, was in charge of Captain J. S. McKean and Lieutenant-Commander John H. Towers. The advisory staff in aeronautics of Chief Naval Constructor D. W. Taylor, the head of the Bureau of Construction and Repair was as follows: Naval Constructor H. C. Richardson, at the Pensacola station; Naval Constructor George C. Westervelt, supervising the construction of the navy aeroplanes; assistant naval Constructor J. C. Hunsacker, in charge of design specifications and contracts; and Dr. Albert Francis Zahm, in charge of wind tunnel and research work.

The following United States Navy officers have been awarded the Aero Club of America's medal of merit:

1914 AWARDS

Lieut. Commander H. C. Mustin, U. S. N., commanding Aeronautic ship *Mississippi* in the Mexican Expedition.

Lieut. R. C. Saufley, U. S. N., Air Pilot and Observer, Mexican Expedition.

Ensign M. L. Stolz, U. S. N., Air Pilot and Observer, Mexican Expedition.

Ensign W. D. La Mont, U. S. N., Air Pilot and Observer, Mexican Expedition.

1915 AWARDS

Lieut. P. N. L. Bellinger, U. S. N. For breaking American Hydroaeroplane Altitude Record. Height attained 10,000 feet.

Lieut. Warren G. Child, U. S. N. In recognition of excellent work in developing machinery for aircraft.

Lieut. Jerome C. Hunsaker, U. S. N. In recognition of his excellent work in aeronautical engineering.

Commander Henry C. Mustin, U. S. N. For being the first to make a flight from the *North Carolina* on the new launching device.

Holden C. Richardson, Naval Constructor, U. S. N. In recognition of achievements in designing aeroplanes and aeroplane floats.

Lieut. R. C. Saufley, U. S. N. For twice breaking American Hydroaeroplane Altitude Record in one year, attaining height of 11,975 feet.

U. S. Naval Experimental Wind Tunnel

The Experimental Wind Tunnel, which the United States Navy Department ėstablished in the Washington Navy Yard at the Experimental Model Basin, where warship models are tested, has a section eight feet square at the point where the models are placed for testing. In addition to the advantage gained by the size, it is possible with the 500-horse-power, motor-driven fan to get wind speeds up to 75 miles an hour.

The tunnel consists of a closed circuit shaped like the link of a chain, as shown in Figure 1. The 500-horse-power top horizontal discharge fan of the corrugated paddle type, with an inlet diameter of 11 feet, 2 inches, and a discharge duct 7 feet, 6 inches by 9 feet, is placed at one end of the link. At the other end, where the air straightens out before flowing through the experimental chamber, are the baffles, which are necessary to remove the eddies and to control the uniformity of the speed. These baffles consist of 64 cells, each 1 foot square and 8 feet long. Each cell is provided with its own damper, so that the velocity of the air in any one section may be controlled. At the experimental chamber in the vicinity where aeroplane wings or models are tested the maximum variation from uniform flow is about 2 per cent. The tunnel is built of wood, with frames spaced about three feet on centers placed outside and sheathed on the inside with 7/8-inch tongued and grooved sheathing laid in two thicknesses in the direction of the air current, and with building paper placed between the two layers. The necessary curvature is obtained by bending the sheathing, the whole of which is blind nailed.

The fan is driven by a 250-volt, 500-horse-power, direct current motor, arranged for operation on the Ward-Leonard system. The motor also has auxiliary field control, so that any desired speed up to about 200 revolutions per minute, which corresponds to a wind speed of 75 miles an hour, may be obtained. At the discharge side of the fan are located 12 pitot tubes which lead to an integrating manometer which gives the average velocity of discharge. This velocity has been calibrated against the velocity obtained at the section in the experimental chamber where the aeroplane or other model is placed, so that any desired velocity may be obtained at that point with precision without having any pitot tubes or obstructions other than the model being tested. In other words, by calibration the velocity of discharge may be found, and this bears a certain constant ratio to the velocity at the experimental section.

Among recent investigations of interest made at the Wind Tunnel was the determination of the coefficient of air friction for various aeroplane and balloon fabrics. Tests have been made on the new dirigible building for the Navy Department and on models of naval aeroplanes both building and projected. A number of tests have also been made for private concerns. In carrying out experiments for private parties the same practice is followed as in the case of tests of ship models; that is, the actual cost of doing the work is charged in each case. On account of the large size of the tunnel it is possible to test comparatively large models.

The Aeronautic Needs of the United States Navy

On February 21, 1916, Captain Mark L. Bristol, the Director of Naval Aeronautics, appeared before the House Committtee on Naval Affairs and stated that the Navy needed $20,000,000 for aeronautics, $13,600,000 of which was required immediately. The fleet needed, he said, 82 aeroplanes, 5 dirigibles, and 41 kite balloons. In addition to the fleet equipment there were needed 120 aeroplanes, 15 dirigibles, and 15 kite balloons to be operated from the fifteen naval shore stations under the patrol system. We know to-day that Captain Bristol's estimate was most conservative and the one general regret is that Congress allowed only $3,500,000 that year, which was not sufficient even to build up the skeleton of a substantial naval air service.

CHAPTER XXV

REGULATIONS RELATING TO ENROLLMENTS IN THE UNITED STATES NAVAL RESERVE FLYING CORPS

CLASS 5.

Naval Reserve Flying Corps

1. Eligibility.

The following citizens of the United States shall be eligible for membership in the Naval Reserve Flying Corps.

(*a*) Officers and student flyers who have been transferred from the Naval Flying Corps to the Naval Reserve Flying Corps.

(*b*) Enlisted men of the Naval Flying Corps transferred under the same condition as enlisted men of the Navy are transferred to the Fleet Naval Reserve.

(*c*) Surplus graduates of the aeronautic school may be commissioned as ensigns in the Naval Reserve Flying Corps and promoted therein under such regulations as may be prescribed by the President.

(*d*) Civilians skilled in the flying of aircraft, or in their design, building, or operation, shall be eligible for membership in the Naval Reserve Flying Corps, United States Naval Reserve Force.

(*e*) Other members of the Naval Reserve Force may be transferred to the Naval Reserve Flying Corps upon qualification for aviation duties.

Class 5 (*a*).—Officers; provisional.

(*a*) Must furnish satisfactory evidence as to character, ability, and citizenship.

(*b*) Must qualify professionally for a provisional rank before the commandant or an officer designated by the commandant for that purpose and physically before a medical officer of the Navy.

Class 5 (*b*).—Officers; confirmed.

(*a*) Former officers of the Naval Flying Corps who have left the service under honorable conditions or officers who may be transferred to the Naval Reserve Flying Corps from the Naval Flying Corps; provisional appointment not necessary.

(*b*) Surplus graduates of the aeronautic school may be commissioned as ensigns in the Naval Reserve Flying Corps; provisional appointment not necessary.

(*c*) After three months' active service an officer may be confirmed in his provisional rank by qualifying professionally before a board of three officers not below the rank of lieutenant commander and physically before a board of two medical officers of the Navy.

Class 5 (*c*).—Men; provisional.

(*a*) Must furnish satisfactory evidence as to character, ability, and citizenship.

(*b*) Must qualify professionally for a provisional rating before an officer designated by the commandant and physically before a medical officer of the Navy.

Class 5 (*d*).—Men; confirmed.

(*a*) Men who have been honorably discharged from the Naval Flying Corps after one or more four-year terms of enlistment in the Navy or after a term of enlistment during minority. No provisional rating required for this class.

(*b*) After three months' active service a member may be confirmed in his provisional rating by qualifying before an officer designated by the commandant for that purpose.

Class 5 (*e*).—Members of this class do not enroll and are not discharged, but are transferred from the regular Naval Flying Corps to the Naval Reserve Flying Corps in the same manner as men transferred to the Fleet Naval Reserve. Their status corresponds more closely to those on the retired list. Any enlisted man of the Naval Flying Corps with 16 years' naval service may, on the authority of the Secretary of the Navy, upon voluntary application on the expiration of his enlistment, if entitled to an honorable discharge, be transferred to the Naval Reserve Flying Corps in the rating in which then serving.

Class 5 (*f*).—Any enlisted man in the Naval Flying Corps with 20 or more years' naval service may be authorized by the Secretary of the Navy, in his discretion, to be transferred to the Naval Reserve Flying Corps in the same manner as in the previous case, except that the transfer may be made at any time during the man's current enlistment.

2. Pay of the Naval Reserve Flying Corps.

Class 5 (*a*).—Officers; provisional.

Annual retainer pay, $12.

(*b*).—Officers; confirmed.

Annual retainer pay, two months' base pay of the corresponding rank or grade in the Navy.

Class 5 (c).—Men; provisional.
Annual retainer pay, $12.

Class 5 (d).—(*a.* and *b.*) Two months' base pay of the corresponding rate in the Navy.

Class 5 (e).—One-third of the base pay they were receiving at the date of transfer, plus all permanent additions thereto.

Class 5 (f).—One-half of the base pay they were receiving at the date of transfer, plus all permanent additions thereto.

NOTES.—(*a*) Members of the Volunteer Naval Reserve enrolled for the Naval Reserve Flying Corps, or for any other class of the Naval Reserve Force for which qualified, receive no retainer pay or uniform gratuity in time of peace. When on active duty they receive the active-service pay of their rank or rating. The only distinction between a Naval Reserve Flying Corps Reservist and a Volunteer Naval Reservist enrolled for duty in the Naval Reserve Flying Corps is the one of retainer pay and uniform gratuity in time of peace.

(*b*) Members of the Naval Reserve Flying Corps who enroll for a term of four years within four months from the date of termination of their last term of enrollment and who shall have performed the minimum amount of active service required during preceding term of enrollment shall, for each enrollment, receive an increase of 25 per cent. of their base retainer pay. (Base retainer pay is two months' base pay of the corresponding rank or rating in which serving.)

(*c*) When actively employed, either under provisional or confirmed rank or rating, the pay of officers and men in the Naval Reserve Flying Corps shall be the same as the pay of officers and men in the Naval Flying Corps on active duty of corresponding rank or rating and of the same length of naval service.

(*d*) The retainer pay is *in addition* to the active-service pay.

(*e*) Officers and men of the Naval Reserve Flying Corps will have their retainer pay accounts carried by the Disbursing Officer, Bureau of Supplies and Accounts, Navy Department, Washington, D. C., and will be paid quarterly by check.

(*f*) Upon first reporting for active duty for training, officers receive a uniform gratuity of $50; men, $30. This uniform gratuity is given for each enrollment. Upon reporting for active service in time of war or national emergency the uniform gratuity is $150 for officers and $60 for men, less any previous uniform gratuity credited during the current enrollment. Should a member sever his connection with the service without compulsion on the part of the Government before the expiration of his term of enrollment the amount so credited shall be deducted from any money that may be or may become due him.

(*g*) Members who shall have completed 20 years of service in the Naval Reserve Force, and who shall have performed the minimum amount of active service required in their class for maintaining efficiency during each term of enrollment, shall, upon their own application, be retired with the rank or rating held by them at the time, and shall receive in lieu of any pay a cash gratuity equal to the total amount of their retainer pay during the last term of enrollment.

(*h*) Pay penalties are given under paragraph 3.

3. DUTIES AND REQUIREMENTS OF THE NAVAL RESERVE FLYING CORPS

(*a*) Three months' active duty each enrollment. This does not apply to class A (*e*) and class 5 (*f*).

The Secretary of the Navy is authorized to assign officers and men to active duty on application. This also applies to men of class 5 (*e*) and class 5 (*f*).

This service may be taken in one or more periods of not less than three weeks.

Penalty for Noncompliance.—If a member fails to perform three months' active service during an enrollment, he shall on reenrollment receive a retainer pay at the rate of $12 per year until such time as he shall have completed the three months' active service during current enrollment.

(*b*) Enrolled members of the Naval Reserve Force shall be subject to the laws and regulations for the government of the Navy only during such time as they may by law be required to serve in the Navy, in accordance with their obligations, and when on active service at their own request, as herein provided, and when employed in authorized travel to and from such active service in the Navy.

(*c*) Make such reports concerning movements and occupations as may be required.

Penalty for Noncompliance.—Retainer pay must be forfeited.

(*d*) An officer or man of the Naval Reserve Flying Corps shall not be an officer or enlisted man in any branch of the military service of the United States or any State thereof, but may accept employment in any other branch of the public service.

4. ENROLLMENTS, TRANSFERS, APPOINTMENTS, PAY ACCOUNTS, DISCIPLINE, ORDERS TO ACTIVE SERVICE, UNIFORM, DISCHARGES, RECORDS OF THE NAVAL RESERVE FLYING CORPS

ENROLLMENTS

(*a*) Officers designated by the commandant of the naval district shall enroll men who are eligible under the rules given. Enrollments shall be for a period of four years.

(*b*) Every officer enrolling a man in the Naval Reserve Flying Corps shall—

(1) Explain to the man that the commandant of the naval district is the man's commanding officer.

(2) Explain that all requests shall be made to the commandant either by letter or in person.

(3) Explain that any change in address must be promptly reported to the commandant and disbursing officer, Bureau of Supplies and Accounts, Navy Department, Washington, D. C.

(4) Make out and forward account cards in triplicate to the disbursing officer.

(*c*) To be given a provisional rating a man must have the technical knowledge of the corresponding rating in the Naval Flying Corps; to be confirmed in a provisional rating a man must, in addition, have a fair knowledge of naval discipline and customs; to be advanced in rating a man must have the technical knowledge of the corresponding rating in the Naval Flying Corps, and a good knowledge of naval customs and methods.

(*d*) After a man is confirmed in his rating he receives retainer pay of class 5 (d). In case he does not perform the minimum active service required in an enrollment, upon reenrollment he shall receive the pay of class 5 (c) until such time as he shall have completed three months' active service.

TRANSFERS

(*e*) Officers and student flyers may, in the discretion of the Secretary of the Navy, be transferred from the Naval Flying Corps to the Naval Reserve Flying Corps.

APPOINTMENTS

(*f*) Former officers of the Naval Flying Corps or graduate of aeronautical schools may make application for enrollment to the Bureau of Navigation, stating briefly his Naval and Naval Flying Corps service. The procedure is as follows:

(1) Applicant applies to the Bureau of Navigation for enrollment.

(2) If the application is approved, the bureau authorizes the applicant to report for medical examination.

(3) If physically qualified, appointment is issued to grade or rank last held in the Navy. Appointment is for four years.

(4) Bureau of Navigation forwards appointment with letter of transmittal and blank form "Acceptance of office and oath of allegiance."

(5) Form is returned to the Bureau of Navigation properly accomplished. Pay and allowance and eligibility for service begin from date of acceptance.

(6) Orders issued to report by letter to commandant of naval district.

(7) Officers are detailed for active service upon their own request, orders for such being issued either by the Bureau of Navigation or the commandant of naval district.

(*g*) An enrolled man eligible for appointment to a provisional rank or grade may be examined professionally by the commandant or an officer designated by the commandant for that purpose and examined physically before a medical officer of the Navy. The physical requirements shall be the same as for officers of the Naval Flying Corps.

(*h*) A civilian eligible for appointment to a provisional rank or grade may likewise be examined.

(*i*) The commandant shall make recommendation to the Bureau of Navigation for appointment to provisional rank or grade. Appointments are issued by the Bureau of Navigation, and upon receipt of "Acceptance and oath of office" the Bureau of Navigation will issue orders to officers to report by letter to commandant of the naval district.

(*j*) Officers must complete not less than three months' active service in a provisional rank or grade to become eligible for confirmation. The commandant is authorized to order such eligible officers as have been satisfactory in their provisional appointment to appear before a board of two medical officers for physical examination, and before a board of three naval officers not below the rank of lieutenant commander, for professional examination or confirmation of provisional appointment. The boards shall conduct examinations in accordance with the department's precepts of October 16, 1916, and subsequent modifications which have been furnished the commandants in blank. Upon the receipt of records of examination in the department the commandant and candidates will be notified of the department's action thereon. The commandant is authorized to appoint supervisory boards in special cases when it is deemed impracticable for the candidate to appear before the regular boards above named.

(*k*) For appointment to a provisional rank the technical requirements shall be the same as those for an officer in the Naval Flying Corps.

(*l*) An officer to be confined must have, in addition to the knowledge required for provisional appointment, a general knowedge of the customs and discipline of the service and a good knowledge of the tactics of the Naval Flying Corps.

PAY ACCOUNTS

(*m*) Account cards are made out in triplicate by the commandants of naval districts, as follows:

(1) Upon enrolling or reenrolling a man in the Naval Reserve Flying Corps.

(2) Upon the receipt of records of the men when first enrolled.

(3) Upon appointment of an officer under his command to the Naval Reserve Flying Corps.
(4) Upon receipt of records of man transferred after 16 or 20 years' service.

(*n*) Account cards are sent to the disbursing officer, Bureau of Supplies and Accounts, Navy Department, Washington, D. C. Account cards may be obtained from the supply officer, navy yard, Washington, D. C.

(*o*) Account cards will be made out in accordance with instructions on the back thereof.

(*p*) Any change in rank or rating or any change that would in any way stop or affect pay shall be reported at once to the disbursing officer.

DISCIPLINE

(*q*) Members of the Naval Reserve Flying Corps when on active service shall be subject to the discipline of the Navy.

RETIREMENTS

(*r*) Enrolled members who have completed 20 years of service in the Naval Reserve Force and who shall have performed the minimum amount of active service required in their class for maintaining efficiency for each term enrolled shall, upon their application, be retired with the rank or rating held by them at the time, and shall receive in lieu of any pay a cash gratuity equal to the total amount of their retainer pay during the last term of their enrollment.

ACTIVE-SERVICE ASSIGNMENTS

(*s*) To order a man to active service upon his own request the following procedure shall be followed:

(1) The commandant of a naval district issues orders, provides transportation and subsistence, and forwards enrollment and health records, with copy of orders to the commanding officer of the ship or station to which the man is ordered.
(2) The commanding officer enters on his enrollment record the date of reporting, the date of detachment, and proficiency marks. The record is handled in the same manner as that of a regular enlisted man.
(3) The commanding officer shall, on completion of the active service training period, as indicated by orders, or as soon thereafter as practicable, detach the man and procure for him transportation and subsistence to his home (or other place, provided this can be done at no greater expense) and forward the enrollment record and health record to the commandant of the naval district.
(4) The pay officer of ship or station will pay members of the Reserve Force on active duty as are paid officers and men of the Navy. (Active service counts from date of detachment.) Mileage of officers ordered to active service shall be paid in same way as to officers of the Navy.

UNIFORM

(*t*) Men in the Naval Reserve Flying Corps shall keep on hand such part of the clothing outfit as may be prescribed.

DISCHARGES

(*u*) In time of peace men shall be discharged upon their own request or by proper authority.

RECORDS

(*v*) An enrollment record shall be kept in the same manner as enlistment record and the health record used shall be the same as the service health record, but shall have written on the face "Naval Reserve Flying Corps," and shall be kept in the same manner as the service health record.

(w) During the period of active service an entry shall be made in the health record to indicate the physical condition of the reservist. Upon completion of active service, fitness reports shall be made out for officers; enrollment records of men shall be marked, and all records shall be returned to the commandant of the naval district.

TRANSFERS TO OTHER CLASSES OF THE NAVAL RESERVE FORCE

(*x*) Members of the Naval Reserve Force may, upon application, be transferred from one class to another for which they are qualified, and may in time of war volunteer for and be assigned to duties prescribed for any class which they may be deemed by their commandant competent to perform.

(*y*) Although the men of class 5 (*e*) and class 5 (*f*) are transferred in the same manner as men of the Fleet Naval Reserve, the entire class 5, "Naval Reserve Flying Corps," will be handled by the commandants of the naval districts. The records, etc., of men in class 5 (*e*) and 5 (*f*) will be sent to the commandant of the naval district instead of the commanding officer of the recruiting district.

L. C. PALMER,
Chief of Bureau.

NAVY DEPARTMENT, BUREAU OF NAVIGATION,
Washington, D. C., November 27, 1916.

The christening of the first seaplane presented to the New York Naval Militia. Miss Olive Whitman, the pretty daughter of Governor Whitman, of New York, holding the bottle of champagne, saying, "I christen thee N. Y. N.-1."

CHAPTER XXVI

NAVAL MILITIA AERONAUTICS

As the naval militia is practically the second line of defense, particularly in connection with the work of patroling the coasts—work which has become of extreme importance since the advent of submarine warfare—it is most important that the naval militia of the twenty-two States and insular possessions which have naval militia organizations should have substantial aëronautic divisions.

The regulations provide that an aeronautic division shall consist of two aeronautic sections and shall be commanded by an officer of not higher rank than lieutenant commander (aeronautic duties only).

(b) An aeronautic section shall consist of 23 enlisted men, and may have 5 officers, as follows:

One lieutenant (aeronautic duties only).

Two lieutenants, junior grade (aeronautic duties only).

Two ensigns (aeronautic duties only).

The enlisted strength may be divided as follows:

Enlisted in Naval Militia as	*Duties performed in aeronautic branch*
1 chief machinist's mate.	Chief aeronautic machinist.
1 machinist's mate, first class.	Aeronautic machinist, first class.
1 machinist's mate, second class.	Aeronautic machinist, second class.
8 electricians, third class (gen.).	Aeronautic machinist, third class.
1 carpenter's mate, second class.	Aeronautic mechanic, second class.
8 carpenter's mates, third class.	Aeronautic mechanic, third class.

1 yeoman, third class.
1 hospital apprentice.

(c) In a locality where there are insufficient men to form an aeronautic section and there already exists an organized deck or engineer division, an officer and not more than 4 enlisted men for aeronautic duty only may be additionally enrolled in such divisions until such time as there is a sufficient number of them to form a separate aeronautic section.

(d) In cases where 4 additional enlisted men of the aeronautic branch are enrolled in a deck or engineer division there will be allowed an additional ensign (aeronautic duties only).

(e) The following additional chief petty officers, petty officers, and other enlisted men of the seamen branch, artificer branch (engineer force), and special branch will be allowed each aeronautical division:

One chief boatswain's mate.
One boatswain's mate, first class.
One yeoman, second class.
One electrician, first class (radio).
One seaman (signalman).

Section 10. (a) The minimum strength of a deck or engineer division or a marine company shall be 40 enlisted men; the minimum strength of an aeronautic section shall be 1 officer and 5 enlisted men.

(b) A deck or engineer division consisting of more than 80 enlisted men, or an aeronautic section of more than 6 officers and 28 enlisted men may be maintained only by permission of the Commanding Officer, Naval Militia.

Applications from men who have had experience in aeronautics as aviators or mechanics are especially welcomed by the naval militia commanders.

HEADQUARTERS OF NAVAL MILITIA ORGANIZATIONS:

State	Mail address
California	Commanding Officer, California Naval Militia, Room 402, Sharon Building, 55 New Montgomery Street, San Francisco, Cal.
Connecticut	Commanding Officer, Connecticut Naval Militia, South Norwalk, Conn.
District of Columbia	Commanding Officer, District of Columbia Naval Militia, Water and O Streets S.W., Washington, D. C.
Florida:	
First Battalion	Commanding Officer, First Battalion, Florida Naval Militia, Key West, Fla.
Second Battalion	Commanding Officer, Second Battalion, Florida Naval Militia, Jacksonville, Fla.
Hawaii	Commanding Officer, Naval Militia of Hawaii, care Executive Chamber, Honolulu, Hawaii.
Illinois	Commanding Officer, Illinois Naval Militia, Steamship Commodore, Chicago, Ill.
Louisiana	Commanding Officer, Louisiana Naval Militia, 326 Camp Street, New Orleans, La.
Maine	Commanding Officer, Maine Naval Militia, 375 Fore Street, Portland, Me.
Maryland	Commanding Officer, Maryland Naval Militia, 500 Continental Building, Baltimore, Md.
Massachusetts	Commanding Officer, Massachusetts Naval Militia, State Armory, Fall River, Mass.
Michigan:	
First Battalion	Commanding Officer, First Battalion, Michigan Naval Militia, 718 Penobscot Building, Detroit, Mich.
Second Battalion	Commanding Officer, Second Battalion, Michigan Naval Militia, Hancock, Mich.
Minnesota	Commanding Officer, Minnesota Naval Militia, 120 North Fifteenth Avenue East, Duluth, Minn.
Missouri	Commanding Officer, Missouri Naval Militia, 709 Leclede Gas Building, St. Louis, Mo.
New Jersey:	
First Battalion	Commanding Officer, First Battalion, New Jersey Naval Militia, U. S. S. Adams, Hoboken, N. J.
Second Battalion	Commanding Officer, Second Battalion, New Jersey Naval Militia, U. S. S. Vixen, Camden, N. J.
New York	Commanding Officer, New York Naval Militia, 2 Rector Street, New York, N. Y.
North Carolina	Commanding Officer, North Carolina Naval Militia, Newbern, N. C.
Ohio:	
First Battalion	Commanding Officer, First Battalion, Ohio Naval Militia, Calvin Building, Toledo, Ohio.
Second Battalion	Commanding Officer, Second Battalion, Ohio Naval Militia, 408 Federal Building, Cleveland, Ohio.
Oregon	Commanding Officer, Oregon Naval Militia, 640 Morgan Building, Portland, Oreg.
Pennsylvania	Commanding Officer, Pennsylvania Naval Militia, 333 Walnut Street, Philadelphia, Pa.
Rhode Island	Commanding Officer, Rhode Island Naval Militia, State Armory, Providence, R. I.
South Carolina	Commanding Officer, South Carolina Naval Militia, Charleston, S. C.
Texas	Commanding Officer, Texas, Naval Militia, care Blum Hardware Co., Galveston, Tex.
Washington	Commanding Officer, Washington Naval Militia, 732 Central Building, Seattle, Wash.

Development of Aeronautics In the Naval Militia

The development of aeronautics in the Naval Militia of the United States was started by the Aero Club of America in 1915–16, and until the

The Aviation Division California Naval Militia. Lieut. Frank Simpson, Jr

end of 1916 the expenses were paid entirely by public contributions.

Appreciating the need of supplying the Naval Militia with trained aviators and seaplanes, to make up for the navy's inability to organize an adequate air service, the Aero Club of America in 1915 took steps to develop aeronautics in the Naval Militia, as part of its extensive campaign for national preparedness. The plan was approved by Secretary of the Navy Daniels, in the following letter to Mr. Hawley, the president of the Aero Club of America:

My dear Mr. Hawley:

Your letter of the 18th ultimo, in regard to a public subscription for aeronautical purposes, was duly received.

I am greatly interested in anything that is being done to assist in the development of aeronautics in this country. I congratulate the Governors of the Aero Club of America on the public spirit which has prompted them to start a public subscription to raise funds to further develop aeronautics in this country.

As you undoubtedly know, I am not allowed legally to consider public subscriptions for the Government's use. It would seem, though, that you could be of great assistance to the Naval Militia at the present time by obtaining aeroplanes for them by popular subscriptions.

If you will apply to Captain Bristol, he will be very glad to assist you in any way that is possible so far as he properly can. By thus conferring with him, you will be able to work, as you have suggested, in harmony with the United States Navy.

Your idea of creating a valuable and efficient aeronautical reserve is an excellent one, and I am sure that you will meet with that measure of success that your efforts deserve.

I desire to thank you and the Governors of the Aero Club of America, so far as the Navy Department is concerned, for the interest taken in this subject.

Sincerely yours,

(Signed) Josephus Daniels,
Secretary of the Navy.

California Naval Militia

The aeronautic section of the California Naval Militia was started in February, 1916, when a contribution of $1200 was made for the purpose through the Aero Club of America. Subsequently, Mr. Glenn L. Martin presented a Martin biplane, and a further contribution of $750 toward defraying the expenses of operating the machine was made to the militia through the Aero Club of America by Mr. Emerson McMillin.

The aeronautic section was attached to the ninth division of the California Naval Militia, and Ensign Frank Simpson, Jr., was put in charge.

The aeronautic section was mustered in February 3, and the four drill periods of that month were devoted to outfitting the enlisted men and to other details connected with the organizing of this section.

On March 2, the roll of the aeronautic section was taken separately for the first time.

On the right is shown the aeroplane presented by Mr. Glenn L. Martin.

The five regular drill periods during the month of March were devoted to instruction in ordnance, discipline, signaling, and instruction in technical aeronautics.

The training continued through the year of 1916, including two weeks of camping with the Second Battalion, N. M. C., on North Island, San Diego Bay, where the members had the opportunity of gaining experience by contact with the United States Army aviators who gave them valuable advice and guidance.

At the close of 1916 when the Navy Department made arrangement for the training of naval militia men at the United States Naval Aeronautic Station at Pensacola, Florida, the following were assigned to take the course of training from the Aeronautic Section, N. M. C.: Lieut. Frank Simpson; Samuel Kroner, H. V. Reynolds, P. S. Ryan, J. G. Weyse.

Connecticut Naval Militia

An aviation section for the Naval Militia of Connecticut formed at Bridgeport in February, 1916, with twenty-three men headed by Ensign John D. Cooper.

Ensign John D. Cooper left Bridgeport, February 28, in charge of the following men, to report to Pensacola, Florida, for three months' training: Warren S. Renolds, chief machinists' mate; Leon S. Moran, machinist, second class; LeRoy Sweeney, electrician, third class; James V. Porto, electrician, third class.

District of Columbia Naval Militia

The first step in organizing the aviation section for the Naval Militia of the District of Columbia was taken when Adjutant-General J. C. Castner of the District of Columbia designated Ensign Dean R. Van Kirk to take the free course of training offered through the Aero Club of America by the Curtiss Aeroplane Company. The expenses of sending Ensign Van Kirk to aviation school to the extent of $200 were defrayed by the National Aeroplane Fund of the Aero Club of America.

When the Navy Department decided to give courses of training to naval militiamen at Pensacola, the following from the Naval Militia of the District of Columbia were sent to take the course: Ensign D. R. Van Kirk, W. H. Boteler, W. R. Garland, A. J. Natho, H. W. Roughly.

Illinois Naval Militia

The Naval Militia of Illinois was presented with the use of a 100-horse-power flying boat by Messrs. A. M. Andrews and Stuart McDonald, in May, 1915. It was officially christened at Chicago on May 22, Miss Mona Dunne, daughter of Governor Dunne of Illinois, acting as sponsor. The ceremony was held at the hangar at the foot of Washington Street, Chicago, and was attended by Governor Dunne, Mayor Thompson, and by the state, militia, and city authorities. During the succeeding summer, training was given to six of the men in handling and taking care of the machine. Training was also given in observation work in connection with the "Isle of Luzon," and the cruises of the Naval Militia.

Illinois Naval Militia

The first steps in organizing an aviation section for the Illinois Naval Militia were taken in May, 1915. The following resolution gives the details of how it started:

WHEREAS, On Saturday, May 22, 1915, the Illinois Naval Militia launched the first hydroaeroplane to be commissioned by a naval reserve organization of this country since the Department of Aeronautics, United States Navy, issued a call for volunteer aviation corps, and

WHEREAS, This machine was placed at the service of the local Naval Militia through the patriotism of Mr. A. M. Andrews, and Mr. Stuart McDonald, citizens of the city, at a considerably outlay on their part, and

WHEREAS, This prompt response to the call of the Navy Department has reflected great credit on the city of Chicago and the State of Illionis and the enterprise of its Naval Militia, be it therefore,

RESOLVED, That this act of patriotism deserves the hearty commendation of this body and that the City Clerk be directed to prepare a letter voicing the sentiments of this body to be signed by the Mayor and forwarded to Messrs. A. M. Andrews and Stuart McDonald and officers of the Illinois Naval Reserve.

STATE OF ILLINOIS, }
COUNTY OF COOK. } *ss.*

I, John Siman, City Clerk of the City of Chicago, do hereby certify that the above and foregoing is a true and correct copy of the certain resolution adopted by the City Council of the City of Chicago on the twenty-fourth (24th) day of May, A.D. 1915.

I do further certify that the original of said resolution is in my custody for safe-keeping and that I am the lawful custodian of same.

In witness whereof, I have hereunto set my hand and affixed the corporate seal of the city of Chicago this ninth (9th) day of June, A.D. 1915.

(Signed) JOHN SIMAN,
City Clerk.

This flying boat was used in connection with the militia cruises and for scouting operations.

Lack of funds prevented the organizing of an aviation division in the summer of 1916.

Maine Naval Militia

The first steps to establish an aviation section in the militia of Maine were taken under the auspices of Rear-Admiral Robert E. Peary in October, 1915. Details of the events which led to starting the movement are given in the chapter on "The Aerial Coast Patrol."

The Chamber of Commerce of Portland, Maine, enthusiastically took up the proposition to have an aeronautic station established near that city. President George L. Crosman, of the Chamber of Commerce, appointed a committee, representing the whole State, to take the matter in charge, composed of: Hon. William M. Ingraham, Portland; Col. Fred. N. Dow, Portland; Hon. E. B. Winslow, Charles F. Flagg, Col. Frederick Hale, Richard Payson, Frank L. Rawson, Lieutenant Reuben K. Dyer, Rear-Admiral Robert E. Peary, Eagle Island, South Harpswell; Hon. Edward W. Hyde, Bath; Hon. Arthur Chapin, Bangor; Col. F. E. Boothby, Waterville; William D. Pennell, Lewiston; Hon. Charles H. Prescott, Saco; Prof. George T. Files, Brunswick; George L. Crosman, President of Chamber of Commerce, W. B. Moore, executive secretary, Chamber of Commerce, members *ex officio*.

A meeting took place on November 5, at Portland, the results of which are told in the following letter:

The Chamber of Commerce
Portland, Maine.

My Dear Mr. Hawley:

You will probably be interested in knowing that the question of the establishment of an aeronautical base in Casco Bay, an Atlantic coast patrol, for the State of Maine, was received with great interest and enthusiasm by about two hundred representative citizens of this city and State at our meeting last night at the Falmouth Hotel.

Our honored guests, Rear Admiral Peary, Henry A. Wise Wood, Henry Woodhouse, and Elmer A. Sperry, delivered interesting and instructive addresses and thoroughly convinced all those present that this is an opportune time for Portland and the State of Maine to take the initiative in this big movement for preparedness.

A committee of about thirty-five representative citizens of the State have been selected to take charge of the campaign for private capital necessary to install this station.

We feel sure of the success of the project, and to you and our guests of last evening, and to the Aero Club of America, we owe a sincere debt of thanks.

We appreciate more than we can express the practical proposition as submitted by you, and without a doubt your plan will be carried through to completion.

The people of Portland and the State of Maine are a unit in this movement and as usual in all progressive projects in this country, Maine leads.

We extend to you now and the members of your organization a hearty invitation to be with us in a very few months at the opening exercises of the first aeronautical base established in the United States.

Very truly yours,
W. B. MOORE,
Executive Secretary.

The committee raised the sum of $10,000 for starting the aeronautic station, $910 of which was contributed by the National Aeroplane Fund of the Aero Club of America.

As an order was about to be placed for the first seaplane in July, 1916, Ex-Senator Charles F. Johnson of Maine advised the committee that a bill had been introduced in both Houses providing for the establishing of a system of Aerial coast patrol stations. Therefore it was advisable for the committee to wait until the Government could establish the station. Unfortunately that measure was not adopted at that session of Congress and is at date of writing being considered by the naval committees of both Houses of Congress.

Burgess seaplane presented to the Massachusetts Militia by the Aero Club of New England.

Massachusetts Naval Militia

Steps to organize an aviation section for the Naval Militia of Massachusetts were first taken in December, 1915, as the result of the combined efforts of the Aero Club of America and the Aero Club of New England, under the personal supervision of Messrs. Godfrey L. Cabot, president of the Aero Club of New England; Norman W. Cabot, G. Richmond Fearing, Greely S. Curtis, Norman Merrill, and others.

Contributions of $2500 and $500 respectively were made through the Aëro Club of America by Mr. T. Jefferson Coolidge for the training of aviators for the Massachusetts Naval Militia in 1915–16. The following letter from Governor Walsh acknowledging receipt of the first contribution shows the interest that was taken in the newly launched movement:

Alan R. Hawley, Esq.,
President Aero Club of America,
297 Madison Avenue, New York City.

My dear President Hawley:

Your letter of December 16, with checks enclosed amounting to $2500, received. As commander-in-chief of the organized military forces of the Commonwealth of Massachusetts, I accept this contribution which an undisclosed patriotic citizen has made for the important work of forming an aviation corps in the militia.

Please convey to the donor and to your organization as well, the thanks of the people of Massachusetts, for I am sure I voice their sentiments when I say to you that our people are most grateful for this evidence of your organization's interest in a great patriotic work. I shall transmit the checks to the official authorized to receive funds for the benefit of the military organization of the State.

Yours very truly,
(Signed) David I. Walsh,
Governor.

In the early part of 1916, the Curtiss Aeroplane Company offered through the Aero Club of America to train an aviator from the militia of each of the forty-eight States, and trustees for the National Aeroplane Fund allowed $40 for the expenses of each man being sent to the school. This sum was increased to $150 per man when the Mexican situation grew critical, to enable the States which did not have an appropriation for this purpose to send officers to take advantage of the free course of training. Ensign Normal Merrill, third deck division, naval battalion, was assigned to take the course of training at Newport News.

Governor Walsh also authorized an aeronautic squad of one officer and four men to be attached to each of the 9th and 10th deck divisions, one of the divisions to operate a seaplane to be given to the State by the Aero Club of New England and the other to operate a private seaplane loaned for the purpose.

The seaplane presented by the Aero Club of New England was officially turned over to the Commonwealth of Massachusetts on November 8, 1916. The ceremony took place at Boston. The presentation speech was made by Godfrey L. Cabot, president of the Aero Club of New England. Governor McCall accepted the machine on behalf of the State, and congratulatory speeches were delivered by Major-General

Leonard A. Wood, U. S. A., Senator Henry Cabot Lodge, former Governor David I. Walsh, and Augustus Post and G. Douglas Wardrop of the Aero Club of America. The invited guests included the members of the Governor's council, the mayors and city councillors of Boston and Cambridge, members of the Massachusetts Legislature, Metropolitan Park Commissioners, United States Army and Naval instructors detailed to duty in New England, and other prominent national guard, army, and navy officers. The seaplane was built by the Burgess Company, and is of the single pontoon "U" type tractor. It is equipped with a Curtiss OXX2, 100 horse-power motor.

When the navy offered to train naval militia aviators at Pensacola, the Massachusetts militia sent the following to take the course of training: Ensigns Godfrey L. Cabot and Norman Merrill; Leon T. Blood, R. E. Self, C. J. Thurlow, Harold Hudson.

At date of writing the aeronautic sections of the Massachusetts Naval Militia are mobilized at Marblehead under the command of Lieutenant James O. Porter. The aviation officers in order of seniority are: Lieutenant Godfrey L. Cabot, aviation aide First Naval District; Lieutenant J. G. F. S. Lincoln, Ensign Norman W. Cabot, Ensign C. L. Flint, Ensign F. S. Allen, Ensign F. S. Amory, Ensign G. Richmond Fearing.

All of these officers have done more or less flying, and all except Lieutenant Lincoln and Ensigns Flint and Fearing have flown as pilots.

The senior officer in his seaplane, the *Lark,* has been patroling Boston Harbor since the declaration of war and taking pupils with him in this work.

The most important advance so far made has been the establishment of the Squantum School for students in which Ensign Cabot and Ensign Fearing have been very active and which was turned over by the State to Capt. W. R. Ruse, Commandant of the First Naval District as representative of the Navy Department on May 18, 1917.

Michigan Naval Militia

Steps to organize the aviation section for the Naval Militia of Michigan were taken in the fall of 1915 by the Aero Club of Michigan, Russell A. Alger, president, with the cooperation of the Aero Club of America. A public subscription was started by the Aero Club of Michigan, and through the generosity of Mr. Emerson McMillin, the trustees of the Aero Club of America contributed a bonus of 10 per cent. of the funds raised by February 1, 1916, which amounted to $11,800. An aeroplane of the L-W-F type, equipped with a Thomas motor, was ordered, which was delivered at the camp of the Michigan Naval Militia at Grayling in July, 1916. Flying took place between the dates of July 20

Miss Mona Dunne christening the Flying Boat presented to the Illinois Naval Militia by Messrs. Andrews and McDonald.

The L. W. F. seaplane presented to the Michigan Militia by the Aero Club of Michigan, which was employed in the manœuvers and which was piloted by H. W. Blakely.

and August 18. On this last date a hurricane came up suddenly and wrecked the machine. In his report of the work of this machine, Mr. Sidney D. Waldon, the treasurer of the Aero Club of Michigan fund, gives the details of this extraordinary storm and mentions how some of the national guardsmen present who had waited and hoped for an aeroplane for so long actually cried over the wreck. The following are excerpts from the report:

The plane was consequently brought into shore and, through the efforts of twenty-five or thirty men, cased up the sand beach and roped to stakes, the machinery and control mechanism being covered with canvas.

When it was seen that the two storms were going to strike camp, additional stakes were driven and ropes passed about the wing struts. A detail of signal corps men was sent to stand by the plane through the storm. The wind at first blew from the west in fitful gusts. This was from off the lake and tended to blow the plane farther on the shore. Suddenly, without warning, the wind veered to the opposite point of the compass and blew a perfect hurricane, accompanied by a terrific downpour, for fifteen minutes.

The intensity of the wind, coupled with the force of the water, leveled the whole camp, broke off large trees, rolled all the heavily loaded transport wagons down hill with their brakes set, and brought a stream of water knee-deep down the slope where the plane was. Some of the stakes pulled loose; others held, but, by holding, hurt the wings. In the intensity of the blow, however, the men were lifted off their feet, everything came loose, the plane was picked up, in the air, whirled around, and landed, bottom side up, well out from shore.

Some of the boys in the signal corps detail were so heartbroken, they blubbered over the wreck; but with the passing of the storm every one pitched in to make the best of a bad job. First of all, they took off all the damaged wings and steering and elevating attachments, then they turned the fuselage right side up and brought it to shore. They then fitted the running gear for land flying, ran the motor to make sure it was all right and then examined the fuselage inside and out.

The state authorities did not have the funds with which to pay for the repairing of the plane, and as no assistance could be obtained from the Federal Government the aeronautic activities of the Michigan Naval Militia were suspended for the time being.

New Jersey Naval Militia

New Jersey was one of the first naval state organizations to make an effort to organize an aeronautic section. In December, 1915, Commander Edward McPeters wrote to the Aero Club of America the following letter:

Gentlemen:

It is desired to bring to the attention of the Aero Club of America the fact that the First Battalion,

Naval Reserve of New Jersey, is organizing an aviation section.

This has received the approval of the Navy Department, and I have appointed Ensign-elect J. Homer Stover to the immediate charge of organizing the section.

The Division of Naval Militia Affairs of the Navy Department has been requested to state what aeronautic equipment the Government would supply for this section.

It has replied that clothing and equipment for the enlisted men in accordance with the funds available would be furnished.

In regard to furnishing an aeroplane or other special equipment it is advised that "steps will be taken to render such assistance as may be possible with the Federal funds available."

There is, however, no definite assurance that an aeroplane, hangar, etc., will be provided before some remote date from that source.

Knowing the aid that the Aero Club of America, with the assistance of the Aeroplane Fund, is giving in the promotion of military aeronautics, it is asked what assistance, if any, and in what form the Aero Club of America will give this organization in its work of obtaining an aeroplane and other necessary equipment for the aeronautic section.

Very respectfully,

EDWARD McC. PETERS,
Commander, N.R.N.J.

The Aero Club of America officials made an appeal on behalf of the New Jersey Naval Reserve, and a hydroaeroplane was presented to the Naval Reserve by Mr. Inglis M. Uppercu, the president of the Aeromarine Plane & Motor Company, who subsequently also presented another training biplane.

Contributions aggregating over $1500 were also made to the Aero Club of America to defray the expenses of the upkeep and operation of the two machines. An aviation camp was established at Keyport, and a number of men were given a limited amount of preliminary training.

When the Navy Department decided to train the naval militiamen at Pensacola, the following were sent from the New Jersey Naval Militia to take the course of training: Ensign W. A. Lee, E. A. Denton, G. MacCreagh, J. C. Rolfe, and F. Prove.

New York Naval Militia

In the early part of 1915 Commander Charles L. Poor, of the New York Naval Militia, wrote to President Alan R. Hawley of the Aero Club of America for cooperation in organizing aviation corps for the New York Naval Militia. His letter read as follows:

First Battalion, N.M.N.Y.
U.S.S. *Granite State,*
U.S.S. *Wasp.* Foot West 97th St.,
New York City.

President of the Aero Club of America.

Sir:—

It is purposed with the cooperation of the Navy Department to organize as a part of this battalion an

One of the two aeroplanes presented to the New Jersey Naval Reserves by Mr. Inglis M. Uppercu, the president of the Aeromarine Plane & Motor Co. The personnel of the Aviation Detachment is shown on the photograph.

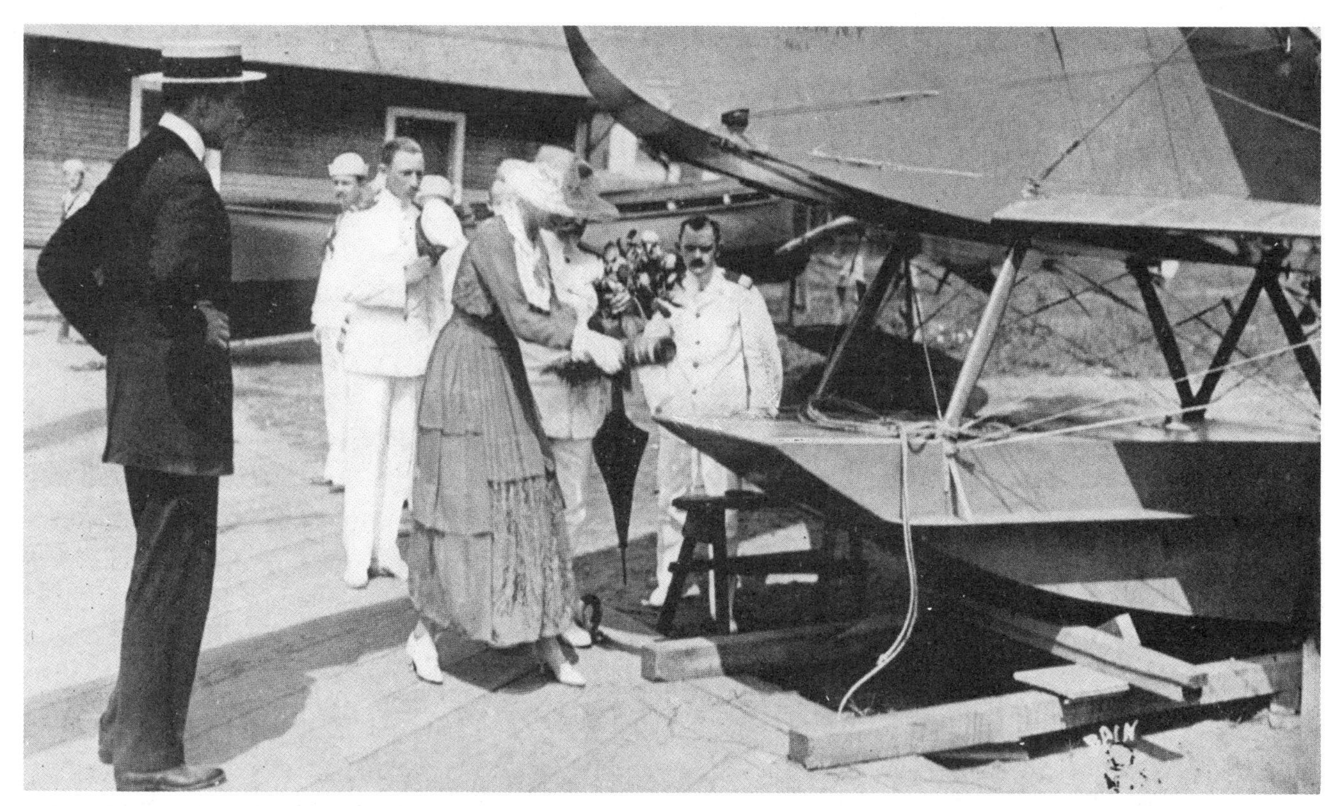

Mrs. Vincent Astor breaking the champagne bottle at the christening of the Second Battalion seaplane, "N. M. N. Y. No. 1." The christening took place on July 1, 1916, at the Second Battalion Armory, at the foot of 52d Street, Brooklyn. Captain E. T. Fitzgerald, commanding officer, Second Battalion, New York Naval Militia, is shown on the right; Ensign Samuel S. Pierce on the left; Mr. Vincent Astor on the extreme left. The seaplane was of the Burgess-Dunne type, equipped with a 140 horse-power Sturtevant motor.

aeronautical squad, or division. It is the plan at first to start in a small way, say with two units, of an officer and six men each, and expand as occasion offers. It is believed that an active and efficient organization can be built up in this way.

It seems logical that the naval militia offers the best medium for those interested in aviation to ally themselves to the national defense and be of service to their country, and the Navy Department is much interested in building up an aviation corps of volunteers through the Naval Militia. It is through the Navy and Naval Reserves in Great Britain that the volunteer aviators have been able to render such active and valuable service.

The members of the aviation units would be enlisted regularly in the Naval Militia, in the regular way for three years, and would have the same obligations and same privileges as the Naval Militia, but their drills and duties would be entirely in connection with their special branch. At such times as the battalion went afloat on a battleship on its regular ten days' or two weeks' practice cruise, it is suggested by the Navy Department that it would be desirable for the aviation squad in lieu thereof to be ordered to the Navy Aviation School for the same period and under the same conditions of pay and so forth, to receive instruction there.

Knowing the strong interest of your members in this science and believing there must be many who would be glad to serve their country in this way, I make this appeal for your interest and cooperation, and ask that you make known this organization to your members.

Yours respectfully,

Charles L. Poor,
Commander N.M.N.Y.
Commanding First Battalion.

An appeal was made by the club on behalf of the militia and as a result, in May, 1915, the First Battalion was presented with a flying boat, and the training for an officer and a mechanic, by Mr. Glenn H. Curtiss. Ensign Lee H. Harris, First Battalion, Naval Militia, New York, and a noncommissioned officer of the First Battalion, were sent to the Curtiss Schools at Buffalo and Hammondsport to take their course of training. The flying boat was christened on November 3, 1915, Miss Olive Whitman, daughter of Governor Whitman of New York, acting as sponsor. The ceremony was attended by the state, city, naval militia, and aeronautical authorities, and was a most impressive event.

The sum of $1250 was also contributed for training aviators for the New York Naval Militia by Mr. T. Jefferson Coolidge.

The aeronautic section of the First Battalion, Naval Militia, New York, was mustered into the service of the State on January 17, 1916. It consisted of one officer and eight men, and the new Curtiss flying boat. The Hudson River, where the headquarters of the Battalion is located, not being a suitable location for a flying school, steps were taken to locate the school elsewhere. Through the courtesy of Mr. Charles Laurence, the loan of a large site of water-front property was secured at Bay Shore, Long Island, on the shores of the Great South Bay, an ideal spot for seaplane work. The progress of this unit, and its difficulties, are told in the official report of the officer in charge, as follows:

On May 13, 1916, the flying boat was carted out to Bay Shore. Here it was assembled by the men of the Aeronautic Section. On May 27, the first flight was made. By this date the strength of the section had been increased to two officers and fourteen men. At this time the aeronautic section had absolutely no equipment, and no funds with which to purchase same. The Department of Naval Militia Affairs at Washington, had informed the Naval Militia of the various states that it would furnish hangars, gasolene, and supplies to States having aeronautic units. Advantage was taken of this offer, and bids secured for a tent hangar; and gasolene, tools, and such necessary equipment requisitioned. Due to various delays the hangar was not received until July 8, 1916. In the meantime, the officers and men of the section had been reporting at Bay Shore every Saturday afternoon and Sunday, and instructions in flying had been in progress.

There had been shipped out to the grounds such equipment as the battalion could furnish, consisting of three small tents, cots, blankets, mess gear, and some tools. Lumber had been secured with which to construct tent floors and runway for launching the flying boat. Upon the arrival of the hangar it was erected, the three small tents pitched, and thus the Aeronautic Station, First Battalion Naval Militia, New York, was established. It should be borne in mind that, as was previously mentioned, the section had no funds with which to carry on its work, or funds to expend in establishing comfortable living quarters for its men. Such expenditures as were absolutely necessary, such as carting, lumber, and other incidental expenses, were met by funds from the battalion, which were limited and needed for meeting its usual expenses. The funds of the battalion were heavily drawn upon in paying the transportation of the men to and from Bay Shore for this week-end work.

In view of the above every endeavor was made to keep the outlay as low as possible, and the actual labor necessary in assembling plane, erecting tents, building platforms, and numerous other necessary duties were performed by the men of the section. The work of trenching in log anchors to hold the hangar was indeed a laborious task. All this work was performed Saturday afternoons and Sundays, and therefore covered a considerable period.

It should be mentioned that the section had added another office to its roster, making a total strength of three officers and fourteen men. On July 15, 1916, the section reported at the station for a fifteen-day tour of Federal duty, while the other members of the battalion were away on their annual cruise. It was intended during this tour of duty to establish a mess and that the entire section should quarter in the tents. This was found to be impracticable, as the quarters were limited and the facilities for cooking and serving proper meals almost out of the question. Therefore, arrangements were made for some of the men to sleep in a near-by boarding house and meals for the entire section provided. This, however, did not prove entirely satisfactory.

During the fifteen-day encampment the weather was anything but favorable for flying. It was possible to

Governor Whitman, of New York, expressing the thanks "of the people of this State" for the flying boat presented to the naval militia of New York.

The officers of the aviation divisions of the First and Second Battalions, New York Naval Militia, at Bay Shore, Long Island. The officers from left to right are: Lieut. J. J. Carey, Ensign Samuel S. Pierce, Ensign Vincent Astor, Lieut. L. H. Harris, Ensign F. E. Wysong, Ensign C. E. Ruttan.

fly but six days during this period. However, during these six days, remarkable progress was made. At the expiration of the tour of duty one hundred flights had been made. Flying was continued during the month of August and into September. Due to the motor needing a general overhauling, which would require considerable time, and the living and working conditions not being satisfactory, as the fall approached, it was deemed advisable to discontinue the station for the season. Up to this time a total of one hundred and forty-four flights had been made, totaling 1694 minutes actually in the air.

As a result of the season's work, four aviators were trained, two of them were actually flying alone in the boat and two were ready for their first flights alone. Others were partly trained. Of these there were two three-fourths finished and two were about one-half through with their course of training.

When the Navy Department offered to train the Naval Militia at Pensacola, in October, 1916, the following were appointed to take the course: Lieuts. Lee H. Harris, and C. E. Rutan; Charles B. Vandy, Elwood H. Neener, George C. Matteson, Charles A. Blanchard.

The aviation section of the second Battalion, New York Naval Militia, had its inception at the tenth annual dinner of the Aëro Club of America, at which Commodore R. P. Forshew, Commanding Officer of the New York Naval Militia and Mr. Vincent Astor decided on a plan of action. A committee was formed consisting of Messrs. Vincent Astor, Aymar Johnson, F. Meredith Blagden, and Charles Laurence, and through their efforts a fund was raised for the purchase of a Burgess-Dunne seaplane. Ensign Vincent Astor had already taken a course in the operation of a seaplane and, with Ensign S. S. Pierce, a veteran aviator, took charge of the aviation section of the Second Battalion—of which Captain E. T. Fitzgerald is the commanding officer. The seaplane was christened on July 1, 1916, at the Second Battalion Armory, at the foot of 52d Street, Brooklyn. Mrs. Vincent Astor acted as sponsor and the ceremony was attended by the militia, state, city, and Aero Club of America authorities. This section camped at Bayshore during the summer with the aviation section of the First Battalion.

Early in 1917 the naval militia authorities, with the cooperation of the Aero Club of America, and the National Special Aid Society, raised funds for the establishing of aeronautic divisions of the militia at the Bayshore station. By June 1 the station was running on a substantial plan. Lieutenant Commander F. R. Lackey was placed in charge of the station which became the training camp for the aviation sections of the First, Second, Third and Fourth Battalions.

The personnel of the aeronautic divisions and sections of the four Battalions on June 15, 1917, was as follows:

AERONAUTIC DIVISION, 1ST BATTALION, N. M. N. Y.

Date of organization, January 8, 1916.

Lieutenant: Harris, Lee H. *Lieutenant (j.g.):* Ruttan, Charles E. *Ensigns:* Wysong, Forrest E.; Wrightsman, Charles B.

Chief Machinists Mates: Peterson, Herman A.; King, Frederick E. *Machinists Mates* 1*st Class:* Roder, Walter L.; Boyd, Theodore P. *Machinists Mate 2d Class:* Lopez, Anthony. *Electricians 3d Class:* Delaney, Charles E.; Douglass, Kingman; Beakirt, Robert H.; Henry, Charles T.; Grant, Frank L.; Farnham, James P.; Dalrymple, Fitzwilliam, Jr.; Madill, Edward J.; Freeman, Frank; McAdoo, William G., Jr.; Reynders, John V. W.; Munson, Curtis B.; Fleischmann, Charles; Blossom, Francis R.; Thomas, John C.; Cummings, James H., Jr.; Evans, George B., Jr. *Carpenters Mate* 1*st Class:* Gorey, Frank M. *Carpenters Mates 2d Class:* McEnroe, Charles J.; Strong, Howard A. *Carpenters Mates 3d Class:* Gordon, Wilmot G.; Laughlin, George M., 3d; Roberts, Charles H. J.; Guest, David P.; Fuller, Roswell H.; Spencer, Dumaresq; Winslow, Alan F.; Berger, Frederick G.; Stanley, Julian C.; Moseley, George C.; Shaffer, Harvey W.; Tevis, Gordon B.; Winter, Wallace, Jr.; Eastman, Julian, Jr.; Taylor, James B.; Requa, Charles P.; Turner, Frank B.; Matthiessen, Conrad H., Jr.; Atkins, Samuel, W. *Boatswains Mate* 1*st Class:* Butler, Thomas J. *Electricians* 1*st Class Radio:* Noble, James K. *Landsmen:* Lamar, Lamartine V.; Humphreys, William H., Jr.; Overton, John W.

AERONAUTIC DIVISION, 2D BATTALION, N. M. N. Y.

Date of organization, May 1, 1916.

Ensigns: Pierce, Samuel S.; Lawrance, Charles L. *Machinists Mate* 1*st Class:* Poor, Roger A. *Machinists Mates 2d Class:* Cusachs, Philip A.; White, Lawrance G.; Breese, James. *Carpenters Mates 2d Class:* Keeler, Joseph; Cogswell, Edward D. *Yeoman 2d Class:* McCormick, Joseph F. *Electricians 3d Class:* Coddington, Dave H.; Eckerson, George D.; Inglis, William C.; Rumpelt, Mortimer R.; Brennan, Edward S.; Quinlan, Joseph F. *Landsmen:* Pollock, Edward S.; Nelson, Clarence O.; Eaton, Alvah H.; Johnson, Albert R.; Lackey, Russell H.; Ansbro, Francis P.; Brennan, Joseph V.; Collins, Reginald; Dollard, Oakly; Harris, Frederick M.; Keith, Frederick W.; Whyte, Dave R.; Burton, James H.; Naylor, Henry R.; Cobb, George W.

AERONAUTIC SECTION, 3D BATTALION, N. M. N. Y.

Ensigns, Garlock, Harold C. J.; Hathaway, Bradford G.; Donnelley, Thorne.

Machinists Mates 2d Class: Blanchard, Charles A. Neener, Elwood H.; Mattison, George C. *Carpenters Mate 3d Class:* Vandy, Charles B. *Seamen* 1*st Class:* Schreiner, Anthony E.; Walter, Steward W. *Apprentice Seamen:* Barry, Edmund C.; Saunders, Donald W.

AERONAUTIC DIVISION, 4TH BATTALION, N. M. N. Y.

Date of organization, May 17, 1917.

Ensigns: Verplanck, J. Bayard; Rutherford, John M.

Chief Machinists Mate: Danner, Edward. *Machinists Mate* 1*st Class:* Smith, Nathaniel. *Yeoman 2d Class:* Newlin, James C., Jr. *Mechanics:* Rogers, Harry; Croft, Leon E.; Hood, John P.; Patterson, John D.; Lyon, R. D.; Patterson, Z. H.; Gates, G. D.; Earle, Jesse B.; Clarke, William J.; Albertson, Nelson; Cullen, John Frank; Brown, David L.; Maier, P. L.; Needham, Charles; Watson, Paul; Jova, John A.; Delahay, Raymond; Swartout, Van Etten; Sittenham, Frederick W.; Pollard, Edward.

Oregon Naval Militia

The Oregon Naval Militia acquired a flying boat in the fall of 1915, but it was damaged in one of the test flights. There being no funds available for aeronautics for the Oregon Naval Militia, and being unable to get assistance from the Navy Department, the Commanding officer appealed to the Aero Club of America, which sent $250 with which to pay for the repairs. The following letter from the commanding officer gives the status of the aviation section of the Oregon Naval Militia at that time:

The 140 horse-power Sturtevant seaplane presented to the Rhode Island Militia through the National Aeroplane Fund starting for a flight.

Portland, Oregon,
November 2, 1915.

FROM: Commanding Officer.
TO: Alan R. Hawley, President, Aero Club of America, 297 Madison Avenue, New York.
SUBJECT: Repairs to aeroplane.
REF.: Your letter November 17, 1915.

1. I beg to acknowledge receipt of your letter of November 17th with the enclosed check in favor of the Oregon Naval Militia in the sum of $250.00, and wish to thank you most sincerely for the assistance you have given our organization. It is not expected that the amount of the repairs now being made will equal that of the check and any excess will be returned.

2. It is expected that two aviators will qualify under the rules of the Aero Club during the next month and that immediately thereafter a full aeronautical section will be organized as a branch of the Oregon Naval Militia.

3. The interest shown in our organization by the Aero Club is of great benefit to us and we shall endeavor to procure funds for a new machine and an aero station along the lines laid down in your various communications.

Very sincerely yours,
(Signed) G. F. BLAIR,
Oregon Naval Militia.

Lack of funds prevented the expansion of the aviation section of the Oregon Naval Militia in 1916.

Pennsylvania Naval Militia

On June 1, 1915, Mr. David H. McCulloch offered the use of his two flying boats to the Naval Militia of Pennsylvania, but the militia could not avail itself of this offer, lacking the funds with which to operate the machines. Subsequently the Aero Club of America sent at its own expense the aviator William S. Luckey to fly for the National Guard of Pennsylvania at Indiana, Pennsylvania, August 7–14, inclusive, who served for that period under orders from Brigadier General Albert J. Logan, commanding second brigade.

Rhode Island Naval Militia

When the Aero Club of America started the National Aeroplane Fund, in the early part of 1915, one of the first contributors was Mr. George I. Scott, a retired banker, of New York and Rhode Island, a member of the club. Finding that the United States Navy's aeronautic program was hopelessly limited by law which allowed it only a small personnel, and by the inadequate appropriations, and realizing the necessity, of establishing aeronautic sections in the naval militia of States having such organizations, he started a subscription for the establishing of an aeronautic section in connection with the militia of Rhode Island. As he was developing the plans, the late Miss Lyra Brown Nickerson, of Providence, Rhode Island on October 3, 1915, subscribed $7500 to the National Aeroplane Fund of the Aero Club of America.

Miss Nickerson's check was transmitted to Brig.-Gen. C. W. Abbott, Jr., of Rhode Island on November 29, 1915, with an offer of $500 additional to go toward defraying the expenses of operating the machine. But in the meantime there was received $11,000 additional in public subscriptions and the $500 was not needed.

The aeronautic section of the Rhode Island Militia was authorized on April 27, 1916. The authorization is reproduced for historic purposes:

STATE OF RHODE ISLAND AND PROVIDENCE PLANTATIONS.

THE ADJUTANT GENERAL'S OFFICE,
PROVIDENCE, April 27, 1916.

GENERAL ORDERS,
No. 13.

I. The provisions of General Orders, No. 153, Navy Department, series 1915, are hereby adopted for the Naval Militia of this State.

II. By authority of Section 34, Chapter 394 of the Public Laws, as amended, and in conformity with General Orders, No. 153, as above, the organization of the Naval Militia, until further orders, will be a battalion, consisting of one Engineer Division, 1st; three Line Divisions, 2nd, 3rd and 4th; and an *Aeronautic Section.*

The battalion headquarters, staff, commissioned, warrant and enlisted and the enlisted personnel of the Divisions will be in accordance with Sections I, II and III, pages 3–14 inclusive. The minimum enlisted strength of a division will be 40, maximum 60. Battalion, Division and Section Commanders will conduct the instruction of their respective commands in such a manner as to fit their enlisted personnel for the examinations prescribed in Pars. 86 to 92, pages 44 to 83 inclusive.

By order of EMERY J. SAN SOUCI,
Lieutenant Governor,
Acting Governor and Commander-in-Chief.
CHARLES W. ABBOT, JR.,
The Adjutant General.

A Sturtevant hydroaeroplane was ordered, which was delivered about the middle of June, 1916, and the aeronautic section of the Rhode Island Naval Militia went to camp on July 17, where the machine was used.

A Curtiss hydroaeroplane was also ordered and delivered to the Rhode Island Militia in the middle of June. But the battery which was to use it went to the Mexican border before the machine arrived, therefore could not use it to train pilots from among its officers.

The status of the aviation section of the Rhode Island Militia at the close of 1916 was given in the following letter from Adjutant Abbott dated December 23, 1915:

The Rhode Island fund is now about $21,500. Twenty-five men, two of whom are qualified aviators, and all of technical training and vocation, have enlisted in one of our Naval Militia Law providing for an aero section and these men will compose it. Instruction in flying has been given to a section of men in Battery A, Field Artillery, R. I. N. G., under the direction of one of its lieutenants, who owns a hydroaeroplane, and is a qualified aviator. Until the Legislature takes action providing for maintenance of our proposed flying plant and personnel we can take no further definite steps, but we are hopeful that Rhode Island will soon be able to do its part for preparedness in that most vital adjunct of defense, aviation.

Sincerely yours,
(SIGNED) CHARLES W. ABBOT, JR.,
Adjutant General.

When the United States Navy arranged for the training of naval militia men at Pensacola, the Rhode Island Naval Militia assigned the following to take the course of training:

Ensign J. K. Park, George B. Proy, Charles J. Whitford, Henry M. Fallon, Raymond N. Estey.

When they completed their course the following were assigned from the Rhode Island Militia to take a course:

Ensign T. J. H. Pierce, W. Biehler, C. A. Dowler, W. G. Fielder, H. H. Walsh.

Upon the declaration of war the aeronautic section of the Rhode Island Militia was mobilized at Marblehead Neck—and at date of writing is "somewhere" in the United States.

Wisconsin Naval Militia

A flying boat was presented to the Wisconsin Naval Militia through the Aero Club of America by Mr. B. R. J. Hassell on September 1, 1915. But the militia could not avail itself of this offer, lacking the funds with which to operate the machine.

An informal snapshot of part of the members of Aerial Coast Patrol Unit No. 1, at West Palm Beach, Florida.

CHAPTER XXVII

AERIAL COAST PATROL

Duties of.—Owing to the trick of circumstances caused by the advent of submarine warfare, aircraft in the great war have hardly been used for the purposes for which they had been assigned before the war; i.e., scouting and spotting for the fleet. They have, instead, been employed very extensively for purposes least anticipated, as follows:

(1) To locate, and assist destroyers, trawlers and submarine chasers in capturing or destroying hostile submarines (seaplanes, dirigibles, and kite balloons used).

(2) To locate submerged mines and assist trawlers in destroying mines (seaplanes, dirigibles, and observation balloons used).

(3) Searching the coasts for submarine bases (seaplanes and dirigibles used).

(4) To convoy troop and merchant ships on coastwise trips (dirigibles used).

(5) To patrol the coasts, holding up and inspecting doubtful ships and convoying them to examining stations (dirigibles used).

(6) Attacking hostile ships and submarines that may show up near the coasts, with torpedoes, bombs and guns (large torpedoplanes and large seaplanes mounting guns used).

(7) Protecting ships at sea and in ports against attack from hostile submarines and battleships (seaplanes and dirgibles used).

(8) Communicating to incoming ships information regarding the location of mines, submarines and the courses to follow to avoid disasters and confusion (seaplanes and dirigibles used).

(9) Serving as the "eyes" of mine planters, minimizing the time required for mine planting (dirigibles and observation balloons used).

(10) Defending and protecting naval bases

and stations from naval and aerial attacks (armed air cruisers and combat planes used). The aerial coast patrols have, therefore, had most of the work to do.

Duties of Coast Patrol Aviators

The duties of aerial coast patrol aviators are many and varied. They are epitomized in the following notices of election of two aviators to the Distinguished Service Order of Great Britain under date of June 22nd, 1916:

Flight Lieut. (Acting Flight Commander) Redford Henry Mulock, R. N. A. S.

In recognition of his services as a pilot at Dunkirk. This officer has been constantly employed at Dunkirk since July, 1915, and has displayed indefatigable zeal and energy. He has on several occasions engaged hostile aeroplanes and seaplanes, and attacked submarines, and has carried out attacks on enemy air stations, and made long-distance reconnaissances.

Lieutenant John Henry Dalbiac, R. M. A.

In recognition of his services as an aeroplane observer at Dunkirk, since February, 1915. During the past year Lieut. Dalbiac has been continually employed in coastal reconnaissances and fighting patrols. The Vice-Admiral Commanding the Dover Patrol, in reporting on the work of the R. N. A. S. at Dunkirk, lays particular emphasis on the good work done by the observers.

The duties of the aviators connected with the Dunkirk station are the intensive duties which aerial coast patrol aviators have to perform under the most severe conditions, since Dunkirk is one of the most exposed stations. But this gives a clear idea of what the duties of the aerial coast patrol are. They have to fight the enemy above, on, and under the water.

How to perform the different duties of the aerial coast patrol is told in detail in different chapters.

The establishing of aerial coast patrol service in the United States is a most popular movement in the field of national preparedness, and thanks to the efforts of the National Aerial Coast Patrol Commission, and the Aero Club of America, a number of units of the aerial coast patrol were established and either operating or under training by June 1, 1917. The story of the founding of the Aerial Coast Patrol Commission is told herewith by Rear-Admiral Robert E. Peary, the chairman of the Aerial Coast Patrol Commission.

THE UNITED STATES AERIAL COAST PATROL

By Rear-Admiral Robert E. Peary,

Chairman of the National Aerial Coast Patrol Commission

History has brought down to us from 600 B.C. Themistocles' dictum, "He who commands the sea, commands all." This dictum was true until the advent of practical aircraft, then it was changed and now we have the new dictum, *"He who commands the air, commands all."*

From time immemorial man has sought to protect himself from savage animals and other equally savage men, by posting watchmen or sentinels. At first a skin-clad, semi-apelike figure, crouching with stone ax in the shadow of a cave mouth where his family or tribe were sleeping. Later armed with spear, or mounted perhaps upon some half wild horse; and still later clad in complete armor, peering from the battlements of some walled city. But the fundamental idea has been the same, that a few could watch against the approach of danger, that the masses might sleep or go about their ordinary occupations in safety, and that the possessions of the individual and the wealth of the community might be protected.

Even the birds and animals have such a system, and perhaps primeval man obtained his first ideas from them. In war time, these watchers were largely increased in number, until around every city and camp there stretched a continuous living cordon of armed men stationed at short intervals of easy communication with each other, each traversing continuously his assigned beat. A living human fence to prevent surprise attack.

Now, I shall try to picture herewith the modern evolution of this idea and tell you of an up-to-date picket line around a great nation; a plan which links Icarus with Wright and Curtiss and Hammond. This is the conception: a continuous picket line of seaplanes off shore around our entire coasts from Eastport, Maine, to Brownsville, Texas, and from San Diego, California, to Camp Flattery, Washington, each machine traveling back and forth—back and forth—over its section or "beat," a winged sentinel, forming a cordon, a continuous line of whirring shuttles, weaving a blanket of protection around the country.

History of the United States Aerial Coast Patrol

In 1914, Captain Virginius E. Clark of the United States Army Aviation Corps published in the "Coast Artillery Journal" an article on 'An Aeroplane Patrol of our North Atlantic Coast from Roanoke Sound to Portland." A review of this article, with a sketch map, was published in a New York paper, attracted my attention and the page on which it appeared went into my scrap book.

In August, 1915, accepting an invitation to address the Portland, Maine, Rotary Club, I spoke on "Preparedness," and suggested to the business men composing my audience, "Might it not be well for you to look into this, with a view to establishing this end of such an Aerial Coast Patrol, and have one of the nation's eyes here in your harbor, at Flag Island, or other suitable place?"

A few days later, the current issue of "Flying" reached me with John Hays Hammond, Jr's., comprehensive plan for an Aero Coast Patrol of our entire coasts—Atlantic and Pacific—by a system comprising forty-four stations.

August 28, 1915, the largest newspaper in Maine printed a half-page story containing extended extracts from the plans of both Clark and Hammond, and a few days later the Portland Chamber of Commerce wrote me, offering its support and assistance.

September 10, 1915, Mr. Henry A. Wise Wood visited me at Eagle Island; we devoted a day to a careful examination of Flag Island, the use of which for a landing station had been tendered to the Aero Club of America a year earlier, and Mr. Wood reported to Mr. Alan R. Hawley, the president of the club, the results of his investigations.

September 13, at my suggestion, Mayor Ingraham, of Portland, called a conference of leading Portland men in his office, to whom the idea of a National Aerial Coast Patrol was presented, and the suggestion made that Portland establish the first station of that system. Mayor Ingraham then assigned the matter to the Portland Chamber of Commerce, whose President, George L. Crosman, appointed a state-wide committee of representative men to take the matter up.

This committee arranged a dinner in Portland on the 5th of November, 1915, at which Maine senators and congressmen, two or three ex-Governors, and the Mayor and other prominent Maine citizens were present. Messrs. Henry A. Wise Wood, Henry Woodhouse, and Elmer A. Sperry came on from New York to lend assistance to the movement. Portland had hoped also to have President Hawley, but illness prevented.

At the dinner, the following telegrams of endorsement were received from the President, the Secretaries of War and Navy, whose attention had been called to the movement by Senators Johnson and Burleigh of Maine:

> I join the Secretary of War and the Secretary of the Navy in feeling a very great interest in the development of aeroplane service in this country, and in hoping that your citizens will meet with entire success in their interesting undertaking.
>
> Woodrow Wilson.

> I assure you I am deeply interested in the development of aeroplane service in this country, and I trust that your citizens will meet with success in their undertaking.
>
> L. M. Garrison.

> Experience has proven the great value of aircraft, and every movement looking to their larger utilization meets with the cordial sympathy of the Navy Administration.
>
> Josephus Daniels.

The Aero Club of America made formal offer of ten per cent. on all sums raised by February 1, and Mr. Curtis of Marblehead offered to build seaplanes for the Maine station at reduced prices, and to educate two aviators.

This meeting impressed upon those present, and through them the entire State, that the matter was not a mere local proposition, but that Maine was inaugurating a great national proposition of vital importance. This dinner was held Friday evening, November 5. On November 12, 1915, at the annual meeting of the Aero Club a resolution was adopted that the Aero Club concentrate its efforts on the Coast Patrol System.

Following this, and thanks to the generosity of Mr. Emerson McMillin, who offered, through the Aero Club of America, to give eleven per cent. for any amounts raised up to $500,000, the trustees of the National Aeroplane Fund of the Aero Club of America wrote to the Governors of all the States and the heads of militia organizations and aero clubs, urging them to raise funds with which to establish stations for the aerial coast patrol on water as well as on land.

From that time on the interest in the aerial coast patrol system grew rapidly.

As public interest increased the interest of Congress was also aroused and senators and congressmen who became interested proposed to realize the plan to establish a substantial aerial coast patrol as part of the national defense program. In July, 1916, Senator Charles F. Johnson and Congressman Julius Kahn introduced bills in the Senate and House of Representatives respectively providing for the establishing of the Aerial Coast Patrol, under the direction of the Navy Department, providing an appropriation of $1,500,000 for same, but the chairman of the committee on naval affairs of the House of Representatives objected, stating that Congress had already given $1,500,000 more for naval aeronautics than had been asked for in the estimates, so there was no progress on this measure.

As the Navy Department has had no facilities for training aviators for the Naval Reserve Flying Corps and could not supply aeroplanes and aeronautic equipment to the Naval Militia, neither of these two organizations were in a position to undertake to train the numerous applicants who applied for training. The National Aerial Coast Patrol Commission has encouraged those who could afford to pay their own expense to form units and prepare themselves to be of utmost use in case of a national emergency—which came.

Aerial Coast Patrol Unit No. 1

On July 3, 1916, Mr. and Mrs. H. P. Davison held a conference with Messrs. John Hays Hammond, Jr., and Henry Woodhouse, regarding the possibilities of their two sons, F. Trubee Davison and H. P. Davison, Jr., taking up aviation training. The value of the aerial coast patrol was discussed and the Davisons decided to form Aerial Coast Patrol Unit No. 1, to consist of twelve men who were selected from among the college friends of the Davisons as follows: F. T. Davison, Robert A. Lovett, John Vorys, John Farwell, 3rd; Albert Ditman, Wellesley Laud Brown, Artemus L. Gates, Erl Gould, Allan Ames, C. D. Wiman, A. D. Sturtevant, H. P. Davison, Jr.

Arrangements were made with the America Transoceanic Company for the training of the unit at Port Washington, and the training began immediately, David H. McCulloch acting as instructor.

In September the unit participated in the manœuvers of the "mosquito" fleet of power craft, which had its headquarters at the Atlantic Yacht Club at Gravesend Bay.

As the report written at the time states, in foggy weather members of the unit went out to locate mines submerged at depths of eighteen feet. Although the water was far from clear and the haze further made conditions difficult, every one of the mines were located. No other craft could have located them under these conditions, and the officer in charge was so impressed that when the manœuvers were over and the aviators had left for their base which is located at Manhasset Bay he called on them to come again to locate the mines, so as to save a lengthy and tedious search for the boats.

On September 9 one of the Curtiss seaplanes of the unit went out to look for the two torpedo boat destroyers representing the enemy attacking fleet. The fog prevented the aviators from seeing further than six miles, and when they were about twenty-five miles outside of New York Harbor they were caught in a thunderstorm. But they located the torpedo boats, unknown to the latter, and reported them to the Atlantic Yacht Club.

The destroyers ***Flusser*** and ***Warrington*** left their anchorage in Gravesend Bay in the morning and proceeded to sea. Their function was to return over a different course in the afternoon, the ***Flusser*** representing an advanced screen of scout vessels and the ***Warrington*** the main body of an attacking fleet whose objective was New York Harbor and the battleships ***Kentucky, New Jersey,*** and ***Maine*** at anchor in the harbor's entrance.

The power craft of the 'mosquito" fleet put out to reconnoiter the approaches of the harbor,

One of the flying boats of Volunteer Aerial Coast Patrol Unit No. 1, photographed from another flying boat of the Unit.

and they were directed in their observations by a seaplane from Volunteer Aerial Coast Patrol Unit No. 1.

Like true seamen, the members of the unit who manned the seaplanes in both locating mines and locating the destroyers do not wish their names given, preferring to have the achievements credited to the unit.

The two aviators who were caught in the storm sighted the two destroyers and plotted their location on their maps when the vessels were sixty miles from Gravesend Bay. In the meantime, there was considerable consternation on board the vessels of the fleet regarding the safety of the aviators.

Thomas W. Slocum's steam yacht, the ***Ranger,*** with Lieutenant A. M. Cohen, U. S. N., on board, was the last of the craft that put out from the Atlantic Yacht Club to observe the seaplane, and she carried the information to the battleship ***New Jersey.*** Meanwhile the ***Dodger II*** with E. S. Willard on board, and other motor craft, sped in search of the air scouts. The ***New Jersey*** was unable to establish radio communication with the destroyers because of static conditions, so that the destroyers could not be advised to assist in the search.

Several steamship masters and others on board of sturdy power craft that were hard put to hold their course in the storm reported that they had seen the seaplane in the storm and it was an awe-inspiring sight to see the craft against a scowling background of copper-colored cloud banks, shot with incessant lightning flashes. But the seaplane went on, though at times in spasmodic jerks, when it met the squally blasts, then it disappeared in the storm haze which hung along the shore.

The aviators had located the torpedo boat destroyers one mile from Fire Island and had landed on the Great South Bay, close by Oakland Island. Immediately upon landing they went to the nearest telephone and succeeded in sending a telephone message, which was relayed by the Quogue Life Saving Station to the Atlantic Yacht Club, reading, "At 3:41 o'clock at a height of 3100 feet, we sighted both destroyers and then put into Oakland Island on account of the squall."

Upon receipt of the official reports showing the valuable work that was done by the Volunteer Aerial Coast Patrol Unit No. 1, the executive committee of the Aero Club of America sent to the unit expressions of appreciation, which said, in part:

Conditions obtaining at the time when one of your seaplanes located the two torpedo destroyers, the *Flusser* and the *Warrington,* make us realize that such a thing could have happened under war conditions and this achievement would have saved New York from being bombarded by the enemy's fleet. Owing to the fact that the United States Navy has only a few aëroplanes and trained aviators, in case of war the naval aviators available would barely be sufficient to operate at one of the naval centers or in connection with one of the squadrons of the fleet. We can conceive of the *Flusser* and *Warrington* representing a raiding squadron of an enemy's fleet dominating a strategical point outside of New York Harbor, and the raiding squadron having succeeded in getting near New York, screened by the fog, and being found by one of the seaplanes of the Volunteer Aerial Coast Patrol Unit No. 1, thereby saving the ships at the entrance of New York Harbor from being destroyed and New York City from being bombarded. Your valuable work in locating mines also deserves commendation.

Throughout the Fall the unit practiced at the New London Navy Yard and experimented in locating mines and submarines.

Upon their return to the Yale University for their studies the members of the unit found that much interest had been aroused among the students, and many applied for admission. Eighteen more were admitted, increasing the unit to thirty.

At the time of Germany's declaration of her intention to prosecute a campaign of ruthless U-boat warfare, the Davisons and all the members and officials of the Aerial Coast Patrol Commission and the Aerial Club of America realized that this policy might mean the involving of this country in war, and that in such a case there would be needed many aviators for coast patrol duties. The thirty members of Volunteer Aerial Coast Patrol Unit No. 1 made plans to leave Yale University and go to Florida to continue their training, so as to be available and to be thoroughly trained in case of emergency.

The report that the unit was to go to Florida for training stirred up tremendous interest among the students at Yale, Princeton, Columbia, Harvard, and other universities, and as a result, the unit was flooded with hundreds of applications from college men anxious to join the unit and to form other units.

Arrangements were made for Unit No. 1 to go to the aviation training camp established by Mr. Rodman Wanamaker, at West Palm Beach. Mr. Lewis S. Thompson agreed to assist the Davisons in this patriotic work, and as Mr. H. P. Davison, Sr., could not himself go to Florida, Mr. Thompson took charge of the general supervision of the unit. As the demand for training was so large, and in view of the approaching emergency, it was decided to give the Unit a military formation, which was done, and the members of the Unit enrolled in the United States Naval Flying Reserve Corps. The camp at West Palm Beach was run on a military basis, the members of the unit getting training in discipline, aerial gunnery, wireless, and general aerial coast patrol work, as well as in the operation of aeroplanes.

Mr. Henry Woodhouse presented to the Unit, through the Aero Club of America, a Lewis aeroplane gun, with 5000 rounds of .333 ammunition, to enable the Unit to get practical experience with an up-to-date aeroplane gun. The plan was to have Unit No. 1 train at West Palm Beach until the weather permitted establishing a camp on Long Island. While the plans for this camp were being considered, the applications for training from college men became so numerous that the patriotic people who were contributing toward the training and equipping of Aerial Coast Patrol Unit No. 1 decided to extend their support so as to increase the number of men to be trained and to establish an aerial coast patrol station at Huntington, Long Island.

The plan was submitted to the Navy Department, and the authorities gave valuable suggestions in carrying it out. Lieutenant (J. G.) Edward O. McDonnell was assigned from the Pensacola Aeronautic Station to supervise the training of the unit.

Aerial Coast Patrol Unit No. 2

Aerial Coast Patrol Unit No. 2 was organized in April, 1917, by Messrs. Ganson G. Depew, Frank Goodyear, with the cooperation of Robert A. Lovett of Unit No. 1, and the National Aerial Coast Patrol Commission. The Unit reported to the Curtiss School at Buffalo for training with the following members: Frank Goodyear, Gannon G. Depew, Ashton T. Hawkins, Ed. De Cernea, Stephen Potter, E. T. Smith, Percival Fuller, Seymour Knox, Clifford Rodman, Winter Meade, Philip Allen, John Joy Schieffelin. The address of the Unit is 165 Summer Street, Buffalo, New York.

Aerial Coast Patrol Unit No. 3

Aerial Coast Patrol Unit No. 3 was organized by David Clinton Backus, a Yale man, in the middle of April, 1917, with the cooperation of the National Aerial Coast Patrol Commission. It reported for training to the Knapp Seaplane station, at Mastic, on the South Shore of Long Island, with twelve members, as follows: Clinton D. Backus, Harold Pumpelly, William J. Connors, Jr., Harold Howe, Thomas Dixon, Jr., Irving Paris, Leslie Macnaughton, Duncan Forbes, William Hamilton Gardner, Austin Feuchtwanger, Stewart Johnson, Joseph Knapp, John Laird, Bruce Campbell. The address of the unit is 205 West Fifty-seventh Street; headquarters, Knapp Seaplane Station, Mastic, Long Island, New York.

Aerial Coast Patrol Unit No. 4

Aerial Coast Patrol Unit No. 4 is being organized at date of writing by Clarence Martin, chief officer of the aviation division, Organization for National Defense, VIII Corps, Columbia University, New York City. Other aerial coast patrol units are under formation at date of writing. One hundred and fifty applications for training had been received, from Columbia students and Alumni. The address of the unit is Room 311, East Hall, Columbia University, City of New York.

Those who went to the West Palm Beach station for training were: F. Trubee Davison and Henry P. Davison, Jr., Robert A. Lovett, Albert J. Ditman, Charles Wiman, Artemus L. Gates, Allen Ames, John V. Farwell, John Vorys, Earl Gould, Wellesly Laud-Brown, Reginald G. Coombe, Oliver B. James, G. Franklin Lawrence, William Rockefeller, Frederick Beach, Kenneth Smith, Kenneth McKleish, Curtis Read, Bartow Read, David Ingalls, A. D. Sturtevant, William Thompson, Frank Lynch, Graham Brush, Henry Landon, Samuel S. Walker, Charles Stewart, Archie McIlwaine.

CHAPTER XXVIII

THE EVOLUTION OF THE SEAPLANE AND THE FLYING BOAT

In broad outlines a hydroaeroplane is a craft having a water borne base in the form of a pontoon, or boat, and the organs of an aeroplane, whereby it operates from the water as a base as distinguished from the land, and is adapted to travel at speed upon the surface of the water, or fly in the air. The boat or pontoon is for these purposes fitted with a hydroplaning bottom so arranged that when under power on the water the craft is lifted to the surface by the combined aeroplane and hydroplane action and skims thereover. It may then either be operated at speed without leaving the water, or, having attained sufficiently high speed to enable the wings to support the entire body, rise from the surface and fly in the air. These two capabilities, cruising at speed and flying, render the hydroaeroplane not only of inestimable value for naval warfare, but extremely attractive for sporting and commercial purposes.

It is, therefore, the hydro that makes the hydroaeroplane—which explains why the present chapter opens with the first experiment with hydroplanes rather than with the first attempt to fit an aeroplane with pontoons to keep it afloat.

The difference in principle between a hydroplane and a pontoon is about the same as that between an aeroplane and an airship. The first cuts through the water in about the same way as an aeroplane cuts through the air, with practically no displacement; while the second displaces the fluid according to its bulk. The first may attain high speed with moderate power, while the latter requires, comparatively, high power for moderate speed. The float used at present is a combination of pontoon and hydroplane; that is, the pontoon is flat-bottomed, usually with an upward slope in front, a combination that works like the hydroplane, causes it to rise to the surface when traveling, so that it can acquire the necessary speed for launching in flight.

The flying boat, which was developed a year after the hydroaeroplane had proven a success, is essentially a boat with wings.

Experiments with hydroplanes were made as early as forty years ago. In 1872 an engineer named Froude, under the auspices of the British Admiralty, experimented with planes sheathed with polished metal, grouped and inclined. This experiment was, however, without result. The writer is indebted to Mr. Orville Wright for pointing out that Comte de Lambert, the well-known aviator engineer, was the inventor of the hydroplane. In Mr. Wright's own words: "Although suggestions of the hydroplane idea had been made years ago, and although Froude had made some experiments without results as far back as 1872, Comte de Lambert was the real inventor of the hydroplane. He was the first to produce a successful one, and all modern hydroplanes are based upon his work. In 1897 Comte de Lambert experimented with a catamaran formed of two narrow floats, to which were attached four transverse planes, whose inclinations could be varied two or three degres. At a speed of ten miles an hour, the floats were lifted entirely out of the water and the machine glided over the surface of the water on the four hydroplanes. Comte de Lambert continued these experiments during the following years up to 1907, and he succeeded in increasing the speed to thirty-four miles an hour."

The first to fit pontoons to an aeroplane was

William Kress's aeroplane fitted with pontoons, 1898–1902.

William Kress, the Austrian engineer and father of Austro-German aviation. His test was made in 1898–1902 as the result of forty years of experimenting with heavier-than-air structures, during which Kress had constructed a number of machines, including a helicopter and an ornithopter. The Kress was a triple monoplane, fitted with two parallel elongated floats, a 30-horse-power motor and two propellers. The two floats were of aluminum and had runners, so that the apparatus could be operated on the snow as well as on the water. The tests took place at the Unter-Tullnerbach docks, Austria, and gave remarkable results—considering that the inventor had to create all except the motor, and was handicapped by the latter on account of its great weight. The machine traveled over the water under limited control at a speed of a fast rowboat. This apparatus came to grief at the close of 1901, adverse wind and a leak in one of the floats sending it to the bottom of the dock. Another experiment in 1902 was not concluded through lack of funds.

In 1905–06, three of the French pioneers of modern aviation—Ernest Archdeacon, Gabriel Voisin, and Louis Bleriot—who experimented with the Chanute-Wright gliders—thought, after wrecking several machines by falls on land, that the water surface, being elastic, would be less dangerous to the operator, so they fitted their gliders with pontoons. The Archdeacon glider fitted with two boat-like floats, was first tried on the River Seine, near Paris, on June 8, 1905. It was towed by a fast motor boat and was piloted by Voisin. It rose to a height of about 50 feet over a distance of 400 feet. The Bleriot glider also towed by a motor boat and piloted by Voisin, proved to have less stability, and on July 18, 1905, being struck by a sudden gust of wind it dove into the water and Voisin, caught in the cage-like affair, was submerged for twenty seconds. The Archdeacon experiment ended at the close of 1905 when Voisin left Mr. Archdeacon and formed partnership with Bleriot. The outcome of the Bleriot-Voisin partnership was two hydroaeroplanes, one consisting of two elliptical-shaped cells, fitted with three pairs of cylindrical, sheet-metal floats, a 24-horse-power motor, and two propellers; the other was a combination of the first and the Archdeacon glider, and was fitted with improved pontoons and two 24-horse-power Antoinette motors. Nearly a year was spent in experiments, but with little results. The principal trouble was no doubt the insufficient motor-power, and the fact that the pontoons were not scientifically constructed.

Louis Bleriot's experiment, 1906.

An early experiment of Gabriel and Charles Voisin, June 8, 1905.

Early in 1905, Professor Enrico Forlanini of Milan, Italy, applied for a patent for a hydroaeroplane of his own invention. That he had a very lucid conception of the advantages of the hydroaeroplane is shown in the patent specification, part of which reads as follows:

"My invention has reference to ships or vessels of that kind which, instead of plowing their way through the water, skim over the surface, thereby offering much less resistance and as a consequence are capable of attaining very much higher speeds.

"Heretofore many attempts to produce an efficient apparatus of the hydroplane type have been made, the majority of them based upon the phenomenon exhibited when a flat object, such as a stone for example, is thrown in such a manner as to glide over the surface of the water, rather than that of obtaining a true hydraulic flight. To this end it has been usual to make use of hydroplanes arranged, for example, in such a manner as wholly or partially to lift the vessel out of contact with the surface of the water when said vessel is propelled.

"The object of my invention is so to improve such devices that their efficiency is greatly increased, and one of the essential features of my invention is that a boat constructed in accordance therewith will be capable not only of skimming over the surface of the water, but may be also used as a flying machine of the aeroplane type, and I have succeeded in constructing an apparatus which has in practice given most satisfactory results." But the inventor did not carry his experiments in that line to a finish, turning instead to the hydroplane, on which he is an authority.

Further experiments with hydroplanes were conducted in different countries, notably in Italy by Crocco and Riccaldoni, in France by Clement Ader, Ricochet, and Bonnemaison; in Switzerland by the Defaux brothers; and to a less extent by others in different countries. But while these experiments contributed each something to the final development of hydros for aeroplanes, the object of the experimenters was not to develop a marine aeroplane.

In 1905, at Saint Helens, Isle of Wight, England, Dr. F. A. Barton and Mr. F. L. Rowson made experiments with a birdlike apparatus fitted with floats. The machine had two wings or main planes set dihedrally, elevating plane in front and an empennage consisting of two planes, both movable with a rudder. It was 36 feet long and 34 feet across the main planes.

The pontoons were 20 feet long, 10 inches wide, 4 inches deep, were made of a light skeleton framework of whitewood, the sides and bottom of which were covered with three layers of mahogany veneer glued together. The whole pontoon was covered with canvas and varnished. Several towed flights were made at sea in September, 1905. During the most successful, the machine rose four feet, then the rope broke, and in the fall the glider was damaged. Failing to secure a suitable motor for the machine the experiments were abandoned.

In 1906–97 experiments with a large glider fitted with pontoons were made by Israel Ludlow, in America. Several experiments were made with this machine towed by the naval tug *Potomac* and the torpedo boat *Gwinn,* at Hampton Roads, during the Jamestown Exposition. But the experiments were discontinued, following an accident which wrecked the glider.

The Wright Experiment

In 1907 the Wrights experimented with hydros intending to develop them for use on their flyer, and thus have a machine which would have

Louis Bleriot's last experiment on water, 1906.

Israel Ludlow's Kite towed by the United States naval tug *Potomac* across Hampton Roads, passes in front of the battleship at anchor.

permitted them always to find good landing places in large streams.

The following quotation from the Dayton "Daily News" for March 21, 1907, shows that the Wright brothers had at the time a thorough idea of the advantages offered by the water as a field for experiments:

"The balustrades of the Third Street Bridge were lined Thursday morning with curious spectators, who were watching the antics of a modern water bird. The banks of the stream around the central point of action were also spotted with onlookers. The object of interest was the hydroplane, which Orville and Wilbur Wright, inventors of the airship, were tampering with in preparation for its initial experimental run.

"Although the inventors, who are being branded as geniuses, would not state the exact purpose of the hydroplane, it was intimated that it is to be used in connection with their airship. If it becomes possible, during a series of experiments, to lessen the weight of the hydroplane, so that it can be attached to the airship, a new and much sought after proposition of aerial navigation will be solved. With the attachment of the hydroplane to an airship, the machine will then be complete, as it can be navigated either in midair or on water.

"The present machine which is uniquely constructed from water boilers, an old gasoline engine and numerous strips of wood and sheet iron, with the water planes of copper, made its sail down the Miami River Thursday morning amid the encouraging cheers of the assembled spectators.

"The machine, if the square lines were produced, would be about twelve by eight feet in dimensions. The vehicle is built on two air floats about twelve feet in length. On these is a platform which supports a light wooden frame. On the bed sits the gasoline engine and there remains room enough for four men. The framework supports the two propeller screws which furnish the motive power.

"The hydroplane is steered by a small rud-

The Wright Brothers' experiment with hydros, 1907.

Becue piloting Fabre hydroaeroplane, 1911.

der, which extends backward from the middle of the crafts. The water planes, which form the basis of the inventors' idea, are situated between the floats and perpendicular to them. The idea is to have these floats so constructed and set that they will just permit the floats to touch the water enough to keep the structure afloat. The basic idea is much the same as that of ball bearings, which is that of decreasing friction. It is claimed that such a craft, when worked out, will be able to attain a speed of from twenty-five to thirty miles an hour. The present machine is motored by a 20-horse-power engine."

Unfortunately during the very night of the day of this experiment, the dam, which retained the water on which the experiments were made, broke, and the Wrights found it impossible to continue the experiment.

The First Flight from the Water

The first to leave the water with a power-driven hydroaeroplane was Henri Fabre, a young French engineer, who, after three years of experiments, constructed a hydroaeroplane of original design fitted with a 50-horse-power Gnome motor, and succeeded in leaving the water, flying and returning to the water without mishap. This first flight took place on March 28, 1910, near Martigues, France; the height reached was about six feet, the distance covered one thousand feet. A better flight was made on May 17, 1910, of about one mile at a height of 30 feet, but on landing the machine was much damaged. Subsequently a number of flights were made; but the machine had many limitations. When at last it was wrecked, at Monaco, during a storm, Fabre discontinued the experiments. Fabre's floats were a great improvement on the pontoons.

Glenn H. Curtiss's Success

Glenn H. Curtiss was very early impressed with the great advantages of an aeroplane capable of operating from the water. His first machine, the well known *June Bug,* built by him as a member of the Aerial Experimental Association, and flown at Hammondsport successfully for the "Scientific American" trophy in 1908, was in the fall of that year equipped with pontoons and operated on Lake Keuka under a new appellation *The Loon.* This machine was a twin pontoon machine equipped with separate hydro surfaces placed underneath the floats, but the experiments were unsuccessful.

In 1909, Mr. Curtiss gave the problem much study. In that year he conceived the form of the machine which was to make him famous the world over. His idea was a machine of the single float type with small floats beneath the tips of the wings combined with hydro surfaces, the former to act by displacement to buoy the wing tips up when the machine was standing still and the latter to act when the boat was operating at speed to impart a stabilizing hydro-lift to the wing tips when the machine should become unbalanced. In this way operation at speed on an even keel was assured. The single float was to be centrally placed and was to be provided with hydro-surfaces beneath it. In the early spring of 1910 this machine was built. The main float was a light but strong

Glenn H. Curtiss piloting his hydroaeroplane at San Diego in 1911

canoe. Metallic hydroplanes were fitted beneath it fore and aft and suitably braced from the keel and the gunwales of the canoe. The wing tip floats were conically capped tin cylinders. The wing tip hydro-surfaces were flexible wooden paddles inclined downwardly and rearwardly. The motor was a standard 4 cylinder, 40 horse-power Curtiss motor. Early in May of 1910 this machine was tried on Lake Keuka with wonderful results. The problem of operating at speed upon the surface of the water was proven to be solved. Time and time again the little craft was operated up and down the lake at varying speeds from the lowest to the highest, turning sharply to right and to left with absolute safety and with perfect poise. The flexible paddles connected with the cylinders under the wing tips maintained the machine always on an even keel. The solution of this problem made possible not only safe operation at speed upon the water, but also alighting upon the water from the air.

About this time, Mr. Curtiss made his world-heralded trip from Albany to New York and won the $10,000 prize of the "New York World." As a preliminary to this he actually equipped a land machine having the usual wheels with pontoons and hydro-surfaces, and starting from the land took an extended flight through the air and alighted safely upon the surface of the water. With the experience of this achievement he equipped his *Hudson Flier* with pontoons and hydro-surfaces so that should occasion have required he could have landed upon the surface of the water. But so well had he prepared for this flight and so capable was the machine with which he flew that

One of the early experiments of Hugh L. Willoughby, 1908–10.

The Voison machine in flight.

there proved to be no necessity for the use of these devices.

But the small 40 horse-power motor of the canoe machine proved to be too weak to attain that speed necessary for flight, and Mr. Curtiss immediately laid plans for a machine with greater power. That summer while he earned money for his experiments by his daring exhibition work on land machines he kept his shop at Hammondsport busy on the construction of the new machine. The new hydroaeroplane was equipped with a pontoon of the modern streamline form, having a broad, flat hydroplaning bottom and a scow-like bow. Not only was this to have less water and air resistance than the adapted canoe of the former machine, but it was to exert a hydro lift through its own bottom instead of through attached hydros used in the canoe machine. It took time to build such machines then, facilities were not great, and experienced help scarce. But the machine was ready in the late fall, and was immediately shipped to San Diego, California, where Mr. Curtiss had found most excellent climate for winter experiment, and a wonderful stretch of water for tests in the Spanish Bight between North Island and Coronado. Within a fortnight after assembly of the new machine was commenced the Curtiss 60 horse-power motor had propelled the craft from the water to the air and safely back again, and the press had acclaimed to the world the advent of the successful hydroaeroplane. The first public flights were made on January 26, 1911. Already a further improved craft was under way, and within a week after the flights of January 26 we find the inventor flying the machine which was to be adopted by the United States and foreign navies. Three weeks after, on February 17, 1911, he introduced the new craft to the United States Navy by flying to and landing alongside of the cruiser *Pennsylvania,* was hoisted aboard by the ship's crew, and when dropped overboard again rose and flew back to shore. Thereafter a number of the Curtiss aviators used hydroaeroplanes for exhibition work, and the navy having purchased a Curtiss hydroaeroplane, naval officers—Lieutenants T. G. Ellyson and J. H. Towers—learned to fly, won their certificates with it, and made scores of flights. During October 17–21 Hugh Robinson flew from Minneapolis to Rock Island, Illinois, 370 miles, carrying and distributing mail along the route. Hundreds of flights had been made by the close of 1911 with one- and two-passenger types, fitted with motors of 60-horse-power and 75-horse-power, and a speed of 50 miles per hour on the water and 60 miles per hour in the air had been attained.

A further development, a machine with a float of pronounced boat shape, was made in the fall, 1911, and was tried on January 10, 1912. The fundamental idea was changed from that of a floating aeroplane to that of a flying boat. The first trial was successful, the craft fitted with a 60-horse-power motor rising from the water with ease, and traveling on the water at a speed close to 50 miles an hour and in the air at one mile a minute.

The first test of seaplanes in a completion took place at Monaco on March 24–31, 1912. It was organized by the International Sporting Club of Monaco, which offered 15,000 francs for prizes to be divided between the three winners: first, 8000 francs; second, 4000 francs; third, 3000 francs. The contest was for starting and alighting from still water and rough water, starting and landing from dry land, and passenger carrying. Points were awarded to the winners of daily contests, and the one getting most points won the prize.

This contest, being the first of its kind and coming at a period when the navies were looking for air crafts for marine service, was well attended by sportsmen and military authorities of different countries. Ideal weather favored the meet from beginning to close, enabling most of the aviators to contest daily.

Seven aviators flying five different types of machines took part and concluded as follows:

Aviator	*Machine*	*Motor*	*Points*
Fischer	Henri Farman biplane	Gnome	112.10
Renaux	Maurice Farman biplane	Renault	100.80
Paulhan	Curtiss biplane	Curtiss	86.30
Robinson	Curtiss biplane	Curtiss	71.90
Caudron	Caudron biplane	Anzani	63.
Benoit	Sanchez Besa biplane	Salmson Unne	50.30
Rugere	Voisin biplane	—	41.75

The meet was a success, but the craft had its disadvantages, principally that of precluding the test on rough water. It would have been interesting to ascertain whether such large machines as the Henri Farman and Maurice Farman, whose large spread made it possible to rise from the water with as many as five passengers, could withstand the heavy sea as well as the Curtiss Triad, which was quite at home midst four-feet breakers, and was the fastest machine in the contest. The importance of this meet was that it brought forth the importance of testing seaplanes in rough water.

By August, 1913, marine flying had become quite popular. In the United States a number of sportsmen, including Robert J. Collier and Harold F. McCormick, owned their own seaplanes, and the United States Navy conducted important experiments, particularly the launching of seaplanes from ships, the details of which are given in another chapter. This country led in efficiency of types of seaplanes, and different foreign governments purchased our flying boats. It must be stated that much of the progress was due to the able efforts of Mr. Henry A. Wise Wood, who fathered "marine flying" from the very beginning, when the

The Burgess hydroaeroplane piloted by Mr. W. Starling Burgess, 1912.

The Curtiss "flying boat," 1911.

hydroaeroplane was considered a freak. He encouraged experiments in marine flying and prophesied exactly what has come to pass. Much credit is also due to Captain W. I. Chambers, U. S. N., for his able work in 1911–13.

France was also progressing rapidly. The list of seaplanes which competed in the Paris-Deauville race and the Deauville meet held in August, 1913, gives the status at marine flying in France at the time.

The Paris-Deauville Race

The race from Paris to Deauville was held on August 24, ten machines, each carrying a pilot and passenger, taking part. The pilots starting were: Weymann, Levasseur, Prevost, Janoir, Molla, Chemet, Rugere, DeMontalent, and Divetain.

The distance from Paris to Deauville, following the Seine, is 330 kilometers. Chemet covered the distance without a stop in 3 hours, 47 minutes, 51⅕ seconds. Three others stopped *en route* and finished as follows: Levasseur, 7 hours, 38 minutes, 15 seconds; Molla, 8 hours, 46 minutes, 11⅕ seconds; Janoir, 10 hours, 11 minutes, 4 seconds. Levasseur was disqualified, however, for not passing the control at Mousseaux.

Of the other contestants, Weymann, who flew very low and actually flew at full speed under bridges, was stopped by trees. In trying to take a short cut over a narrow branch

The hydroaeroplanes which participated in the first contest at Monaco—1912.

of the river at Saint-Pierre-du Vouvray, Divetain had to quit at Elbeuf, his floats having been damaged in landing on a rough shore. De Montalent fell near Rouen, his heavy machine being caught in *remous* at a low altitude.

The results of this race were not up to the standard established by French aviators—who in the year past had made many flights of between 500 and 1000 miles a day. A 500-mile flight was no longer recorded by the press, and an 800-mile flight, of which half a dozen had been made in the past month, was given but a few lines. The reason for this was that the contestants had to follow the Seine, which is tortuous and often narrow and lined with trees; and to pass each control, of which there was one at every bend of the river, they found it necessary to fly low, and thereby go through the *remous*. This was the cause of the Montalent disaster, and the reason for the victory of the light machines.

The Deauville meeting, like the Paris-Deauville race, was attended by excellent weather, which favored in many ways the light-built, low-powered machines, except during part of two days, the 29th and 31st, when the sea was heavy, once with a gale, then so rough that the destroyers, which had been on the scene of operation each day, remained in the harbor until the wind had abated.

Development Between 1914–17

An idea of the tremendous development that has taken place since 1914 may be gained from the following specifications: issued by the United States Navy Department in August, 1916 (schedule 39), which includes the terms of bid for aeroplanes for the United States Navy opened September 5, 1916.

CLASS 181.—(*Req'n 3, Office of Naval Aeronautics, C. and R. and S. E.—App'n: "Aviation, 1917"—Sch. 39.*)

To be delivered at the Navy Aëronautic Station, PENSACOLA, FLA., WITHIN . . . . DAYS after date of contract.

Bidders will insert in the above blank space the shortest time within which they can guarantee delivery,

The Martin aero yacht in flight.

subject to penalty for delay as provided in Form A. Other conditions being nearly equal, the time of delivery will be considered in making award.

Stock Classification No. 65

Bids are desired for furnishing aeroplanes, complete, and power plants, complete, as follows:

Aeroplane.—Includes the aeroplane proper, exclusive of power plant items, and in order for flight, and as per the following specifications, and includes, in addition, a launching truck and the necessary shipping crates.

Power Plant.—Includes motors, propellers, radiators, starting devices, gasoline and oil tanks, piping, controls, gasoline and oil gages, wireless outfit, power transmission system, tachometers, and the necessary shipping crates, etc., in order for flight, and as per the following specifications:

BID A.—*On the basis of furnishing 3 aeroplanes and power plants.*

Aeroplanes . . . . each
Power plants . . . . each

BID B.—*On the basis of furnishing 6 aeroplanes and power plants.*

Aeroplanes . . . . each
Power plants . . . . each

BID C.—*On the basis of furnishing 9 aeroplanes and power plants.*

Aeroplanes . . . . each
Power plants . . . . each

BID D.—*On the basis of furnishing 12 aeroplanes and power plants.*

Aeroplanes . . . . each
Power plants . . . . each

Bids to be itemized as above; unit prices to govern.

NOTE.—*Bids on this schedule should quote prices only, and should be forwarded in the usual manner to the Bureau of Supplies and Accounts. All drawings, blue prints, and descriptive matter illustrating the apparatus it is proposed to furnish, or amplifying the specifications, should be placed in a separate envelope and forwarded to the Office of Naval Aeronautics, Navy Department, Washington, D. C., in time to be received before the hour fixed for opening the bids. All such envelopes delivered late will not be opened and the bids will not be considered. These envelopes should be marked "Confidential," and should bear the name of the bidder, class and schedule numbers, and date of opening. After award of contract the drawings, etc., submitted by unsuccessful bidders will be returned.*

SPECIFICATIONS

SPECIAL CONDITIONS

Aeroplanes having characteristics differing from those specified will be considered, provided the differences are clearly noted in the specifications proposed, and provided the design proposed has sufficient merit to warrant such consideration.

It is desired to obtain the performance required on the lowest weight and power consistent with the other requirements of the specifications. What is wanted is a handy school aeroplane, embodying to the greatest degree practicable the qualities specified without the use of excessive power for the purpose. The performance outlined in these specifications may be departed from moderately without prejudice to a design, but is intended as a close guide to the type desired.

NOTE.—The following specifications have been purposely made of a very general character, stating broadly the results which it is desired to obtain. Bidders are requested to state in particular detail the methods they propose to employ to attain these results, in order that their proposals may receive intelligent consideration.

The department reserves the right to reject any or all proposals under these conditions. Awards will be based on the merit of the design, the design and manufacturing abilities and facilities of the bidders, the completeness with which information is supplied, and the price and time of delivery. It is particularly desirable that bidders shall estimate exactly deliveries which can be met.

GENERAL REQUIREMENTS

To be a naval aeroplane of the two-place, tractor, biplane type, and to conform in general to the detailed requirements of the following specifications:

PERFORMANCE

Maximum Speed.—Not less than 52.1 knots (60 miles per hour) nor more than 60.7 knots (70 miles per hour).

Minimum Speed.—Not more than 34.7 knots (40 miles per hour).

Climb.—Two thousand five hundred feet in first 10 minutes from surface.

Getaway.—Not over 34.7 knots (40 miles per hour).

Landing.—Not over 34.7 knots (40 miles per hour).

Radius.—Four hours, full power.

Fly in wind of 30.4 knots (35 miles per hour).

Drift in wind of 21.7 knots (25 miles per hour).

Getaway and land in wind of 21.7 knots (25 miles per hour).

1913 type of Wright flying boat.

Mr. Robert J. Collier, and Walter Brookins about to land by the Atlantic Fleet flagship *Washington*. Mr. Collier went on board of the ship and invited Rear Admiral Hugo Osterhaus to attend the Aero Show, May 9, 1912. Burgess hydroaeroplane.

DESIGN AND CONSTRUCTION

Throughout the construction of all aeroplanes and parts contracted for, designated inspectors shall have complete and free access to the shops, plans, specifications and records of tests of material involved in the construction of same. Where required, supplementary tests of materials and parts shall be made as directed by the inspector.

All parts to be of first class workmanship, material and design.

Alterations of parts, plans or material shall not be made without approval of the Navy Department.

Improvements developing between the dates of the contract and the completion of the machine shall be incorporated in the machine, if approved by the department, which shall determine finally the change of cost under the contract, if any is involved.

In the same manner any other changes or alterations ordered by the department after signing of the contract shall be considered and the change in cost determined.

Parts which appear defective in design, material or workmanship will be rejected and satisfactorily replaced unless the objection to their use is removed by a satisfactory demonstration of fitness. Aluminum shall not be used in the structural parts where strength is involved.

Protection from weather and salt spray shall be provided for all parts by the use of approved paints, varnishes, shellac, covers or metal plating, or by the use of non-corrosive material.

Portable covers for the cockpits and power plants shall be furnished with each aeroplane.

All interior woodwork will be given efficient protection against moisture. Particular care must be exercised to prevent access of moisture at faying surfaces, to end grain, and at butts, scarfs and joints. Such protection must be applied before final assembly of parts.

The color scheme will be natural finish or as approved.

The wing section used should be chosen with a view to efficiency and stability, and its characteristics in these respects must be known from laboratory tests.

The wings shall be readily and quickly removed or attached.

The control surfaces shall be of such proportions as to give positive control when flying at slow speed. They are to be capable of operation by either pilot unassisted or in conjunction.

Duplicate control leads are required to ailerons and rudders. The duplicate leads to follow as nearly as practicable different lines from those of the principal leads.

Means for hoisting shall be provided in the form of a fixed eye for a shackle over the top plane, and as nearly over the center of gravity when afloat as is practicable. The means of attachment shall be thorough and permanent and distribute the load to suitable members.

All parts shall be thoroughly trussed to withstand the launching impulse on the catapult. As this will depend upon the point of attachment and the float arrangement, prospective bidders should at once submit general arrangement plans to the department which will approximately indicate the point of attachment or

The Short folding wings seaplane being lifted out of the hold of a seaplane carrier.

any minor rearrangement of substructure required.

The floats may be of any type which will meet the requirements as to seaworthiness. They shall be substantially built to withstand the service intended, and to be stream lined as much as practicable without involving elaborate construction. They shall be divided into water-tight compartments, provided with approved means for inspection and drainage while afloat and when resting on launching trucks. The floats shall be provided with towing cleats, of approved form, securely fastened to the float.

Transportation trucks of approved design are required with each aeroplane.

The following instruments of approved type are to be furnished and installed on the instrument boards:

Air-speed meter, tachometer.*

Longitudinal inclinometer, oil gage *; air gage.*

And the following instruments supplied by the Government shall also be installed or provided for:

Altimeter, compass.*

Those items marked with * are to be installed in pilot's cockpit only.

A suitable gasoline gage, visible from the pilot's seat, to be installed on main gas tank.

POWER PLANT

To be suited to the requirements of the machine and to be provided with self-starters so fitted and installed that the motor may be started from the pilot's seat.

The carbureter shall be provided with a means for heating and with a successful means for muffling to prevent fire in case of a blowback from the engine. Provision shall also be made to prevent any danger in case of fire should the machine turn upside down.

Double independent ignition and double magnetos shall be used.

The motor shall be protected from moisture and spray.

The ignition and auxiliary circuits must be thoroughly protected from short circuits from spray, to insure against failure of the motor from this source. All aluminum parts are to be given protection against the effects of salt water.

All oil piping to be annealed.

The gas leads to reserve tanks, the control leads and the carbureter adjusting rod shall be provided with suitable and safe and ready couplings where these connections have to be frequently broken.

A positive system of pumping gasoline from the reserve tanks to the service tank shall be provided unless gravity feed or pressure feed from all tanks is used.

Gas, water and oil service pipes will be protected against vibration.

A positive means of cutting off the gas at the service tank shall be readily accessible from either seat.

At least one reliable method of stopping the motor shall be provided, to be operated from either seat.

Fuel tank capacity for at least four hours' flying at full power shall be provided. Fuel tanks are to stand an internal pressure of five pounds per square inch and shall be divided by swash-plate bulkheads. The heads of the tanks are to be so formed as to prevent crystallization due to vibration.

The service feed tank shall have a capacity for at least one-half an hour's flight, and shall be so fitted as to prevent danger from fire in case the machine should turn upside down.

All couplings and fittings in the gasoline line are to be thoroughly sweated on.

So far as practicable the entire power plant should be assembled as a unit on a good rugged foundation, which can be readily removed or replaced with a minimum disturbance of connections, controls and other structural fittings.

The motor shall, if practicable, be so installed as to permit of dropping the lower crankcase without the removal of the motor from its foundations.

A complete set of power plant tools to be supplied with each machine.

PROPELLERS

The propellers shall be suited to the requirements of the motor and machine, and their efficiency should exceed 70 per cent. They should be efficiently protected from the action of spray and broken water. The hub face plates shall be thoroughly interconnected independently of the propeller bolts. A safety fitting shall be provided, so that in case the propeller bolts carry away, the propeller cannot come off the hub.

MOTOR TESTS

Before installation, one motor, if of a design new to the department, to be selected, shall be put through the complete set of tests in succession as described herein. These tests shall take place at the navy yard, Washington, D. C.

Mr. Rodman Wanamaker's flyer, *America,* the first twin motored seaplane, taxying on Lake Keuka, Hammondsport, N. Y.

Rodman Wanamaker's Flyer *America* in flight at Hammondsport, N. Y.

Test A.—A run on the block to determine the maximum brake horse-power and the revolutions necessary to deliver the rated horsepower, to be followed by the calibration run for determining the b.h.p.r.p.m. curve.

Test B.—Two five-hour runs of the motor with calibrated moulinet or propeller at rated power. After the five-hour runs the motor shall be disassembled and the motor and the auxiliary parts shall be weighed. It will then be carefully examined and conditions within noted, particular attention being paid to the amount of wear and of carbon deposit. If the above tests and inspections are satisfactory, the motor shall be reassembled and given an additional one-hour run, without any adjustments or replacements during same, and during which observation shall be made in exactly the same manner as in the previous five-hour run. All other motors to be given tests at the factory the same as test B, except that there shall be one five-hour run only to be made with propeller calibrated at the Washington navy yard.

During the above trials records of the revolutions obtained and the corresponding power developed shall be made every fifteen minutes, together with notes as to the general action of the motor while running. The engine shall be thoroughly balanced, and vibration shall be a minimum. Oil and gasoline consumption shall be measured for each of the above trial runs and notes made as to the temperature of the circulating water at the inlet and outlet. No adjustments or replacements are to be made during the above trials.

Definition.—Full load comprises the aeroplane complete in order for flight and, in addition, fuel and oil for four hours' flight at full power and 375 pounds for pilots, instruments and equipment.

TRIALS

Demonstration Trials.—Before entering the prescribed acceptance tests each aeroplane shall be set up and flown by a representative of the builder at such place as shall be agreed upon. During these trials the full load shall be carried, and it shall be demonstrated to the satisfaction of the inspector that the aëroplane is capable of meeting the requirements. Defective features, if any, developing in the course of these trials shall be corrected and demonstrated to have been overcome.

Acceptance Trials.—If the demonstration trials have shown that the aeroplane is capable of entering its acceptance trials with a reasonable assurance of meeting the requirements, it will be given acceptance trials at the Navy Aeronautic Station, Pensacola, Fla., at such times as may be agreed upon in the contract. In the case of duplicated machines, the demonstration trials will be all that are required, and they shall demonstrate that each additional aeroplane can perform consistently with the original of the type.

SEAWORTHINESS

If the weather affords an opportunity, the aeroplane must ride at anchor or adrift in a 21.7-knot (25-mile) wind in Pensacola Harbor without danger of capsizing; otherwise stability shall be demonstrated to the satisfaction of the inspector.

Adrift, it should head into the wind.

Underway at low speeds it should steer readily.

With Full Wind.—The aeroplane shall get away in a calm in smooth water in not over 1500 feet (from a start with the motor idling at not over 25 per cent. of the revolutions for full speed).

It shall be capable of alighting and getting away in a 21.7 knot (25-mile) wind in Pensacola Harbor.

It should be capable of landing at high speed without danger of nosing rudder.

It should begin planing at not over 21.7 knots (25 miles) in rough water.

It should not capsize on a skidding landing or when running with wind abeam at high speed on the surface.

The floats should have a skid-form profile and a sufficiently easy bow to allow of plowing through a moderate sea without undue pounding or wetness.

AIRWORTHINESS

To have efficient longitudinal, lateral and directional stability in strong and rarified winds up to 30.4 knots (35 miles per hour) and to be capable of banking steeply without danger.

Longitudinal control shall be such as to enable recovery after a steep glide and to enable the machine to readily assume the gliding attitude in case power should fail while climbing.

Proposed aeroplanes shall have initial or natural lateral, longitudinal and directional stability in flight, such that moderate variations from the neutral attitude shall produce, positive righting moments without introducing oscillations of increasing amplitudes; any special arrangement of the wings or control surfaces for the purpose shall be clearly described, together with the effects produced.

Inherent or natural stability will be demonstrated by steadying the aeroplane on a path, straight or curved, and then holding the control in a fixed position. Under these conditions the aeroplane should continue to hold its path and trim for an appreciable period without requiring correction or assuming a dangerous attitude.

With Full Load.—A maximum speed of not less than 52.1 knots (60 miles per hour) and not over 60.7 knots (70 miles per hour) is required.

A minimum speed of not over 34.7 knots (40 miles per hour) is required.

A climb of 2500 feet in ten minutes from leaving the surface is required.

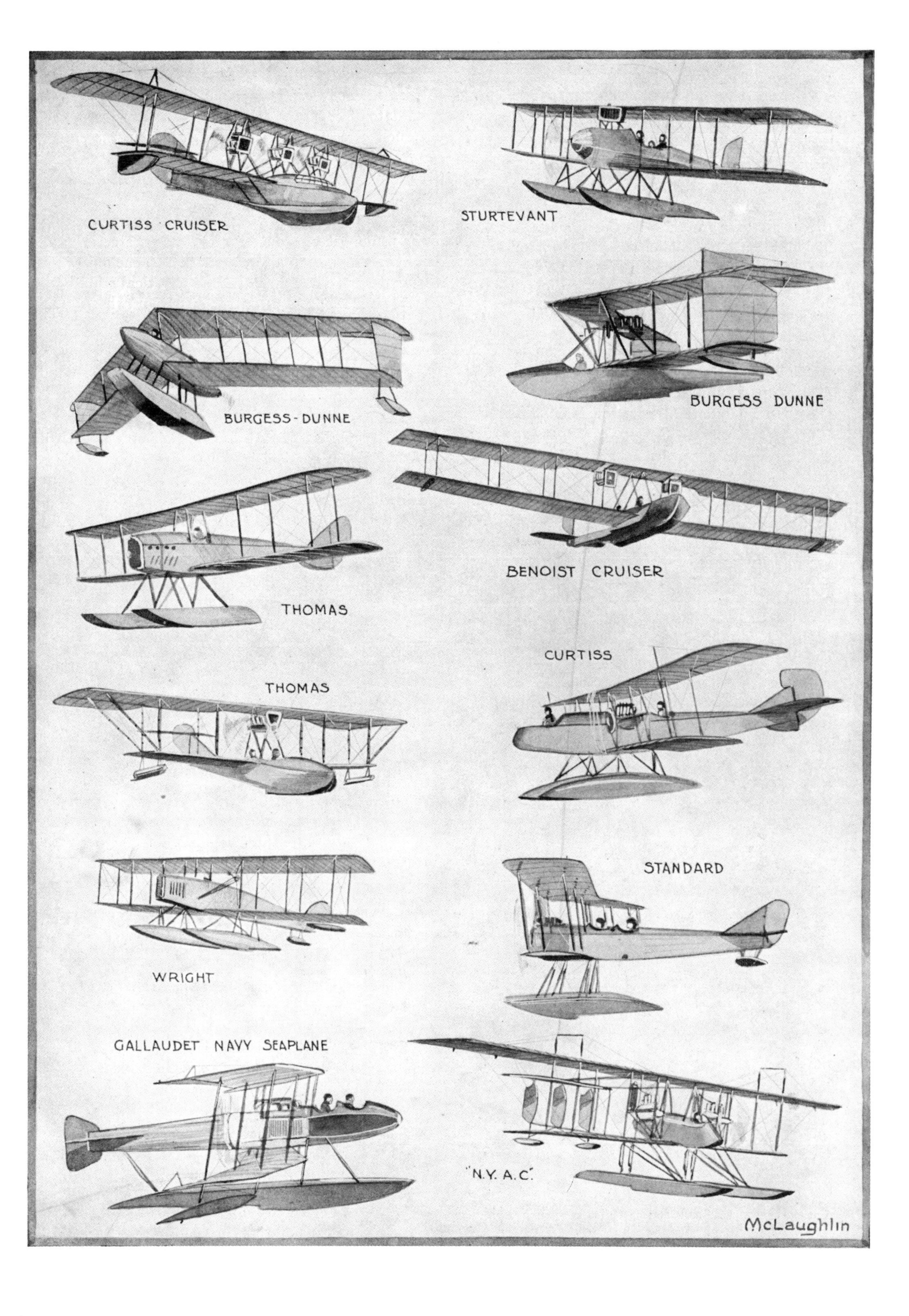
CURTISS CRUISER
STURTEVANT
BURGESS-DUNNE
BURGESS DUNNE
THOMAS
BENOIST CRUISER
CURTISS
THOMAS
STANDARD
WRIGHT
GALLAUDET NAVY SEAPLANE
"N.Y. A.C."
McLaughlin

DATA REQUIRED WITH PROPOSALS

Bidders will submit the following data, in *duplicate*, with their proposals:

1. General arrangement plans of aeroplane.
2. Profile of wing section, with characteristics, stating source of data.
3. Stress diagram for the wings and body.
4. Tabular list of maximum, fiber stresses on truss members, together with strength of materials involved, and factor of safety indicated.

NOTE.—A factor of safety of not less than 7 is required.

5. Horsepower curves for full load, horizontal flight, to show (*a*) wings e.h.p.; (*b*) head resistance e.h.p.; (*c*) total e.h.p. required; (*d*) total e.h.p. available. All curves based on knots (1 knot equals 6080 feet).
6. *Motor Particulars.*—General arrangement plans, specifications, b.h.p. r.p.m. curve, guaranteed fuel and oil consumption, and weight schedule as follows: (*a*) Motor empty, with ignition system and carburetor; (*b*) radiator; (*c*) cooling water; (*d*) propeller; (*e*) starter complete with tank or battery.

Additional information in the form of plans, photographs, catalogues, or other descriptive matter may be submitted if desired.

Spare Parts.—A complete table of spare parts, together with the prices at which the bidder is prepared to furnish each part throughout a period of one year from date of contract, is to be submitted with the bid, but such prices are *not to be included in the cast of the machine* as referred to above. The following is a list of spare parts on which prices are to be quoted as per data required:

(*a*) *Aeroplane Parts.*—Floats, each; upper wing; lower wing; set of struts for one wing; steering rudder, stay wires, each size; turn-buckles, each size; diving rudders; horizontal stabilizers; vertical stabilizers; keel planes; ailerons; control levers or wheels, each type; and items of outfit, etc.

(*b*) *Power-plant Parts.*—Include motor; propeller; crank shaft; cam shaft; starting device; cylinders; valves; pistons; spark plugs; carburetors; magnetos, etc.

(*c*) *Aeroplane Tools.*—Include complete set of socket wrenches; spanner wrenches; open-end wrenches; and any special tools necessary for all sizes of bolts and fittings used in the aeroplane; to be itemized.

(*d*) *Power Plant Tools.*—Include complete set of socket wrenches; spanner wrenches; open-end wrenches; and any special tools necessary for the assembly of the power plant and its auxiliaries; to be itemized.

NOTE.—*It is expected that spare parts to the value of 10 per cent. of the contract will be required with delivery of the aeroplanes. Such additional spare parts as may be required during the year following the date of contract; to be delivered within. . . . . .days after receipt of order.*

Bidders must insert in the above blank space the shortest time within which they can make delivery.

Delivery.—Upon the satisfactory completion of the specified trials each machine will be put in first-class order and defects corrected. It will then be approved for acceptance.

It is desired to obtain these machines at the earliest practicable date.

Payments.—Payments will not be made until satisfactory plans corrected from work are supplied to the *Navy Department, Washington, D. C.*, in triplicate as follows (1 Van Dyke (cloth) and 2 prints each):

General arrangement plans, scale 1 inch to the foot.

Detail plan of planes, scale 3 inches to the foot.

Detail plan of floats, scale 3 inches to the foot.

General arrangement plans of power plant.

Detail plans of power plant.

Detail plans of propellers, and a complete set of corrected data as required with proposals, and in addition the complete data obtained in the trials of the power plant and of the machine, complete.

Recent types of seaplanes and flying boats which are prominent on the various European battle fronts. Courtesy London *Aeroplane*

Table of the Leading American Hydroaeroplanes, Seaplanes and Flying Boats

MAKER	Type	Wing Span Upper	Wing Span Lower	Chord	Gap	Total Length	Floats—No. and Arrangement	Motor	Speed Range Mi. Per Hr.	Net Weight (lbs.)
Aeromarine	Model M-1	37′–0″	33′–0″	6′–3″–	6′–6″	25′–6″	Twin float	Aeromarine 100 h.p.	78–42	1200
Aeromarine	Model 700	51′–0″	45′–0″	6′–3″	6′–0″	27′–0″	Twin float	Aeromarine 90 h.p.	65–45	1800
Benoist Aircraft Co.	G-17 Flying Boat	75′–0″	65′–0″	5′–0″	6′–0″	.....	Boat hull	(2) Roberts—100 h.p.	70–45	2900
Benoist Aircraft Co.	F-17 Flying Boat	45′–0″	41′–0″	5′–3″	6′–0″	27′–0″	Boat hull	Roberts—100 h.p.	65–45	1600
Burgess Co.	Sportsman's Seaplane	46′–0″	46′–0″	5′–0″	5′–3″	24′–6″	1 main float	Curtiss—90 h.p.	71–40	1750
Burgess Co.	Navy Dunne Seaplane	45′–0″	43′–0″	5′–0″	5′–3″	25′–0″	1 main float	Thomas—140 h.p.	82–50	2400
Burgess Co.	Type "U" Seaplane	46′–9″	38′–3″	6′–3″	6′–0″	30′–5″	1 main float	Curtiss—90 h.p.	68–40	1799
Christofferson	Flying Boat	47′–7″	34′–0″	5′–6″	5′–9″	.....	Boat hull	....	.....	....
Cooper Aeroplane Co.	Training Tractor	33′–0″	22′–0″	4′–6″	4′–6″	.....	Twin float	Frederickson—70 h.p.	45–35	500
Curtiss Aeroplane Co.	Model F Flying Boat	45′–2″	35′–0″	5′–2″	5′–11″	28′–0″	Boat hull	Curtiss—90 h.p.	65–45	1440
Curtiss Aeroplane Co.	N-9 Hydroaeroplane	53′–4″	43′–0″	5′–0″	5′–0″	29′–10″	1 main float	Curtiss—100 h.p.	70–45	1900
Curtiss Aeroplane Co.	Twin JN Hydro	52′–9″	43′–1″	4′–11″	5′–2″	29′–4″	Twin float	(2) Curtiss—100 h.p.	85–48	2110
Curtiss Aeroplane Co.	Type H-12 Flying Boat	92′–9″	66′–11″	7′–0″	8′–0″	46′–6″	Boat hull	(2) Curtiss—200 h.p.	85–55	5945
Flint Aircraft Mfg. Co.	....	....	....	....	....	....	....	....	....	....
Gallaudet Aircraft Co.	Navy Seaplane	48′–0″	48′–0″	7′–0″	7′–7″	33′–0″	25′ Pontoon	(2) Duesenberg—150 h.p.	100–	4000
General Aeroplane Co.	Verville Flying Boat	38′–0″	36′–0″	5′–0″	6′–0″	27′–0″	Boat hull	Curtiss—100 h.p., or Maximotor	70–42	1450
General Aeroplane Co.	Hydro Pusher	40′–0″	34′–0″	5′–0″	5′–0″	25′–10″	Twin float	Curtiss—100 h.p.	84–46	1375
Heinrich Corp.	Twin Tractor	48′–0″	48′–0″	5′–0″	5′–3″	28′–6″	Twin float	(2) Aëromarine—90 h.p.	85–50	....
Janney Aircraft Co.	....	....	....	....	....	....	....	....	....	....
Lawrence-Lewis Co.	Model A 1	30′–0″	.....	6′–0″	4′–8″	25′–0″	Boat hull	Hall-Scott—135 h.p.	.....	1750
Lawrence-Lewis Co.	Model B 1	42′–0″	.....	7′–6″	5′–8″	29′–0″	Boat hull	Hall-Scott—135 h.p.	.....	2200
L. W. F. Engineering Co.	Seaplane	42′–0″	34′–0″	6′–0″	6′–6″	28′–0″	Twin float	Thomas—135 h.p.	.....	....
N. Y. Aero Construction Co.	N. Y. A. C. Hydro	73′–0″	44′–0″	7′–6″ top 6′–6″ bot.	6′–9″	37′–0″	Twin float	(2) Aëromarine—90 h.p.	78–45	2750
Pacific Aero Products Co.	Tractor Seaplane	52′–0″	43′–6″	5′–9″	6′–6″	27′–9″	Two 10′ float	Hall-Scott—135 h.p.	.....	2100
Shaw Aeroplane Co.	Flying Boat	42′–4″	34′–4″	5′–6″	5′–6″	23′–5″	Boat hull	....	.....	....
Standard Aero Corp.	Model H-3	40′–1″	40′–1″	6′–6″	6′–6″	27′–0″	Twin float	Hall-Scott—135 h.p.	84–46	2700
Standard Aero Corp.	Model D, Twin	62′–4″	49′–0″	7′–0″	7′–0″	33′–6″	Twin float	(2) Hall-Scott—140 h.p.	.....	3300
Sturtevant Aeroplane Co.	S-4 Seaplane	49′–6″	49′–6″	7′–0″	6′–9″	28′–0″	Twin float	Sturtevant—140 h.p.	.....	2025
Thomas-Morse Aircraft Corp.	H-S Navy Hydro	48′–6″	37′–0″	5′–3″	5′–0″	29′–0″	Twin float	Thomas—135 h.p.	82–47	1800
Thomas-Morse Aircraft Corp.	B-5 Flying Boat	50′–0″	38′–0″	6′–0″	6′–0″	31′–6″	Boat hull	Thomas—135 h.p.	76–48	2000
Thomas-Morse Aircraft Corp.	SH-4 Navy Hydro	.....	.....	....	....	.....	1 main float	.... 100 h.p.	65–40	....
United Eastern Aero Co.	Navy Pursuit Scout	26′–4″	.....	2′–6″	2′–10″	24′–7″	1 main float	Gen. Vehicle Gnome—100 h.p.	110–45	1150
United Eastern Aero Co.	School Hydro	42′–0″	37′–0″	6′–0″	6′–0″	29′–6″	1 main float	Hall-Scott—90 h.p.	80–45	1950
Wright-Martin Aircraft Corp.	Martin "S" Seaplane	51′–8″	51′–8″	....	....	28′–0″	Single float	Hall-Scott—135 h.p.	75–40	2300
Wright-Martin Aircraft Corp.	Martin Model R	50′–8″	36′–10″	5′–6″	6′–0″	26′–8″	Twin float	Hall-Scott—135 h.p.	86–47	1915
Wright-Martin Aircraft Corp.	Wright Seaplane	.....	.....	....	....	.....	Twin float	....	.....	....

The above plans and information should be furnished with the delivery of the first machine, and if this is done the payment for each machine will be made upon delivery as specified.

The next development of basic importance in the evolution of seaplanes was the development of the multiple-motored seaplane, first the flying boat, then the hydroaeroplane. The first twin-motored flying boat was the Wanamaker flyer *America,* ordered from the Curtiss Aeroplane Company by Mr. Rodman Wanamaker in the early part of 1914, for the purpose of making a trans-atlantic flight. The *America* was a flying boat, the hull of white cedar, 32 feet long and with a beam of 4 feet. The upper wings had a 72-foot span and the lower 46. It was equipped with two 100 horse-power Curtiss motors. It was figured out that the machine could lift easily the 5000 pounds of weight which would include: fuel, 2000 pounds, made up of 300 gallons of gasoline and 25 gallons of oil; hull, 550; pilot and assistant, 300; motors, wings, tanks, and supplies, 2150. The two oil tanks were suspended above the engines; the gasoline tanks, five in number, were arranged in the hull, and the gasoline was forced up from the tanks to the motors above the operators' heads.

The *America* was christened on June 22, 1914, and a number of flights were made with it on Lake Keuka, N. Y. As the final tests were being made war was declared and Lieut. John C. Porte, of the British Royal Naval Flying Corps, who was to pilot it, had to return to England for service. The *America* was acquired by the British Admiralty and was used for many months for patrol duty, and then as a training aeroplane.

An important contribution, making it possible to stow seaplanes on board of a ship in the minimum space, was the development by the Short Brothers, the British aircraft manufacturers of the folding wings seaplane.

Another valuable development was the Galaudet seaplane with a new stream line system of propulsion. The connection between the front and rear portions of the body and the mounting of the propeller of the Gallaudet seaplane is so rigid and strong that the chances of propeller failure are reduced to a minimum The part of the propeller nearest the center, which absorbs power without producing any propulsive effect, is inside the body, so that the power usually lost in this portion is expanded with the heat near the tips. The result is estimated to be an increased propulsive efficiency of 10 per cent. to 20 per cent., according to the relative diameters of body and propeller.

In 1916–17 the number of twin-motored seaplanes increased, and there were also built larger seaplanes equipped with three and four motors. One of the Curtiss seaplanes equipped with three motors is shown elsewhere in this book. The larger air cruiser, also built by the Curtiss Aeroplane Company, when sent to England, was equipped with 4, 250 horse-power motors, and during the tests carried 3500 pounds of crew and equipment. This type of air cruiser opens new problems, requiring mechanical piloting, and special arrangements for housing it. The first problem can be solved by the employment of the Sperry automatic pilot; the second would probably be solved by making the hull more substantial so that the machine can be kept in the water all the time. Once these two problems have been solved, larger air cruisers will be built.

There is no visible limitation as to the size which air cruisers can be built, and as the weight and fuel consumption of motors decreases the capacity for carrying useful load increases, permitting the building of stronger hulls to withstand the roughest conditions. In discussing the subject of torpedoplanes with the writer, Mr. Glenn H. Curtiss stated that he was ready to construct a large torpedoplane capable of carrying a half dozen torpedoes and landing at sea and riding the sea for days. This would make it possible for an air cruiser while on patrol duty to fly out 500 miles and remain at sea on the lookout for hostile ships, only rising occasionally to scan the sea for enemy ships, saving its fuel—which would eliminate the expenditure of fuel involved in returning to the base each night.

A Naval Zeppelin dirigible.

CHAPTER XXIX

NAVAL DIRIGIBLES

One often hears expressed the opinion that the Zeppelin has been a failure, which is usually followed by a general condemnation of the dirigibles as instruments of war. These are hasty conclusions, dealing only with one aspect of the subject—with the Zeppelin as an instrument of destruction. In this respect the critics are right, since the expectation was that Zeppelins would destroy cities by dropping tons of explosives on them.

This notion was established before 1912, when the aeroplane was limited in its performances. The reason Zeppelins have not destroyed cities is that swift, powerful aeroplanes have been developed which are a match for a Zeppelin whenever they can find the airship in the immense dark sky, at night. During the day it is too dangerous for Zeppelins to venture on a raid. Theoretically, they would be successful by employing a formidable escort of fighting aeroplanes, but the number of aeroplanes would have to be large, more than equal to the number of aeroplanes which the Allies can send up to fight them, and that Germany has not been able to do. So the Zeppelins only operate in the long, dark winter nights.

Though they are prevented from doing much damage by aeroplanes and anti-aircraft guns, the Zeppelins, like the submarines, are successful in keeping from the front tens of thousands of men who are part of the aircraft and anti-aircraft defenses, and would be at the fronts otherwise.

For coast patrol and as an auxiliary of the fleet, the Zeppelin is very efficient indeed. For submarine hunting, convoying ships, and patrolling ship channels, a single Zeppelin can easily do the work of fifty aeroplanes, and can do work which no aeroplane can do at present.

Germany alone was prepared in the line of large dirigibles at the beginning of the war. Her preparedness in Zeppelins, of which she had about forty, represented the result of the work of Count Ferdinand von Zeppelin since 1900, and the investment of probably over one hundred million dollars. Since then Zeppelins have been produced almost as fast as the U-boats have been produced, and have been used extensively in naval operations and coast patrol work.

The reader can get full details of the hundreds of dirigibles built by different countries from "D'Orcy's Airship Manual" (published by the Century Company, New York, price

$3.50), and the author will not attempt to supplement D'Orcy's exhaustive work.

Germany's Naval Airships

The following information regarding Germany's latest naval airships is reprinted from "The Aeroplane," London, England, and "Aerial Age Weekly," New York City, N. Y.:

The Zeppelin brought down in Essex recently was from 650 feet to 680 feet in length, and measuring 72 feet across its largest diameter, the vessel was of stream-line form, with a blunt, rounded nose and a tail that tapered off to a sharp point. The framework was made of longitudinal lattice-work girders, connected together at intervals by circumferential lattice-work ties, all made of an aluminium alloy resembling duraluminum. The whole was braced together and stiffened by a system of wires, arrangements being provided by which they could be tightened up when required. The weight of the framework is reckoned to be about nine tons, or barely a fifth of the total of fifty tons attributed to the airship complete with engines, fuel, guns and crew. There were twenty-four balloonets arranged within the framework, and the hydrogen capacity was 2,000,000 cubic feet.

A cat-walk, an arched passage with a footway nine inches wide, running along the keel, enabled the crew, which consisted of twenty-two men, to move about the ship and get from one gondola to another. This footway was covered with wood, a material which, however, was evidently avoided as much as possible in the construction of the ship. The gondolas, made of aluminium alloy, were four in number; one was placed forward on the center line, two were amidships, one on each side, and the fourth was aft, again on the center line.

The vessel was propelled—at a speed, it is thought, of about sixty miles an hour in still air—by means of six Mayback-Mercedes gasoline engines of 240 horsepower each, or 1440 horse-power in all. Each had six vertical cylinders with overhead valves and water cooling, and weighed about 1000 pounds. They were connected each to a propeller shaft through a clutch and change-speed gear, and also to a dynamo used either for lighting or for furnishing power to the wireless installation. One of these engines with its propeller was placed at the back of the large forward gondola; two were in the amidships gondolas, and three were in the aft gondola. In the last case one of the propellers was in the center line of the ship, and the shafts of the other two were stayed out, one on either side. With the object of minimizing air-resistance the stays were provided with a light but strong casing of two- or three-ply wood, shaped in stream-line form. The gasoline tanks had a capacity of 2000 gallons, and the propeller shafts were carried in ball bearings. The date, July 14, 1916, marked on one of them, is thought to indicate the date of the launching or commissioning of the vessel.

Forward of the engine room of the forward gondola, but separated from it by a small air space, was first the wireless operator's cabin, and then the commander's room. The latter was the navigating platform, and in it were concentrated the controls of the elevators and rudder at the stern, the arrangement for equalizing the levels in the gasoline and water tanks, the engine-room telegraphs, and the switchboard of the electrical gear for releasing the bombs. Provision was made for carrying sixty of the latter in a compartment, amidships, and there was a sliding shutter, worked from the commander's cabin, which was withdrawn to allow them to fall freely. Nine machine guns were carried. Two of these, of 0.5 inch bore, were mounted on the top of the vessel, and six, of smaller caliber, were placed in the gondolas—two in the forward, one each in the amidships ones, and two in the aft one. The ninth was carried in the tail.

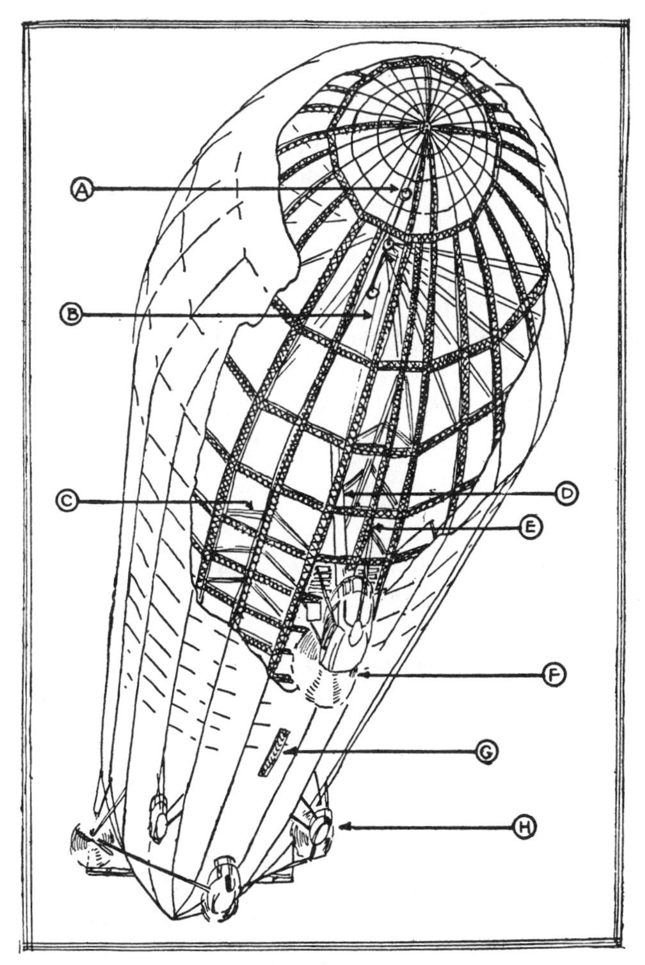

Sketch of frame of the Zeppelin brought down in Essex, England. It was between 650 and 680 feet long, 72 feet diameter, had 24 balloonets with hydrogen capacity of 2,000,000 cubic feet, six engines of 240 horse-power each.

As regards the Zeppelin airships of the naval class, the least one can say about the information which follows hereafter is that it is as accurate as any other that has filtered through neutral and other channels.

The following are the outstanding facts regarding this list: All matter in sections 1, 2, 3, 4 is absolutely the genuine stuff; the airships of classes A, B, C have been described in detail by many Swiss and Italian magazines, although often with errors, and have here been corrected according to data given the writer by a neutral who just returned from Germany. The existence of class D is beyond doubt, although opinions differ regarding their volume and power plant. Finally, there is undoubtedly a larger type than the *L.33*, for a power plant of seven or even eight motors has actually been seen on recent airships by various observers. I would not bet a cent that the *L.33* class actually displace 54,000 cubic meters, for although I may be mistaken, I personally believe they are rather between 45,000 and 50,000 cubic meters. I repeat I do not vouch for the correctness of the displacement of classes D, E, F, but I most emphatically do vouch for the linear dimensions given.

Incidentally, one might remark that Count Zeppelin himself announced in 1913 the increases in size for which calculations had been made, and gave the master diameters of the types to be as 17.5 meters, 20 meters, 22 meters, 25 meters, and 30 meters.

And one more thing, if you put down 250 grammes (i.e., about 0.5 pounds) as the fuel and oil consumption (including water) per horse-power hour, and multiply according to power plant, you will find out there has never been, nor will there be for a time, a Zeppelin having an endurance of 75 or 100 hours at full speed, as has been said about the naval airships. For a 40-hours endurance with a power plant of 1000 horse-power would alone necessitate carrying ten tons of fuel, always assuming Maybach fuel consumption.

Naval Airship Division

Headquarters: Cuxhaven.—The eventual strength of the Naval Airship Division (Marine-Luftschiff-Abteilung) was laid down in the four years' building program authorized by the Reichstag on October 13, 1913, providing $2,750,000 for the construction of 10 airships of the rigid type, $3,500,000 for the construction of two first-class air harbors and several subsidiary ones, all fitted with revolving sheds; and $2,500,000 for the maintenance of the *matériel.*

Two airship squadrons, each to consist of four active and one reserve units, were to be organized and stationed at the first class air harbors of Heligoland and Tondern, respectively. Second-class air harbors were either to be created or already existed at Emden, Wilhelmshafen, Cuxhaven, Hamburg, Kiel, Wismar, Rostock, and Königsberg. In Hamburg and Königsberg private hydrogen generating plants are available. This program was to be completed by January 1, 1918.

At the outbreak of the Great War the strength of the Naval Airship Division consisted of one airship (*L.3*) commissioned and four (*L.4–L.7*) building; the *L.3*, *L.5*, and *L.6* were of the Zeppelin type, the *L.4* and *L.7* of the Schutte-Lanz type. In addition three air-liners of the German Airship Navigation Co. (Delag) of Frankfurt, had been chartered to serve as training airships; these were the *Sachsen, Hansa,* and *Viktoria-Luise.*

The airships of the Naval Airship Division all bear the mark L. (Luftschiff) regardless of their type, and are numbered currently.

Class A

This class represent the highest development of orthodox Zeppelin design. The hull, made of aluminium trellis-work, is in the form of a cylinder with ogival ends, the bow being but slightly blunter than the stern; a triangular keel fitted as a gangway connects the two open cars and contains in the middle a closed cabin. Cross-wire monoplane fins and elevator, multiplane rudder.

L.3 (May 11, 1914), *L.5* (November, 1914), *L.6* (December, 1914).

L.8 (February, 1915), *L.9* (March 12, 1915).

Builders: Zeppelin Works, Friedrichshafen.

Length: 158 meters (521.4 feet).

Beam: 16.6 meters (54.8 feet).

Height: 18.9 meters (62.4 feet).

Fineness ratio: 9.5 to 1.

Volume: 27,000 cubic meters (953,000 cubic feet).

Total lift: 29.7 tons.

Useful lift: 8 tons.

Compartments: 18.

Propelling apparatus: Four 210 horse-power 6-cylinder Maybach engines (equals 840 horse-power) mounted in twin-units on each car and driving through shaft and bevel gear transmission four 2-bladed side propellers. Fuel consumption, 0.240 kilograms (0.528 pounds) per horse-power-hour, 1200 revolutions per minute.

Full speed: 80 kilometers (50 miles).

Full speed endurance: 26 hours.

Maximum altitude: 2500 meters (8250 feet).

Complement: 16.

Armament: (1) Four Maxims mounted on the cars and firing broadsides; (2) 1.5 tons of explosives.

Notes.—The *L.3* stranded during a storm on February 17, 1915, near Esbjerg; the crew were saved and interned, but the airship was a total loss. The *L.5* was destroyed on June 7, 1915, in the airship shed of Evere by Flight Sub-Lieuts. J. P. Wilson and J. S. Mills, R. N. A. S. The *L.6* was set on fire and destroyed in mid-air on the same day near Ghent (Gand) by the late Flight Sub-Lieut. R. Warneford, R. N.

A. S. All the crew perished. The *L.8*, while returning from a raid on England, was engaged by Flight Commander Bigsworth, R. N. A. S., on May 17, 1915, near the Belgian coast, and so damaged that she was wrecked on landing near Tirlemont. All of the crew are reported to have thereby been killed. The *L.10*, while returning from a raid on England on August 10, 1915, was engaged and wrecked off Ostende by the Dunkirk squadron of the R. N. A. S.

The *L.9* is thus the only survivor of this class.

CLASS B

This class is characterized by an ellipsoidal hull built up of laminated wood girders forming a closely meshed lattice-work which is kept under tension by wire stays. There are five cars, four of which are crosswise hung from the hull by cables and constitute the engine-rooms; the fifth car is rigidly connected with the hull near the bow, and serves as the navigation room. Cruciform "fin-and-flap" tail. No keel.

L.4 (January, 1915), *L.7* (February, 1915), *L.?* (1915), *L.21* (1915).

Builders: Schutte-Lanz Works, Rheinau, near Mannheim.

Length: 165 meters (544.5 feet).

Beam: 18.4 meters (60.7 feet).

Height: 21 meters (69.3 feet).

Fineness ratio: 8.9 to 1.

Volume: 30,000 cubic meters (1,059,000 cubic feet).

Total lift: 33 tons.

Useful lift: 14 tons.

Compartments: 7.

Propelling apparatus: Four 240 horse-power Mercedes engines (equals 960 horse-power) mounted singly on each engine-car, and driving through a clutch one 2-bladed rear propeller each. Fuel consumption: 0.240 kilograms (.528 pounds) per horse-power hour.

Full speed: 85 kilometers (53 miles).

Full speed endurance: 26 hours.

Maximum altitude: 2500 meters (8250 feet).

Complement: 16.

Armament: (1) Five Maxims, mounted one on each car; (2) 1.5 tons of explosives.

Notes.—The *L.4* foundered in a storm on February 17, 1915, off Esbjerg; four of the crew were lost with the airship, the remainder were interned. The *L.7* was shot down by H. M. ships *Galatea* and *Phaeton*, on May 4, 1916, off the Schleswig coast. Seven of the crew were rescued and made prisoners. The *L.21* was set on fire and destroyed with all aboard on September 3, 1916, near Cuffley, by Lieutenant William L. Robinson, V. C., R. F. C.

CLASS C

The airships of this class resemble generally speaking, those of Class A, except for the stern, which is pointed, and carries in lieu of multiplane-tail surfaces a cruciform "fin-and-flap" tail. Cars and gangway like on Class A. Hull frame: aluminium lattice work.

Fighting a battleplane from the platform of a Zeppelin. (Courtesy of "Illustrated London News.")

L.11 (July 10, 1915), *L.12* to *L.19* (August to November, 1915).

Builders: Zeppelin Works, Friedrichshafen and Potsdam.

Length: 160 meters (528 feet).

Beam: 17.5 meters (57.8 feet).

Height: 19 meters (62.7 feet).

Fineness ratio: 9.1 to 1.

Volume: 30,000 cubic meters (1,059,000 cubic feet).

Total lift: 33 tons.

Useful lift: 10 tons.

Compartments: 18.

Propelling apparatus: Five 210 horse-power 6-cylinder Maybach engines (equals 1050 horse-power), two of which are mounted in the bow-car and drive two 2-bladed side-propellers; the stern-car houses three engines, two of which drive two side-propellers, while the third engine drives through a clutch transmission a rear propeller.

Full speed: 85 kilometers (53 miles).

Full speed endurance: 26 hours.

Maximum altitude: 3500 meters (11,550 feet).

Complement: 16.

Armament: (1) Four Maxims mounted on the cars and one on the roof, near the bow; (2) 2 tons of bombs.

Notes.—The *L.15* was damaged, while raiding England, by anti-aircraft guns, and by Lieut. A. de W.

Brandon, R. F. C., and came down on April 1, 1916, in the mouth of the Thames, near Kentish Knock, where the crew scuttled the airship and surrendered. The *L.18* caught fire and blew up on November 17, 1915, in the airship dock at Tondern. The *L.19* was damaged by anti-aircraft guns during a raid on England, and subsequently foundered on February 2, 1916, in the North Sea with the entire crew.

CLASS D

This class differs from the one preceding, essentially in that the hull has no keel, the gangway being enclosed in the bottom. The cars are entirely closed and of streamline shape; they are, on account of the third engine each carries, much longer than heretofore. Central cabin in the bottom of the hull.

L.20–L.29 (November, 1915–April, 1916).

Builders: Zeppelin Works, Friedrichshafen and Rheinau.

Length: 170 meters (560 feet).

Beam: 20 meters (66 feet).

Fineness ratio: 8.5 to 1.

Volume: 35,000 cubic meters (1,235,000 cubic feet).

Total lift: 38.5 tons.

Useful lift: 13 tons.

Compartments: 19.

Propelling apparatus: Six 210 horse-power 6-cylinder Maybach engines, mounted in triple units on each car; each unit drives through shaft and bevel-gear transmission two side-propellers, and through a clutch one rear-propeller.

Full speed: 95 kilometers (59 miles).

Full speed endurance: 30 hours.

Maximum altitude: 4000 meters (13,200 feet).

Complement: 18.

Armament: (1) Two large bore machine-guns: one on the roof, near the bow, and one on the bow-car, mounted forwards; (2) six Maxims: two each on the cars and cabin, all firing broadsides; (3) 2.5 tons of bombs.

Notes.—The *L.20* stranded on May 3, 1916, near Stavanger, having run out of fuel while homeward bound from a raid on Scotland. The crew were interned and the airship was blown up by the Norwegian authorities as a measure of precaution. The *L.22* is reported to have been wrecked in the early part of December, 1915, at Tondern.

CLASS E

This class, often referred to as the Super-Zeppelin class, may be regarded as a product of the combined Zeppelin and Schutte-Lanz designs, for although the hull frame is made of aluminium trellis work, the crosswise mounting of the cars is strongly reminiscent of Class B. No central cabin, but bomb emplacement amidships; gangway like on Class D. Four cars.

Severely damaged by British light cruisers and finally destroyed by a British submarine: Zeppelin "17," wrecked and on fire in the North Sea.

L.30 (May 29, 1916), *L.31* (June 17, 1916), *L.32* (June, 1916), *L.33* (July 14, 1916), *L.34* (August, 1916).

Builders: Zeppelin Works, Friedrichshafen and Mannheim.

Length: 207 meters (680 feet).

Beam: 22 meters (72 feet).

Fineness ratio: 9.4 to 1.

Volume: 54,000 cubic meters (1,906,200 cubic feet).

Total lift: 59.4 tons.

Useful lift: 19.4 tons.

Compartments: 19.

Propelling apparatus: Six 250 horse-power 6-cylinder Maybach engines; three are mounted singly on the bow-car and in two central side cars ("power eggs") and drive through a clutch one 2-bladed rear-propeller of 8.35 meters (17.5 feet) diameter each; the three remaining engines are mounted in a triple-unit on the stern car and drive two side propellers and one rear-propeller of the same diameter as above.

Full speed: 105 meters (65 miles).

Full speed endurance: 35 hours.

Maximum altitude: 5000 meters (16,500 feet).

Complement: 20–22.

Armament: (1) Two large bore machine-guns carried side by side on roof, near the bow, on collapsible tripods; one similar gun is mounted on the roof, near the stern; (2) six Maxims, viz., two each on the bow and stern-cars, and one each on the side-cars, all firing broadsides; (3) sixty bombs aggregating 3.5 tons.

Notes.—The *L.31* was shot down by A.-A. guns and aviators while raiding London on October 2, 1916, and was destroyed with all aboard at Potter's Bar. The *L.32* was shot down during a raid on London by A.-A. guns and aviators on September 24, 1916, and fell in Essex. All of the crew were killed. The *L.33* was damaged by A.-A. guns and aviators while raiding London on September 24, 1916; landed in Essex, and was blown up by commander. Crew surrendered.

CLASS F

These airships are very similar in general design to those of the preceding class; there are, however, seven or even eight engines to the power plant, with an equal numbr of propellers. Four cars.

Builders: Zeppelin Works, Friedrichshafen and Rheinau.

Length: 235 meters (775.5 feet).

Beam: 25 meters (82.5 feet).

Fineness ratio: 9.4 to 1.

Volume: 70,000 cubic meters (2,471,000 cubic feet).

Total lift: 77 tons.

Useful lift: 28 tons.

Compartments: (?)

Propelling apparatus: Seven 250 horse-power Maybach engines; mounted in twin-units on the bow-car, singly on two central side-cars ("power eggs"), and in a triple-unit on the stern-car. There are, consequently, six side-propellers, four of which are mounted on outriggers, and one stern-propeller. On a later model of this class the power plant consists of eight engines, there being three engines in the bow-car instead of two, so that one stern-screw is added to the propelling apparatus.

Full speed: 110 kilometers (68 miles).

Full speed endurance: 40 hours.

Complement: 20–22.

Maximum altitude: 5000 meters (16,500 feet).

Armament: (1) Four large bore machine-guns mounted in pairs on the roof near the bow and near the stern; (2) six Maxims, mounted in pairs on the

The Zeppelin "L. Z. 6" on the stocks, strong aluminum frame 490 feet long.

The "D. N. 1," the first United States Naval Dirigible, during one of its test trips.

bow and stern-cars, and singly on the side-cars; (3) 4 tons of explosives.

Notes.—One airship of this class is said to have been destroyed on September 22, 1916, in the airship dock of Rheinau by French aviators. Another airship of this class appears to have been wrecked by the storm on November 21, 1916, near Mayence, while en route from Friedrichshafen to Wilhelmshafen.

Two naval airships (numbers unknown) were shot down during a raid on November 28–29, 1916; one off Durham by aviators of the R. F. C., and one off Norfolk by aviators of the R. N. A. S. Another naval airship, whose number is unknown, apparently foundered off Sylts on September 3, 1916, while homeward bound from a raid on England. (Witnessed by numerous vessels.)

A small coast patrolled dirigible photographed while passing over one of the ships "somewhere in Europe."

One can hardly read the above and of the valuable services rendered to Germany of the Zeppelins without feeling like exclaiming, as Hon. Arthur J. Balfour did in the House of Commons on February 16, 1916:

"England's greatest error before the war was made when the Government decided not to develop the 'rigid lighter-than-air' machine.

"I am sorry that we did not develop that type of vessel, not so much for aggression and defense as for maritime and other scouting. Such airships might have played for us an important part. Certainly Germany has had an advantage in possessing them."

Mr. Balfour amplified this statement later, by saying:

"It is extremely desirable that we should have lighter-than-air machines . . . in order to sup-

Floating hangar of the United States Navy dirigible "D. N. 1."

plement the efforts of our fleet by machines which, in many respects and in favorable weather, are far more effective than the swiftest destroyer or the most powerful cruiser. Therefore, we have done and we are doing our best to develop the lighter-than-air machines."

Coming from a British first lord of the admiralty, these words are particularly significant, so significant, indeed, that they make all comment superfluous, excepting that one can never urge too often the necessity of the United States building large dirigibles for naval purposes.

One of the Coast Patrol dirigibles at Salonika which are used extensively in connection with naval operations.

The "Blimps," "Coast Patrol Airships," "Submarine Spotters"

While Great Britain has reasons for regretting that she did not develop large dirigibles, she also has reasons for being pleased with the work of her small dirigibles, which are, at times, called "submarine spotters" or coast patrol airships, because of their remarkable work in spotting submarines, and in coast patrol, but are generally known as "Blimps."

The Allies have hundreds of these small dirigibles and have used them in all their naval campaigns. The details of construction are practically the same as the details of the sixteen dirigibles of this type ordered by the United States Navy in March, 1917, and the performances obtained, only a little less than are expected under the United States Navy's specifications, which are as follows.

Memoranda:

CHAPTER XXX

SPECIFICATIONS FOR UNITED STATES NAVY SCOUTING TYPE DIRIGIBLES

Specifications for scouting type dirigibles were issued by the United States Navy Department in February, 1917, and prices asked on one, two, four and eight dirigibles, and an equivalent number of power plants, to be delivered at any point to be designated by the Government on the Atlantic or Gulf coasts of the United States. Final acceptance tests to be completed within 120 days after date of contract or bureau order, the bids to be opened at 10 A. M. on March 6. (See U. S. Navy Aeronautics.)

The complete specifications follow:

SPECIFICATIONS

(SPECIAL CONDITIONS WHICH MODIFY FORM A)

The following specifications, terms and special conditions are understood to apply to each dirigible contracted for:

The department having adopted, as foundation for this contract, the plans and specifications herein below guaranteed by the contractor to be capable of certain performances, assumes no responsibility with reference thereto, and will consider any changes therein suggested by the contractor, and will feel it to be its duty to permit changes therein so long as the requirements of the contract, plans and specifications remain substantially the same. The contractor shall submit from time to time as required the detailed drawings, plans, and specifications of the vessel for the examination and approval of the department, and all drawings shall have the approval of the department before the work is begun, but the department shall not be held responsible for any failure of the vessel to fulfil any of the contract requirements unless such failure be due to the adoption and use of any design or material not specifically covered by the specifications and made on the request or demand of the department for which the contractor disclaims responsibility at the time of such demand or request by the department. The permission of the department to make changes suggested by the contractor shall not place the responsibility therefor upon the department.

The contractor will, at his own risk and expense, construct, in accordance with the aforesaid drawings, plans, and specifications, including duly authorized changes therein, one dirigible, to be provided with fittings, equipment, machinery, devices, appliances, and appurtenances complete in all respects, and in accordance with detailed plans and specifications to be prepared in accordance with the aforesaid drawings, plans, specifications, and current practice at date such drawings are submitted for approval by the contractor and submitted to the department for its examination and approval from time to time, as may be necessary, before the work is begun.

The prices quoted above are not to include transportation charges from the contractor's works to point of delivery. When the destination of each dirigible is determined, the contract price will be modified to include the transportation charges.

Guarantee.—The contractors must agree to guarantee the dirigble for three (3) months from date of delivery. Any defects which may develop within that time which can be attributed to faulty workmanship, material, or design must be made good by the contractors without cost to the Government.

The contractors shall guarantee to protect the Government against any patent infringement; the Government also having the right to use with the dirigible any and all of the patented features controlled by the manufacturer of the dirigible without any obligation on the part of the Government.

Payments.—Payments shall be made by the department upon bills submitted in quadruplicate and certified by the inspectors of the department in such form as the department may direct, as follows:

Fifty per cent. of the contract price of the dirigible when—

(*a*) The envelop and ballonets have been satisfactorily air tested.

(*b*) All parts have been satisfactorily assembled at the works of the contractors.

(*c*) All parts have been satisfactorily tested at the works of the contractors.

(*d*) The complete dirigible has been prepared and placed on board for shipment to destination and transportation company's receipt delivered to the inspector.

Forty per cent. of the contract price when the dirigible has been accepted.

Ten per cent. of the contract price or such part thereof as may remain at the expiration of three months as is due the contractors after remedying by

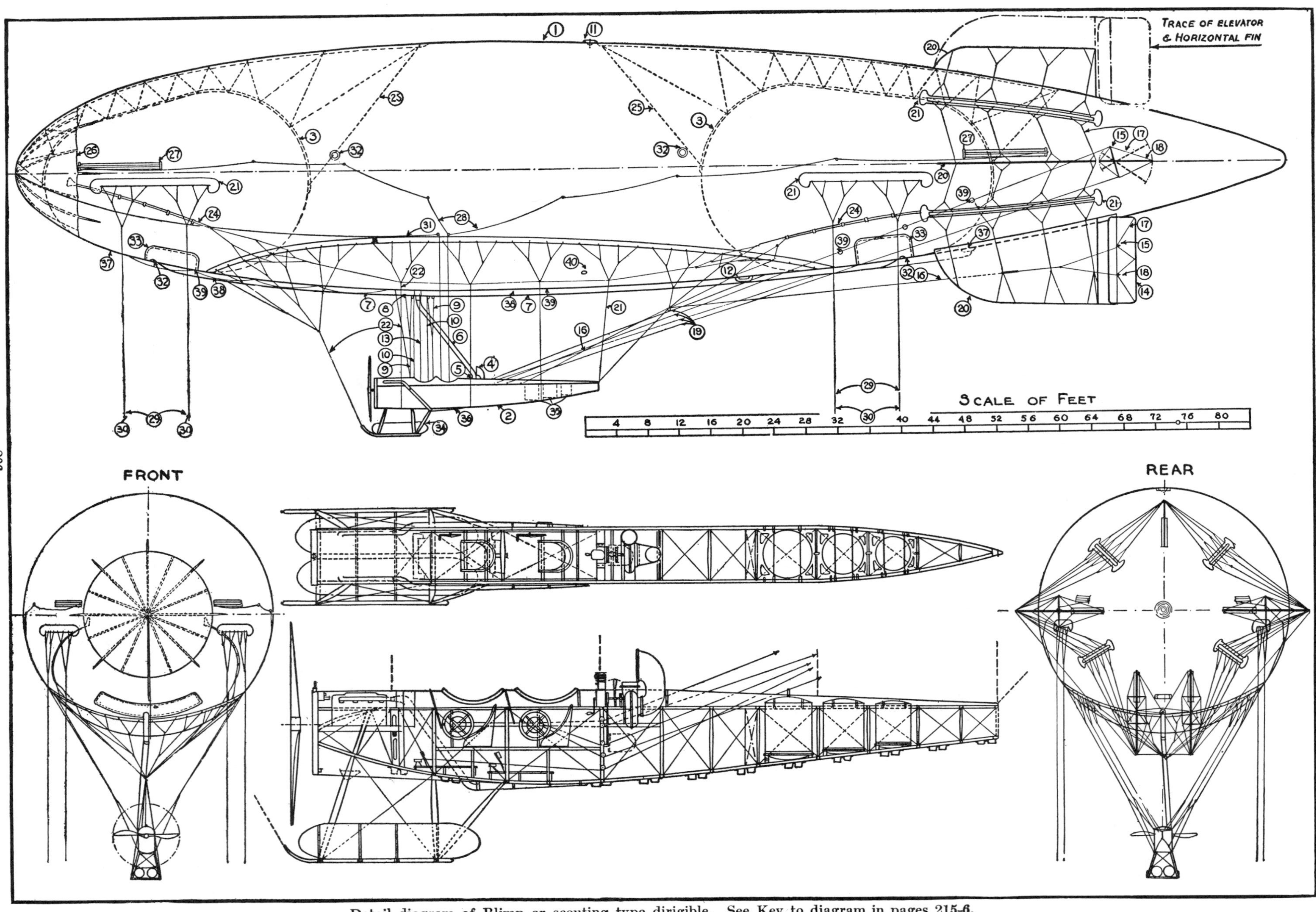

Detail diagram of Blimp or scouting type dirigible. See Key to diagram in pages 215-6.

the Government of such defects as they may be responsible for.

The contractor agrees to refund to the Government one-half of the payment of 50 per cent. of the contract price above referred to, in case a dirigible after trial shall be rejected.

The bond on the contract will, therefore, be 50 per cent. of the contract price, the liability being limited to 25 per cent. of the contract price for satisfactory performance of the contract, and 25 per cent. of the contract price for refund of the amount specified above.

Insurance.—The dirigible, after being delivered to the transportation company, shall be insured by the contractors at their expense in such form as the Secretary of the Navy may designate, against risks of transportation, fire, and damage, except such damage as may occur during actual trials free from the ground.

Bonus and Penalty.—The dirigible is desired to be ready for its trials within three months from date of contract. If the dirigible satisfactorily completes its acceptance trials before the expiration of four months, a bonus of $100 a day will be paid for each day so saved up to 30 days. Liquidated damages at the rate of $100 per day will accrue for each day beyond the four months period until the ship is given satisfactory trials, subject to remission for any of the causes stated in paragraph 10 of the conditions of Form A.

No extension of time hereinabove allowed for completion shall operate to extend the premium period fixed above, except such extensions as are approved by the department to correspond to delays in trials or delivery is prevented or delayed due to unsuitable weather conditions, failure of the department to provide a trial board promptly, or other causes for which the contractor is not responsible.

The total bare weight shall not exceed the figure given in the weight statement marked "Penalty weight," and any overweight shall render the ship liable to rejection. A bonus of $5 per pound will be paid for each pound of weight as listed above which the contractor can show has been saved. Items of outfit or equipment, such as instruments, fire extinguishers, etc., which may be added or omitted at the department's request shall not be counted in computing the bonus earned. Furthermore, if the department should later decide that any fin, tank, or other feature of the design is unnecessary, the weight saved by its omission shall be subtracted from the "penalty weight" above before bonus is computed.

Uncompleted Work.—The dirigible when delivered to the Government must be as nearly complete in accordance with the contract as possible, but should there be any work not completed by the contractors it will be done by the Government at the contractor's expense.

Inspection.—The work of construction shall be at all times open to inspectors of the department and their assistants, and every facility shall be afforded such inspectors for the prosecution of their work.

Inspectors may peremptorily reject any inferior workmanship or material or forbid the use thereof. Should the contractors question such decision, they shall have the right of appeal to the department whose decision shall be final.

The contractors shall furnish under the contract, without additional cost, such samples of material and information as to the quality thereof and manner of using same as may be required, together with any assistance necessary in testing or handling materials for the purpose of inspection or test.

Suitable offices, conveniences, and necessities for the use of the inspector or his assistants during the con-

A close view of one of the blimps. A bomb may be seen attached to the undercarriage.

struction of the dirigible shall be furnished under the contract.

Correspondence.—All correspondence with the department shall be forwarded through the inspector in duplicate. Letters involving claims for delay and appeals to the department shall be forwarded through the inspector in triplicate.

Weighing of Material.—The contractors shall furnish under the contract competent weighmasters, who shall make daily returns of weights to the inspector in such manner as the inspector may prescribe.

The contractors shall furnish the necessary scales and other apparatus for making correct return of weights. Great care shall be taken that all scales are properly adjusted and that no estimated or calculated weights are returned.

All material of every description placed on or attached to the dirigible shall be weighted, together with all material of every description which after being weighed and placed on or attached to the dirigible are removed; the weight and description of the part weighed in all cases to be reported to the inspector.

Where material is assembled before being weighed the center of gravity of such assembly shall be ascertained. The center of gravity of each part or group of parts entering into or attached to the dirigible shall be reported in relation to the nose of the dirigible and from the axis of the envelope.

Trials.—The acceptance trials shall be conducted by the contractor, who shall furnish all necessary gas, fuel, or other material and the operating personnel. The contractor may select good weather for the trials in which the wind velocity at a height of 100 feet from the ground is less than 15 miles per hour. The department expects to have ready sheds for housing the dirigible at the point of delivery and it is agreed that the contractor shall have the use of such shed from which to conduct the trials. In case trials are held from a Government shed, the Government agrees to provide hydrogen gas at a cost to the contractor of 1 cent per cubic foot, and also labor for handling the dirigible on the ground under the supervision of the contractor.

The contractor may conduct the acceptance trials at a place selected by him, furnishing his own gas, labor, shed, and all facilities. When trials are completed the dirigible is to be created and prepared for shipment to destination, where it shall be erected and inflated by the Government under the supervision of the contractor.

The following trials shall be conducted, the procedure to be followed to be settled by agreement before such trials are undertaken:

(*a*) *Endurance Trial.*—With full load at start consisting of the bare weight listed in the weight statement and a useful load of 2 men, full supply of fuel and oil and such ballast as the contractor may choose to carry, the ship is to be driven through the air continuously at full power for three hours.

(*b*) *Speed Trial.*—With any convenient load the ship is to be driven at a full speed through the air over a measured course at least 10 miles in length.

(*c*) *Manœuvering Trial.*—With any convenient load, the ship is to ascend to a height of 6,000 feet in 10 minutes, to descend from this height in 15 minutes, and to be so manœuvered as to demonstrate that she is completely under control in three dimensions both at full speed and at cruising speed.

The dirigible shall, in addition, fulfil on said trials the following conditions:

(1) The working of all the machinery, devices, appliances, and appurtenances, and all parts thereof, shall be to the satisfaction of the department.

(2) The dirigible shall be found in all its parts to be strong and well built and in strict conformity with the contract, drawings, plans and specifications, and duly authorized changes therein. During the time of these trials and thereafter until delivery of the dirigible all machinery and equipment shall be kept in an efficient and operating condition by the contractor.

If, upon the trials and tests required by this contract, the dirigible shall fail to fulfil the requirements and conditions hereof, the contractor shall be entitled to make further trials, sufficient in number to demonstrate its capabilities: *Provided,* That the number of trials shall be determined and limited by the depart-

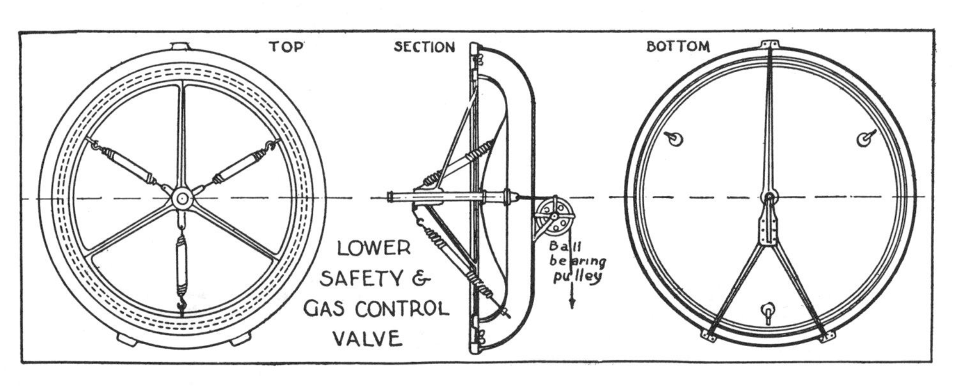

SAFETY AND GAS CONTROL VALVE

Gas valves are set to automatically release when pressure at bottom is 1 inch of water. The safety valve is located on top of the envelope.

The manœuver valve is located under the envelope and operates as a safety valve also, but may be opened by means of a line to the pilot's seat.

Openings in valves are 20 inches in diameter. Tightness is secured by spring loading which brings the valve against a rubber seat on the valve frame.

ment and that all the expenses of all trials and tests of the dirigible prior to conditional acceptance shall be borne by the contractor.

Such tests and trials shall be witnessed and reported upon by a board appointed by the department in order to determine whether or not the dirigible as completed satisfies the requirements. The board shall have power to prescribe the precautions and methods necessary to insure accurate results, but shall have no power to direct the methods of operation of the dirigible on trial, and the board shall take into account and report upon the performance of the dirigible and her qualifications and capabilities with respect to the demonstration by tests and trials, as aforesaid, of special features not specifically covered in the tests and trials prescribed. If, at and upon the trials and tests before mentioned, the foregoing requirements and conditions shall be fulfilled, and if the dirigible is built in accordance with the requirements of the contract, plans, and specifications, and duly authorized changes, and is complete in every respect and ready for delivery to the department, then, and in such case, the dirigible shall be conditionally accepted; but if the speed and other qualities attained by the dirigble on trial shall fall below the guaranties and requirements aforesaid, but not below certain minimum requirements hereinafter set forth, deductions from the price shall be made in accordance with the following provision, viz:

Every effort shall be made to attain the designed speed of 45 miles per hour, but if the speed on trial is below 35 miles per hour the dirigible may be rejected.

Accounts.—During the construction of the dirigible detail cost accounts shall be kept and such accounts, in condensed form, shall be submitted to the department upon completion of the contract. These accounts shall be at all times open to the inspection of the department's representatives. It is agreed by the department that all cost data supplied by the contractor is to be considered strictly confidential.

THE CONSTRUCTION OF A NONRIGID NAVAL DIRIGIBLE FOR COAST PATROL

This specification contemplates a nonrigid self-propelled dirigible or airship designed for use in connection with coast or harbor patrol. It is intended that the dirigible shall be operated from a base on shore, but that it shall be possible for it to rest upon the surface of the water in good weather.

The airship shall consist of a nonrigid envelope made of rubberized fabric and containing hydrogen under sufficient pressure to maintain the rigidity of the envelope. There shall be attached to the envelope vertical or horizontal fins and vertical and horizontal rudders, mooring line, rip panels, manœuvering and safety valves, ballonets or internal air sacks with means for their inflation.

Beneath the envelope and supported thereby is carried upon a suspension a car or body containing the power plant, fuel, ballast, personnel, radio, etc.

The envelope fully inflated has a displacement of about 77,000 cubic feet, corresponding to a gross buoyancy of 5,275 pounds when inflated with hydrogen of good commercial purity and under normal conditions of barometer and temperature. Under these conditions the life is reckoned at 0.068 pound per cubic foot, at 15° C. and 760 mm.

The length of the envelope is 160 feet and the maximum diameter 31.5 feet; maximum width over tail fins, 36.2 feet; the center of buoyancy is 69.2 feet from the nose; the height over all is 50 feet; horsepower of motor, 100; horsepower of blower engine, 2; maximum safe attitude, 7,500 feet.

Designed maximum speed at an altitude of 600 feet, 45 miles per hour; endurance at full power, 10 hours; cruising speed, 35 miles per hour; endurance at cruising speed, 16 hours.

Capacity of tanks, 100 gallons, 600 pounds.

Total volume of both ballonets, 19,250 cubic feet.

Reserve ballast tank in car, 300 pounds of water.

Trimming tanks attached to envelope: Forward, 40 pounds of water; after, 50 pounds of water.

The following weight statement is furnished for the guidance of the contractors and is subject to such changes in distribution as are found necessary during development, but the total weight of all items is not to be exceeded:

Item	Weight
Envelope (including seams, nose piece, and doublings only for ballonet attachment, tail group, grab ropes, and ballonet suspension, also belly band)	1,177
Gas valves and sight holes	36
Air ducts, air valves, and air manifold	54
Suspension, wire cable, crow's feet, grab ropes	35
Ballonets, including ballonet suspension	350
Fins and rudders, including their bracing	480
Running rigging for steering and rip panels, and parts of nose rope on envelope	43
Car:	
Structure (including tank weights)	321
Engine, mufflers, radiator and water, propellor and hub	568
Blower engine and blower	100
Starting crank for main engine	25
Lighting cells, wiring and lamps	30
Landing gear and floats	58
Miscellaneous fittings, including jack stay, stirrups, and nose rope	57
Total "penalty weight"	3,334

Useful load:

Pilot and observer	320
Instruments	100
Radio	250
Fuel and oil	670
Water ballast (including 90 pounds for trimming)	390
Sandbag ballast	211
	1,941

The following plans are furnished as a part of these specifications. In case of any discrepancy between the plans and specifications, however, the specifications shall prevail:

(1) General arrangement plan.

(2) Detail sheet showing type construction of miscellaneous parts.

The contractors shall develop from the plans and specifications such working plans as may be necessary for the construction of the dirigible. Before undertaking work or ordering material from such plans the plans must first receive the approval of the inspector, and all work done on the dirigible must be strictly in accordance with such approved plans.

The inspector shall be furnished with a copy of all requisitions for material. In case it is desired that such material shall be inspected at the place of manufacture, the inspector shall be so advised at the time of placing the order and furnished with such additional copies of the requisition as he may require. Should it not be convenient to inspect the material at the place of manufacture, the contractors will be so advised and the material will be inspected after delivery at the contractors' works.

Within one month after the delivery of the dirigible the inspector shall be furnished with complete plans, in the form of cloth tracings in ink with blue prints in triplicate, of all parts of the dirigible, corrected to show the work as it was actually installed at the time of delivery.

Parts which are partially or completely assembled before installation shall be photographed and two (2) prints from a negative (6 by 8 inches) shall be supplied the inspector. The object of this requirement is to obtain photographs of fin and rudder construction, valves, suspension car, blower installation, power-plant installation, etc., to form a record of progress. With the final plans the contractor shall deliver to the inspector 12 copies of a descriptive booklet giving complete description of material, construction, operation, testing, care, and maintenance of the completed dirigible and each of its principal parts and accessories.

Workmanship.—The workmanship throughout shall be of the most thorough character and suitable for the purpose intended and satisfactory to the inspector.

Aluminum shall not be used for important strength members, nor shall any strength member depend for its strength upon brazing, welding or soldering.

ARTICLES OF EQUIPMENT AND OUTFIT TO BE FURNISHED AND INSTALLED BY THE CONTRACTORS UNDER THE CONTRACT

Tachometer.
Gasoline air pressure gage.
Circulating water and lubricating oil thermometers.
Oil pressure gage.
Longitudinal inclinometer.
Map boards.
Mooring rope.

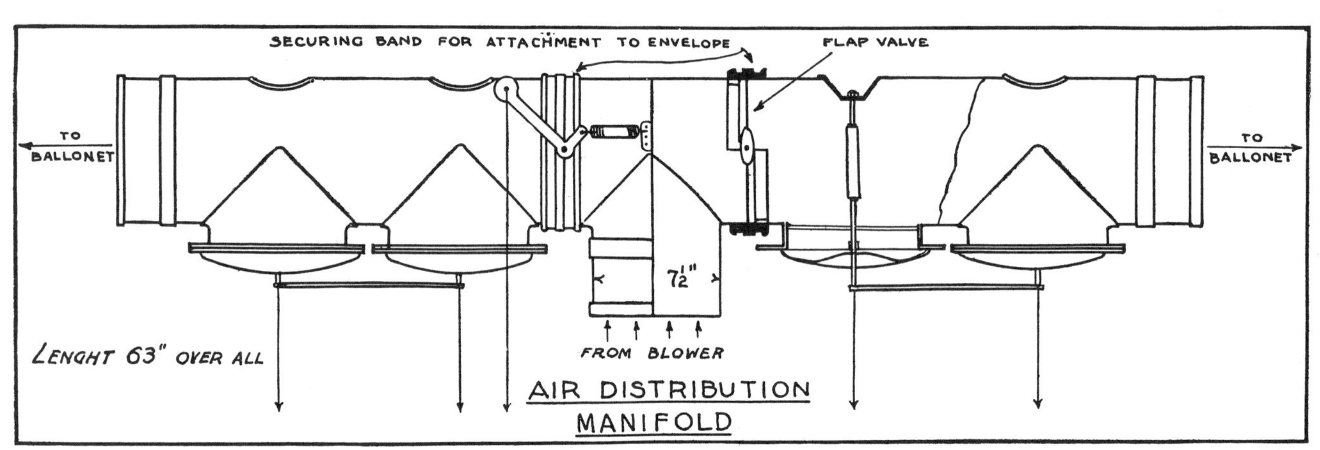

AIR DISTRIBUTION MANIFOLD

The air distribution manifold is provided with four relief valves, each with an opening 9 inches in diameter, which are spring loaded to release air at ¾ inch of water pressure.

Pipe from blower to distribution valve is 7½ inches in diameter, and the wind pipes from distributor to ballonets are of balloon fabric, 8 inches in diameter. Air from the blower can be drected to either or both ballonets by flap valve dampers operated by cords from the pilot's seat.

All parts of the manifold and valves are of aluminum with the exception of pins, guide rods and springs. Springs are of phosphor bronze, and their adjustment device is gas tight.

Two gas pressure manometers.
One ballonet air pressure manometer.

ARTICLES OF EQUIPMENT AND OUTFIT TO BE FURNISHED BY THE GOVERNMENT AND INSTALLED BY THE CONTRACTORS UNDER THE CONTRACT

Altimeter, statoscope, compass, air speed meter, fire extinguisher (chemical sprinkler type). Search light.

Protective Coatings.—The envelope shall be made of rubberized fabric as a protective coating. The car shall have woodwork sandpapered and coated with heavy spar varnish, "Valspar" or its equivalent in quality. Wire cable shall be galvanized and solid wire used in the body construction shall be tinned. Steel bolts and pins shall be coppered and nickel plated. Clips of sheet steel shall be hot galvanized or coppered and nickel plated or copper plated and covered with baked enamel.

The fabric covering the car, fins and rudders, shall be coated with at least five (5) applications of nitrocellulose, and finally coated with "Valspar" varnish or its equivalent.

Portable canvas coverings for cockpits and power plant shall be furnished.

FABRIC REQUIREMENTS

Factor of Safety.—The contractor must be able to show that the fabric factor of safety under normal running conditions for any part of the dirigible exceeds 8. The strength to be taken as a basis, to be found by the methods given below.

Construction of Fabric.—All fabric used in the envelope or ballonets to contain two or more plies of cloth, one of which is to be laid on a bias of 45 degrees. Sufficient rubber of proper quality shall be placed in the fabric to meet the requirements as to diffusion and weather-resisting properties, which are given below. The protective coating on the outside of the envelope shall weigh at least 0.4 ounce per square yard, and on the inside 0.2 ounce per square yard.

No fabric in the balloon is to weigh over 12 ounces per square yard, and no fabric is to test less than 40 pounds per inch in the direction of any of the threads, either bias or straight, test to be made as described below.

Fabric in the top of the balloon shall show an average strength over 60 pounds per inch for the four different directions of threads; or a strength over 100 pounds per inch in either warp or filler if the two plies are doubled straight together for a sample test.

The fabric which is intended to separate air and gas shall have an average hydrogen diffusion not greater than nine liters per square meter per 24 hours, according to the procedure given below, and under a pressure of 30 millimeters of water.

Samples of the finished fabric shall be submitted to the naval inspector for approval before starting construction.

Method of Weight Testing.—The figure for the weight of fabric shall be obtained by getting the net weight of an entire roll and dividing by its superficial area in square yards.

Strength Test.—Dimensions of samples, 2 inches wide by 6 inches long. Cut 12 samples from each roll, including 3 each from the warp and filling of the straight ply, and 3 each from the warp and filling of the bias ply. Place in the testing machine with 3 inches initial gap between jaws. The speed of separation of jaws will be 12 inches per minute. Use jaws 2 inches wide. A number of specimen seams shall also be tested by this method, enough to prove to the inspector that they are over 100 per cent. efficient.

Diffusion.—The following conditions are assumed when getting the final result for hydrogen diffusion:

1. Temperature 15° C.
2. Pressure on the air side of fabric to be atmos-

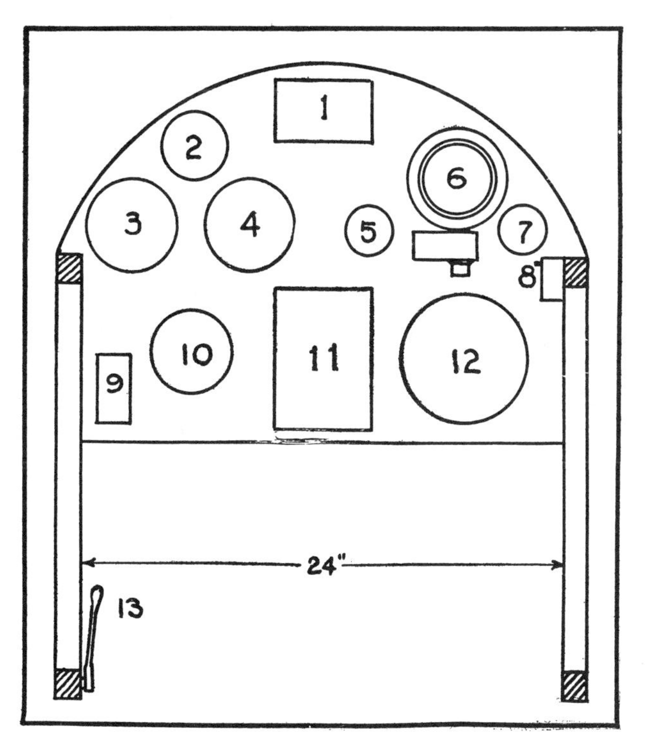

INSTRUMENT BOARD ARRANGEMENTS

In scouting dirigibles as specified by the Navy Department the forward instrument board is to be equipped with instruments arranged as shown in the accompanying drawing. They are as follows:

1. Compass, 2. Thermometer, 3. Altimeter, 4. Statoscope, 5. Air Pressure Gage, 6. Tachometer, 7. Oil Pressure Gage, 8. Switch, 9. Longitudinal Inclinometer, 10. Clock, 11. Two Manometers to indicate Hydrogen pressure, 12. Air Speed Indicator, 13. Throttle.

The board in the aft part of the car is for instruments to be used in connection with a radio set. Both boards are 24 inches wide and 18 inches in over-all height, with the upper edge curved at a 14-inch radius.

A British Blimp dirigible entering its hangar near Salonika.

pheric with pressure difference of 30 millimetres of water.

3. Current of air to be maintained on the air side.

4. Fabric to "soak" in the hydrogen for at least 5 hours before starting test.

5. Run the test itself at least 2 hours and correct the result to apply to 24-hour period.

6. Keep the air stream perfectly uniform (about 5 bubbles per second through sulphuric acid).

7. Apparatus and method to be the same in principle as that used by the Goodyear Tire & Rubber Co., described by R. A. D. Preston in "India Rubber World," November 1, 1914, page 70, in which a constant pressure of hydrogen is maintained on one side of a test piece of fabric, while air is passed over the other side, taking with it the hydrogen which diffuses through the fabric, which hydrogen is burned to water and weighed.

Instead of running the test at 15° C., which is the standard, it may be run at any fixed temperature up to 20° C. by reducing the final result in the ratio of 15 divided by T where T is the temperature actually used. If it is desired to run at a greater difference in pressure than the standard 30 millimetres, this may be done for any pressure up to 50 millimetres by multiplying the result by $\sqrt{\frac{30}{P}}$ where P is the pressure actually used. A diffusion sample will be cut from each roll within 9 feet of one end.

Inspection.—All fabric going into the balloon shall be run over a light to detect pinholes and flaws in the threads, any such places being subsequently either proved of sufficient strength and tightness or patched. Inspector may require such tests of diffusion and strength on various parts of the fabric and seams as he may judge necessary to prove the quality required. Every seam will be inspected when completed.

Model Test.—Each contractor for one or more dirigibles shall construct a model of linear dimensions one-thirtieth the size of the balloon, made of identical fabric, provided with suitable suspension disposed in similar fashion to that on the full-size balloon. This model to be inflated (upside down) with water to correspond with the full-size balloon when inflated to five times its normal running pressure. This will mean on the model a head of water at the top equal to 12½ feet. Leave pressure on for 10 minutes.

Tests on Completed Envelope.—Each envelope when completed shall be blown partially full of air and all fabric inspected from the inside against a light to detect small leaks. After inflation with hydrogen at normal operating pressure, the envelope must show a leakage of gas less than 1 per cent. per day, with all valves and accessories in place. Envelope, with ballonets in place, after inflation with hydrogen shall be inflated to a pressure of 3 inches of water at lowest point and held there for one minute. This is three times the normal pressure for flight, but no defects shall be developed by this proof test.

Guaranty.—The contractor shall guarantee the completed envelope for three months' continuous exposure to the weather or its equivalent, provided this does not occupy more than one year's time from date of acceptance. Under a pressure test at the end of this time the envelope shall resist without injury a pressure of 2 inches of water at the bottom. In case the envelope fails through deterioration to give the specified length of service, it shall be replaced by the manufacturer at the request of the Government and the manufacturer shall receive compensation for the service rendered by the first envelope pro rated on an arbitrary value of $12,000. Reasonable care shall be taken to keep the envelope out of the sun when not in use, and the fabric shall be dry and cool before packing. Hydrogen gas used shall be free from injurious chemicals or excessive moisture.

Tests shall be made of all fittings, so far as practicable, after completion and before installing in place. This testing shall include air and gas valves, gasoline tanks, oil tanks, blower engine, blower, and instruments; also means for controlling the trimming, ballast, and emergency discharge of fuel tanks. Samples of all wire, cable, rope, fabric, typical seams, patches, etc., shall be broken in a testing machine.

Such other tests shall be made as are necessary to determine that the requirements of the specifications have been fulfilled and that all parts and installations are suitable for the purpose intended.

Mooring Eye.—The nose shall be reinforced as shown on the plans by means of a doubling patch of fabric similar to that in the envelope and wood battens about 3/8 inch by 3 inches in section equally spaced, converging at the nose, every alternate batten to extend beyond the doubling 3 feet. An eye shall be provided in the nose for the attachment of a mooring line. This eye will consist of a 1¾-inch metal thimble secured in an eye splice of a four-strand manila rope. The rope will be unlaid and the eight strands let in beneath the doubling patch, equally disposed radially, the ends of the strands frayed and the whole set in cement. The breaking strength of the mooring attachment shall be about 5,000 pounds.

Rip Panels, four in number, shall be located approximately as shown with a rip cord, dyed red, run from each through light agate guides to the pilot. By pulling the rip cord weak stops securing the end of the panel are to break and tear open the panels for rapid deflation in an emergency.

Grab Ropes, eight in number, shall be secured to patches on envelopes as shown. These to be 1½-inch circumference manila rope secured to crows'-feet of ¾-inch flax signal halyard stuff (braided of 3-ply thread in 8 strands of flax twine; breaking strength, 800 pounds).

Ballonets are located as shown. Their combined volume is 25 per cent. of the total volume of the envelope. The relative sizes of ballonets are to be adjusted so that their displacements give equal moments about the center of buoyancy of the envelope when completely inflated. The ballonets are to be fitted with a suspension band of fabric running around the plane of symmetry to which a light flax suspension is to be attached. This suspension holds the ballonet in place and prevents shifting fore and aft when flabby. Sight holes of transparent material are to be placed in the envelope for inspection of this suspension.

The lower part of the ballonet is to be secured to a patch of doubled envelope fabric sewed and cemented to the envelope. This patch when removed carries with it the ballonet and windpipe connection. A sight hole shall be located in the ballonet patch for inspection of inside of ballonet.

The *safety valve* will be located on top of the envelope. Construction of the valve is shown on detail sheet. This valve is to automatically open outwards when pressure in envelope at lowest point exceeds 1 inch of water.

The *manœuver valve* will be located under the envelope and will operate as a safety valve also, but may be opened by means of a line to pilot's seat.

These valves are to be tested for tightness of seat and the spring loading adjusted as stated on the detail plan before installation.

The *belly band* or *suspension band* is to be a heavy fold of canvas running round the lower part of the envelope, bearing at intervals hardwood toggles for the crow's-feet of the car suspension. This band shall be securely sewed and cemented to the envelope in a manner which on test is shown to develop the full strength of the latter fabric.

Fins.—Stability of route is assisted by two horizontal fins and three vertical fins as shown. These fins are to be 170 square feet each in area except the vertical fin shown on top of the envelope, which will be 80 square feet in area, and are to be made up of a light structure of steel tubing and wood with internal wire bracing and covered with aeroplane linen treated with five coats of dope and varnished to give a smooth, taut surface. The fins are to be braced by wires with turnbuckles, and crows'-feet to doublings on the envelope. Weight of fins is to be kept down to one-half pound per square foot. Doubling patches shall be fitted on envelope to secure butt edges of fins by lacing.

Rudders.—Two horizontal rudders, each 70 square feet in area, and two vertical rudders, each 35 square feet in area, are to be provided. Both the horizontal and vertical rudders are to be balanced and to work in ball bearings. These rudders are to be securely trussed and operated by nonconducting leads (flexible cable of flax line) passing through agate guide rings, or ball bearing bronze sheaves as indicated on the plans. Rudder operating leads are to be designed to give the least number of turns so as to reduce friction to a minimum. Fins and rudders are to be readily detachable for shipment or stowage.

Insulation.—The car is to be electrically insulated from the envelope and no valve or other operating leads shall be of continuous wire. All metal parts in car are to be electrically connected. Metal parts of valves and their seats, wherever located, shall be electrically connected.

Mooring Line.—A mooring line consisting of 200 feet of 2½-inch manila rope is to be provided for carrying coiled in the car. The fixed part shall be stopped to the envelope and the end secured to mooring eye in the nose.

Steering Controls.—All steering controls shall be in duplicate and interconnected. The vertical rudders shall be operated by means of foot bars located as shown on the plans, the foot bars to be fitted with foot rests arranged to pivot so as to remain normal to operator's thrusts. The leads are to be fastened to the ends of the foot bar by a suitable fitting. The horizontal rudders are to be controlled by means of hand-wheels located on the right hand side of the car as indicated. The hand-wheels will be arranged to operate the leads by means of sprocket chain and sprockets in accordance with the plans. Both hand-wheels are to be provided with an effective band brake so that the horizontal rudders may be locked in position by means of an arrangement for increasing the tension in the brake band.

Tanks shall be provided and arranged substantially as shown on general arrangements plan, sufficient for a 10-hour supply of fuel and oil at full power for the motor.

The main fuel tanks shall be interconnected and so arranged that fuel may be taken from any tank or combination of tanks. Valves shall be operated from the rear seat and shall be quick acting. Fuel tanks shall be of noncorrosive material. If copper tanks are used, they shall be tinned on the inside. Tanks shall, before installation, be tested to an internal pressure of 5 pounds per square inch and must show no permanent deformation. Tanks shall be provided with swash-plate bulkheads. A small plate shall be soldered to each tank, where it may be readily seen, showing weight of empty tank and capacity in gallons. Each tank shall be fitted with an effective gage to show amount of fuel in tank, and, so far as practicable, visible from rear seat. The three main tanks shall be connected by a length of 1-inch pipe run under them, with quick-acting shut-off valves in the three branches to the tanks. A fourth valve is to be provided for rapid discharge of gasoline to lighten the dirigible in an emergency. Each tank shall have a handhole for filling and cleaning.

Ballast.—The water-ballast tank shall be located in the body and be of 300 pounds capacity, and provided with means for rapid discharge from the pilot's seat. The tank is to be made of waterproof fabric and shall be tight when full.

In addition to trimming the dirigible by manipulation of horizontal rudders or shifting air between ballonets, small water containers are to be placed near the bow and stern of the envelope, fitted with spring loaded valves, and means whereby such valves may be pulled open by the pilot. The forward tank shall contain 40 pounds of water and the after tank 50 pounds of water, and these tanks shall be located at the points shown in the general arrangement plan. The forward tank should be a fabric tube placed along a meridian with a valve in its lower after end. The after tank shall be as flat as possible and secured in a fabric pocket under the envelope and between the lower vertical fins.

Tubes shall lead discharge from each tank to main ballast tank.

Blower System.—A 2-horse-power, 900 revolutions per minute engine of motorcycle type is to be arranged to drive through a 2 to 1 gearing a multivane blower of a capacity of 600 cubic feet per minute at 1800 revolutions per minute against a head of 2 inches of water. A crank is to be provided for starting the motor which is to be accessible from the rear seat. (See Radio.)

A wind scoop is to be arranged to supply air to the blower, which discharges through suitable dampers into one or both ballonets. The wind pipes to ballonets are of balloon fabric and are to be 8 inches minimum diameter. The air manifold is to be made of sheet aluminum and provided with relief valves spring loaded to release air at three-fourths inch of water pressure.

Precautions similar to those stipulated for the main engine shall be provided for the blower engine to prevent flame from fire or exhaust.

Radio.—The radio outfit will be supplied by the Government and will be installed by the contractor in the space reserved and indicated on the plans. It will weigh about 250 pounds and will have a generator to be driven by the blower engine through a silent chain and 3 to 1 increasing speed gear. A friction clutch, to be designed so as to reduce weight to a minimum, is to be provided by the contractor so that the blower engine may be run idle or engage either blower or generator.

Car.—The car or body is of standard aeroplane type consisting of a rigid rectangular girder of spruce trussed with wire. The motor and radiator are to be mounted forward, with a sheet steel fire bulkhead behind them; next the pilot with all controls and instruments; next the observer with duplicate controls and radio key; next the blower and radio outfit; next a fabric tank for water ballast; and last the gasoline tanks for 10 hours at full power. Reserve oil and gravity gasoline tanks may be mounted near the motor as shown on the plans. The car is to be inclosed with aeroplane linen except over the motor where the covering shall be of sheet aluminum. The motor compartment shall be well ventilated and the bottom perforated to prevent accumulation of gasoline in case of leakage.

The department will supply and the contractor install a sprinkler type of chemical fire extinguisher with nozzles located near carburetor.

The car shall have a jackstay along top and stirrups under bottom, as shown, to enable a man to reach motor or gasoline tanks in the air.

The maximum propeller diameter permissible is 8 feet 6 inches.

Suspension.—The car is to be suspended from the envelope by means of galvanized-wire cables with breaking strength of 2700 pounds, arranged as shown. The cables are to be fitted with means for adjusting their lengths to equalize the load. The cables are connected to the suspension band by crows'-feet of braided flax, 3/4-inch signal halyard stuff, as specified elsewhere. To carry the car when the dirigible is inclined the upper ends of suspension cables are connected by a fore and aft stay as shown.

The suspension cables shall be connected to the car by hooks or bolts arranged to permit ready detachments.

Power Plant.—Includes engine, propeller, radiator, starting device, gasoline and oil tanks, piping, controls, gasoline and oil gages, pressure gages, thermometer, power-transmission system, tachometer, necessary shipping crates, etc., in order for flight, and as per the following specifications:

The engine shall be a standard Curtiss OXX-3, 100-horsepower aviation engine, or a Hall-Scott A-7-A, 100-horsepower aviation engine.

The engine shall be provided with an effective starting device, so fitted and installed that engine may be easily started from the front seat. If hand starting is used, a booster coil will be provided.

All oil pipes shall be annealed.

The fuel leads to fuel tanks, the control leads, and the carburetor adjusting rod shall be provided with suitable, safe, and ready couplings where these connections have to be frequently broken.

All couplings and fittings in the gasoline lines shall be thoroughly sweated on.

Gasoline shall be supplied from the reserve tanks to the service tank by pressure feed.

Fuel, water, and oil service pipes will be protected against vibration.

Oil shall be supplied to the engine pump from the reserve oil tank by means of a hand pump operated from the forward seat.

As far as practicable, the entire power plant should be assembled as a unit on a rugged foundation, and should be capable of being readily removed or replaced with a minimum disturbance of connections, controls, or structural fittings.

A complete set of power-plant tools shall be furnished with each engine.

Tachometers of approved type shall be installed on the pilot's instrument board (Reliance No. D-18 with internal counter, or equal).

A thermometer showing the temperature of the circulating water discharge, and one showing temperature of lubricating oil in crank case, shall be mounted on the instrument board (Foxboro transmission type thermometer, or equal).

Each engine shall be fitted with the Bureau of Steam Engineering standard name plate for aeronautical engines.

The air intake opening to carburetor shall be provided with a safety wire gage sphere of generous size to quench flame in case of backfire.

The motor shall exhaust into an effective muffler at each side so arranged as to cool the gases and to preclude possibility of flame at exit. Tests will be run to prove the effectiveness of the muffler as a flame quencher and if necessary wire gauze may be placed over outlet slot.

Engine Tests.—Each engine shall, before shipment from the factory, be subjected to a one-hour full-power run on propeller. During this run revolutions shall be accurately determined and recorded. Report of all engine tests shall be forwarded to the Bureau of Steam Engineering.

With each dirigible the contractor shall deliver the following:

1 ground cloth, 170 by 35 feet.
100 sand bags holding about 40 pounds of sand each.
1 filling balloon or service gas reservoir of 700 cubic feet capacity.
5 screw stakes.
1 extra mooring rope.
200 feet filling tube, 6 inches in diameter, with connecting sleeve.
1 roll of each weight of fabric used.
1 fabric stitcher.
1 fabric roller.
5 gallons of cement.
100 feet of each size wire or cable used.
100 feet of each size rope used.
1 spare gas safety valve, complete.
1 spare manœuver valve, complete.
1 spare gas pressure manometer.
1 set of each size bolts, sheaves, guides, shackles, thimbles, turn-buckles, toggles, or other miscellaneous fittings used—list to be approved by the naval inspector.

KEY TO DIAGRAM

1 Gas Envelope
2 Car
3 Ballonet
4 Blower Intake Pipe
5 Engine for Blower
6 Main Air Discharge Pipe
7 Air Pipe to Ballonet
8 Air Manifold
9 Operating Cord, Ballonet Exhaust Valve
10 Operating Cord, Butterfly Valve
11 Valve, Pressure Relief
12 Valve, Manœuver Gas
13 Operating Cord, Manœuver Gas Valve
14 Rudder—Twin
15 King Post
16 Leads, Steering Gear
17 Bracing Wire
18 Elevator
19 Leads for Elevator
20 Fin, Stabilizing

21 Doubling Patch
22 Car Suspension
23 Belly Band
24 Webbing
25 Ballonet Suspension
26 Nose Reinforcement
27 Rip Panel
28 Rip Cord
29 Grab Ropes
30 Weights
31 Mooring Rope
32 Sight Holes
33 Patch for Removing Ballonet
34 Kapok Floats
35 Fuel Tanks
36 Muffler
37 Trimming Tanks
38 Operating Cords, Trimming Tanks
39 Guides for Operating Cords
40 Filling Hole and Doubling Patch

One of the photographs showing the extent of employment of "Blimps" by Great Britain.

The Navy's kite balloon (Goodyear Type) at the Pensacola, Florida, Aviation station.

CHAPTER XXXI

CONSTRUCTION AND OPERATION OF KITE BALLOONS

BY RALPH H. UPSON

The kite balloon now used in such large numbers in Europe is a combination of two principles, both long known to aeronautical students—the man-lifting kite and the captive balloon. A strong wind was required to fly the man-lifting kite balloon, and an ordinary captive balloon could be flown only in comparative calm. The kite balloon synchronizes the best in both and takes care of either condition.

The purpose of this chapter is to impart to the aviation section of our army and navy working knowledge of the handling of this type of aircraft.

The apparatus is essentially an elongated balloon, which is always kept in an inclined position like a kite. Thus the wind tends to lift it by blowing against the under side. This counteracts the contrary tendency of a wind to blow the balloon over toward the ground.

The outstanding features of the kite balloon are shown in Fig. 1. The shape of the gas bag is modified in such fashion that it has a minimum resistance consistent with other requirements. The steering bag is replaced by an air funnel, *C,* which is the only inflated protuberance of the balloon. This carried the keel, *B,* the tail cup, F, and has a value at its lower end through which air enters to make good any deficiency in pressure. In this way a comparatively small hole will supply all the air necessary to keep the balloon well inflated.

The keel, like the funnel, is non-rigid, being supported entirely by a proper balance of forces. Its great advantage over the old steering bag is the fact that it presents a double concave surface to the wind so that it is held in the wind by a positive pressure on both sides at once. In this way any tendency to yaw is stopped almost before it starts. On the other hand, the old type of steering bag will allow considerable deflection or swing of the balloon before its corrective influence is felt. In addition to this the keel has only a small fraction of the resistance of the steering bag, is lighter in weight, and is simpler in practical use.

The side fins indicated by *A* in Fig. 1 are so

shaped and disposed that they help in the stability as well as in the kite effect of the balloon.

The function of the tail cups, *F*, has been reduced almost wholly to one of dampening the motion of the balloon in a gusty wind. Contrary to expectations, it was found that the form of construction of the tail cup was a considerable factor in its stability.

The resistance was further cut down by making the balloon more nearly self-contained, and eliminating superfluous cordage and protuberances.

In other balloons it was found that the gas leakage through the valve and various appendices was commonly many times greater than all other sources of leakage combined. The valve was therefore designed especially to prevent leakage. Its size is more than ample for all requirements of captive- or free-balloon use. The air valves have also been made tighter and more efficient, thereby cutting down the quantity of air required to keep the balloon properly inflated.

For maximum efficiency, the inclination of the balloon to the air must be calculated with considerable exactness. If it is inclined too much, it carries the balloon over and puts an undue strain on the cable. On the other hand, if the balloon is too nearly horizontal, the lift of the wind is lost. There is a certain best inclination for each balloon at which it is perfectly steady, the pull on the cable is near a minimum and the altitude a maximum. In order to preserve this proper angle for different conditions of use, the basket is made adjustable so that it can be moved forward or back when necessary. In a 30-mile wind, there should be no motion apparent in the basket; the cable tension at the balloon should be less than 1000 pounds and an altitude of 4000 feet readily attainable.

Location of Kite-Balloon Aerodrome

1. Select a field where there are no large obstructions within at least 200 feet or subtending an angle of more than 20 degrees with some central point; at the bottom of a shallow valley or depression is best. Clear the ground of all sharp objects liable to puncture the balloon. The main pulley (8 to 10 inches diameter) should be located at the central point unless the usual wind is from some particular direction.

2. The winch should have a drop of at least 8 inches in diameter, or 10 inches if the balloon is to be pulled up and down many times a day. It should be placed about 100 feet from the pulley in a direction to windward of the average wind. Anchor it firmly in position to withstand a straight pull of 3000 pounds. This can be best obtained by bolting two longitudinal timbers about 12 feet by 10 inches by 2 inches on the bottom. Bury these timbers, from 1 to 2 feet deep. It is important to see that the cable pulls exactly at right angles to the drum.

3. The anchor post carrying the main pulley consists of four standard screw stakes, placed pyramid fashion with the eyes screwed down on a level with the ground. A maximum total pull of 4000 pounds, both vertically and horizontally, should be allowed for. If the ground is not suitable for the use of stakes, a wooden post about 4 inches by 6 inches by 5 feet, with cross bracing of timber, must be used.

Laying Out

1. Fill about fifty bags three quarters full of dry sand (tied shut). Spread out the ground cloth on a leeward side of the pulley with the narrow edge about five feet from the pulley. Place two or three sandbags on the windward edge and peg down the corners.

2. Unroll the balloon, starting one foot from the pulley end of the ground cloth. When unrolled it should be like top up and head into the wind.

3. Any one required to walk on the balloon should either take his shoes off or protect them by tying sandbags over his feet. Spread a cover over the balloon whenever possible to protect it from the sun.

4. Tie the nose rope to the main pulley post. Pull the other ropes out to the side. Put a screw stake or small dead man opposite each manœuvering rope at the edge of the ground cloth (three on each side). Give each rope one-half hitch through the eye of its stake and tie

FIG. 1—OUTSTANDING FEATURES OF MODERN KITE BALLOONS.
A—Fins; B—Keel; C—Funnel; D—Holding Cable; E—Manœuvering Ropes; F—Tail Cups; G—Appendix; H—Air Entrance Valve; J—Gas Valve; K—Basket; L—Main Bridle; M—Suspension Groups.

with a slip knot. Drive two wooden stakes (about 2 inches by 1 inch by 2 feet) on each side about 20 feet from the balloon, to which are attached the central and rear-suspension groups by means of scrap rope tied to the concentration rings. Arrange a few sandbags to hook into the suspension.

5. Thread the main cable through the pulley and attach to the bridle. Tie a conspicuous rag 8 feet below the upper splice of the cable. Lock the winch leaving the cable just slack.

6. Put the slack of the valve cord inside its gland to within a foot of the loop. Tie the gland about 4 inches from the end, and then tie it again through the loop with breakable cord about 2 inches from the top of gland. Note: The cord (not the gland) should come loose at a pull of about 40 pounds.

7. Push the rip cord up into the appendix and tie a smooth weight of about 1 pound to the lower end. Pull the appendix out to the side next the gas supply, keeping out all twists.

8. See that the valve seats are tight all round. It should start to open with 30 pounds' pull. Pack the stuffing boxes tight. Disconnect the valve cord (white) from inside the valve hold and attach to the valve. Screw in the valve and put on the cover in a vertical position. Tighten up the screws again after a few hours. Note: The stuffing boxes together with a bridle attachment on the balloon diaphragm are provided in case it is desired to use an automatic wire for discharge of gas. This has been the prevailing practice in Europe, but after exhaustive tests we recommend in preference the method hereinafter described.

9. See that the manholes are properly (securely) closed. The gas manhole must be fixed to pull out from below. The lacing over the air manhole may be left till later.

10. All preparations will have to be made with unusual care if there is much wind. If possible, postpone inflation if the wind exceeds 20 miles per hour.

Inflation (by Bottles)

1. Lay 12 bottles side by side in a row with the nozzles about one foot apart and pointed up. Lay the manifold above (on its stand) and put one man on each of the six branches. Attach the manifold at every other bottle. Each man after discharging his bottle takes the next full one. As fast as the bottles are emptied they are replaced with full ones by other men, two on each bottle. Put the caps back on the empty bottles and mark to distinguish the empty bottles from the full ones. In discharging a bottle, open the valve slowly as far as it will go. After the gas has apparently stopped it will start again after a few seconds. Leave the connection on till after it has done this twice. Then give the word and all quickly shift to the next bottle. The whole operation should not take more than three minutes for each set of six bottles. In cold weather a bottle must be opened more slowly to prevent "freezing up."

2. Let out on the manœuvering ropes as the balloon inflates. The fabric should be kept just tight enough to prevent flapping and caving in from the wind. Hang sandbags in the suspension if necessary. Do not exceed two bags on each suspension patch.

3. Before the balloon is full, let a little air blow into the ballonet. Then test the safety valves by filling the balloon with gas. When the air is all forced out put still more gas pressure in and see that the balloon pressure indicator works properly. This should open after the air is all out and when the fabric is just tight enough to sound like a drum when hit with a pencil. Try the gas valve to see that it opens and seats properly. See that the small cord running down from the gas manhole is securely tied into the suspension by slip knots.

4. Let the balloon up about five feet from the ground. Disconnect the inflation tube. Let the rip cord out and put back enough slack inside to bring the upper loop about one foot below the opening. Tie a breakable cord through this loop to the appendix opening cord. Tie another breakable cord to the second appendix opening cord through the second loop and through the toggle on the balloon at the front of the keel. This should bring the appendix flat up against the balloon. Each closing cord should pull loose at about 40 pounds. Look in the window and see that the inside cords are in order.

5. Attach the basket and pull the dampening (lower) rope around until the toggle is centered over the basket. See that the tackle is free from twists. (See special instructions on folding the parachute.) Connect up the observing and signaling apparatus. Put one bag of sand (tied up) in the basket. Tie the end of the tail cup adjusting rope to the concentration point above the basket. Bring down the valve and rip cords through their respective rings on the upper ropes, and tie just within reach.

6. Attach three tail cups to loops closest to the balloon. See that the tail cup bridle pulls free without twisting.

Ascension

1. All men at their posts (ground position), observer in the basket. Release the holding ropes from the concentration rings and remove all bags. Untie all manœuvering ropes at the stakes and pull taut. All men pull ropes from position near their respective screw stakes. Then wait for a lull in the wind, and at the command "play out" the balloon is let up gradually.

2. The winch is kept locked until all the load is transferred to the cable. If the command to shift is not given sooner, the air position is taken gradually (without command) as the forward ropes go out of reach. Then the ropes are pulled out of the stakes and as far as possible to the sides. The object of this is to prevent yawing or side pitching as the balloon ascends.

3. Let the cable out until the basket just clears the ground. Then at the command "Drop Basket," the men at the basket let go

H.M.S. *Canning,* one of many balloon ships of the British Navy. (See chapters on "Kite-Balloon Ships and Hunting Submarines with Aircraft.")

and scatter to the help of any posts that need them most. The cable is played out, and the men at the rear ropes follow, just enough to keep the balloon from pitching. The ropes must be played out in this way to the very end (fast if there is much wind). Then at the command "Ready," "Let Go," every one let go at once. If instead of "Let Go," the command "Slack" is given, the ropes may subsequently be let go, without further command, as they go out of reach. Simultaneous with this command the cable is let out fast for about 100 feet and slowed up very gradually to a normal speed. Until this point has been reached all men stay under their ropes as closely as possible so as to seize them again if necessary (when the command is given).

4. Do not let the average speed of ascent exceed 500 feet per minute unless the valves are set to open easier. When the balloon is sent up for the first time, it is better to limit the speed to 200 feet per minute.

5. The man in the basket must watch the pressure indicator while ascending. If it opens, let out gas until it closes again. After once attaining the working altitude, the pressure may be disregarded unless the sun suddenly comes out warm.

6. Experience is the best guide to the proper angle of tilt for various conditions. This can be varied by shifting the basket, either in the air or on the ground. At the proper angle of tilt, the balloon is steady, the angle of cable is near a maximum, and the tension in the cable near a minimum (about 600 pounds at 30 miles per hour.) Fifteen degrees is the usual standard when riding nearly full at normal altitude. The basket tackle needs very little attention after the proper setting has been once ascertained. The balloon will automatically adjust itself through a considerable range of wind and altitude. The tail cups must be set so there is sufficient play in the bridle to meet all conditions. Keep the tail cup adjusting rope just tight.

7. Some device for measuring the pull on the cable is very desirable. A properly calibrated scales, midway between the wind and the pulley, will answer the purpose if allowance is added for the weight of cable in the air. The tension at the balloon should not be allowed to exceed 3000 pounds.

Pulling Down

1. If the wind is strong and has changed abruptly in direction, it is best to move the ground cloth and stakes to correspond. All men at their posts (air position), follow the ropes wherever they go, and seize them near the end when the command is given.

2. The main principle in holding the balloon and basket at first is to dampen or check the lateral movement, but not absolutely to prevent it. The men should move slowly with the balloon and keep it headed into the average wind. The cable is reeled in fast until the basket is about 10 feet from the ground. Then wait till the basket has stopped swinging before bringing it to the ground.

3. The basket man grasps the basket firmly as soon as the top edge comes within reach (without command). If it is merely wished to change passengers, ascension is made again immediately without pulling down farther.

4. For putting in gas or anchoring, however, men take ground position, a signal is given to the engine, and the cable is reeled in. The ropes in the meantime are pulled in correspondingly. The chief of section or corporal must signal the engine driver to stop while there is still 10 feet clearance of scale. Great care must be taken in this not to run too far and break the cable.

5. When the men at either post 4 or 6 get within reach of their own screw stake, the rope is given one complete turn through the eye and held until the opposite grasp can be attached. The chief of each group then pull out the slack while all others pull down from above the ring. All pull down as rapidly as possible, unless one side or the other is ordered to slow up to get the balloon level. If there are not enough men to pull it all the way down at once, tie the ropes to prevent them from slipping, then concentrate all the men in succession on each rope until the balloon is down within reach. Then hook the bags into the suspension and if necessary tie the concentration rings to stakes. Disconnect the fins at the bottom and furl them up. See that the main cable has one or two feet of slack in it. The balloon is now in a position to receive gas and it can be left this way in a light wind.

6. If the balloon sways around much it must be brought flat and tight onto the ground where it catches the wind less and is also less readily seen from above. Plenty of sand bags should be hooked high into the suspension (two in each patch or in the splice below). If left in a strong wind the end of the rip cord should be brought out and tied with plenty of slack to one of the stakes. It is always worth while to cover the balloon if it is to be left anchored for any appreciable time. Emergency anchor straps may then be thrown over cover and all.

Photograph taken from one of the kite balloons protecting Venice from attacks by the Austrian ships on the Adriatic.

Observation

1. It is not the purpose here to go into the details of observation. We will assume the following fundamentals:

The target can be seen from the balloon.
The ballon can be seen from the gun.

2. The observer in the balloon takes:
   (a) The height above ground by barometer.
   (b) The distance and direction of target (by position finder).
   (c) Correction data on subsequent firing.
3. The observer at the gun takes:
   (a) The distance and direction of balloon (by balloon sight).
   (b) The final corrected firing data (by mechanical plotter).
4. Observer at the base of cable takes:
   (a) Zero barometer reading.
   (b) Correcting, for windage and temperature.

5. All data should be taken as near simultaneously as possible. Whenever there is time, at least one check reading should be taken, and more than that if the height of the balloon has

substantially changed. A height reading is taken before and after each set of position readings.

6. For transmission of data from the balloon a combination wireless telephone and telegraph is immensely superior to ordinary telephone for all altitudes. Written messages and diagrams may be sent down the cable by means of a simple carrier device that can be readily installed.

7. If it is required to make quantitative observations in a very strong wind, some form of wind-shield is desirable, and for very cold weather an enclosure for the entire car and some form of heater. There should always be a means of quick exit, however.

8. If the balloon is full, watch the gas pressure during any considerable influence. (See Paragraph F. under "Ascension.")

9. Whenever the wind exceeds 40 miles per hour, the balloon should be watched carefully and brought down as soon as reasonably possible. It is not necessary always to carry an anemometer, however, as it is possible with experience to judge the wind quite closely. If the balloon is tilting to a considerable angle the tension of the cable can be relieved by shifting the basket forward. This is especially applicable when descending from a great height.

Free balloon just leaving for a trip with pilot instructor and students at an aeronautic station "somewhere in America."

Inflating a free balloon for instruction at a United States naval training aeronautic school "somewhere in America." A course in free ballooning is an essential part of a course in operating kite balloons.

10. It is best policy for the observer to wear his safety belt constantly so that he can jump into the parachute and jump at a moment's notice. When nearing the ground in a parachute drop, it is best to detach the parachute from the safety belt again. One can then let go immediately on striking the ground, and roll over unimpeded.

Use as Free Balloon

1. The ordinary position of the appendix in the air is shown by the dotted lines in Fig. B (balloon captive).

2. If the balloon breaks away and it is wished to stay with it, the first thing to do is to pull open the appendix by pulling on the rip-cord. This is one step further than is shown by the full lines in Fig. 1. There are then three more safety cords on the inside which guard against accidentally pulling open the rip panel. At the same time break the string on the valve cord gland and pull the valve open hard. (A man's entire might on the valve cord is not too much.) Hold it open until the balloon just starts down.

Standard hydrogen gas balloon cylinders designed to contain 200 cubic feet of hydrogen at 1800 pounds per square inch.

3. When nearing the ground check the descent if necessary by throwing out the ballast and dispensable equipment. Break the remaining safety strings on the rip-cord. Pull out the rip-panel all the way when about 50 feet from the ground.

4. The above operations cannot be handled properly without some previous free-balloon experience.

5. There are no extra instruments required, but a short piece of very light silk ribbon will be found convenient for gaging the vertical motion through the air.

Inspection and Repair

1. *Ropes.*—All ropes must be systematically inspected every day of actual use and worn parts replaced. When the main cable starts to wear, it may be kept in use by reversing, end for end.

2. *Leakage.*—It should not be necessary to replenish the gas to the extent of more than 300 cubic feet per day average when new, or 600 cubic feet per day at any time. If the leakage exceeds this (with proper allowance for temperature change and valve operation) the first thing to examine is the valve and its connections. See that it seats absolutely tight and is set open at the proper time. Examine the staffing boxes, the appendix, manhole and valve gland, which may also be sources of serious leakage if not properly secured. One hole 1/16 inches in diameter will let out more gas than escapes by diffusion through the entire fabric of the balloon. The greatest care must also be taken to prevent air getting in. Gas can always be replaced, but the only way to get air out is to deflate. With good management this should not be necessary for at least two months.

3. *Gas Replenishment.*—This should be done at some regular time every day. Economy in gas is largely dependent on the judgment that is used in putting it in. There is rarely any need of filling the balloon full on the ground. Allow about 1000 cubic feet of air in the ballonet for each 1000 feet of altitude to be attained. (The total capacity of the ballonet is 5000 cubic feet.) An allowance for expansion

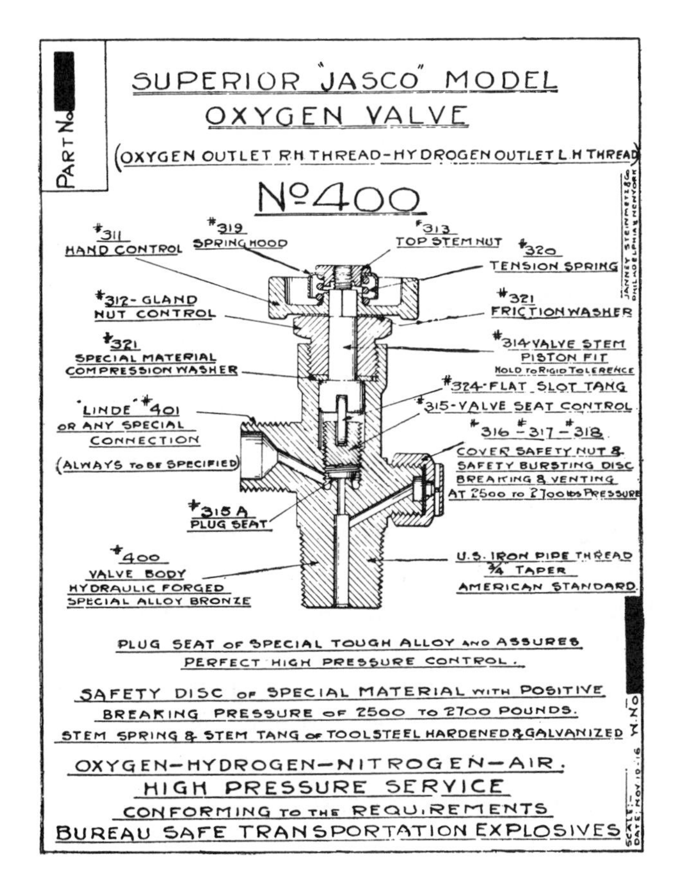

from the heat can be made if the gas is put in during the night (maximum 2000 cubic feet of air). A small cord running down through the ballonet serves as an air gage. This should be calibrated for some particular set of conditions.

4. *Fabric.*—The balloon fabric should be examined frequently and carefully for holes and deep scratches, especially where it comes in contact with the ground and where the ropes are liable to rub. If any are discovered, cut a patch to fit, wash both surfaces with gasoline on a cloth. Apply two light coats of cement (C35 tire cement will do if there is no balloon cement available) both to the balloon and to the patch, allow each to dry about five minutes, press down the patch and roll it down hard against some hard smooth surface. Especial care should be taken to make the edges tight. To detect pin holes and other small leaks inflate about one quarter full of air and get inside. Examine the fabric, block by block; by this means small holes are readily seen against the light. This should be done after every deflation.

5. *To Reinsert the Rip-Panel.*—Get two quarts of Goodyear balloon cement. Clean the old cement off of the edges with gasoline. Allow the surface to dry, then apply a coat of cement to each surface with a brush. Allow to dry five minutes and apply another coat. Let dry for five minutes more before sticking together. During the process keep the cemented surface out of the wind and sun as much as possible. Start at the bottom of the panel and work up toward the top along one side, being careful not to scratch the top edge out of place. Roll down hard. The other side of the panel in the same way. Dry the panel at the top and see that it pulls off properly. Cement up tight again. When finished there should be a 2½ inches lap cemented all around the panel, except for the extreme end, where it is a little wider and brought up a blunt point about 1 inch below the stick. Cover the seems on the outside with two coats of cement 1¾ inches wide. Lay down our specially prepared taping over this and roll it down hard. Allow to stand for several hours before rolling up.

6. *To Clean.*—If any oil gets on the surface, rub at once with gasoline. Ordinary dirt may be removed with water. If it is wished to paint the balloon, a special paint must be used, which will not injure the rubber.

Packing Up and Carrying

1. *Deflation.*—Pull gas manhole wide open and let the balloon rise slightly at the rear.

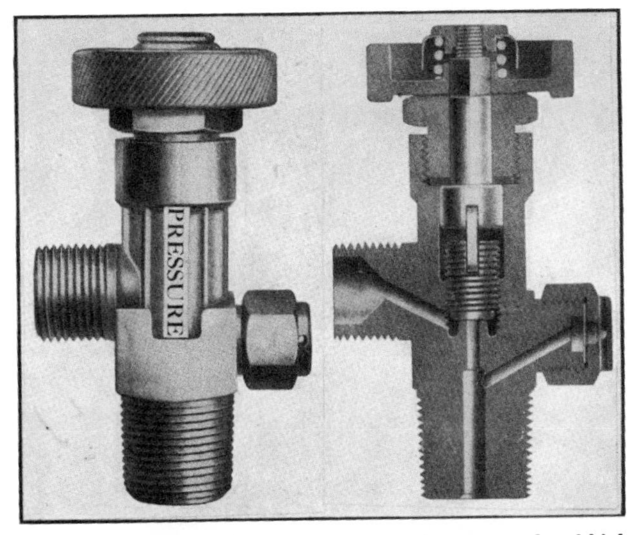

The packing nut, spring nut, stem and plug are made of high-grade brass. Stem spring and stem tang are tool steel, hardened and galvanized, thus preventing any possible corrosion. Plug seat is a special, soft, tough alloy and assures a perfect seat.

Any small quantity of gas that is trapped at the front can be let out through the valve. (Never use the rip-panel except in an emergency.) Open the air manhole as soon as it gets within reach. After the air is expelled there may be a residue of gas against the ballonet diaphragm at the rear. This may be worked back through the air manhole. Be very careful to see that it is securely closed again before leaving.

2. *Folding.*—Remove the valve carefully and put it away. Tie the valve cords to the loop inside. Be careful that there is nothing on the balloon that will cut the fabric. If the rip-panel has been used, pull off the fold tape from the edge before folding. Lay balloon with the rip-panel up. Pull the bottom fabric smooth so that both sides of the suspension are near the outer edge. See that the air funnel is laid flat and as far toward the front as possible. Grasp the balloon at each side and fold one third of the way across. This will bring the suspension patches on top and all loose ropes can be pulled over onto the top of the balloon.

3. *Stowage.*—Be sure the fabric and cordage is dry before packing. Keep covered in a dark room or den of moderate temperature. Sunlight has a destructive influence and the balloon should not be left in the sun any more than necessary.

4. *To Fold Parachute.*—Place the top ring of parachute as smoothly as possible in the bottom of the case. Pass a light breakable cord directly across the top of this ring, from one side of the case to the other, so that it will hold the ring from coming out and also from twisting. Hang the case up by its handle from some high point, if possible so the entire fabric of parachute will hang clear. Pull all the opening parachutes free, pull all the ropes into the center, and the fabric in pleats to the outside. Place a piece of light cardboard in between the ropes at a point about two feet below the bottom of the cloth. This cardboard is notched at the edge to receive the ropes and serves to keep them from getting tangled together. Now starting at the top of the parachute pull it down evenly on all sides, and tie it every six feet with a short piece of cord. When you get down within six feet of the bottom bring the small opening parachute up to the outside and tie another cord around near the top of the opening parachute.

The United States Navy has not, at date of writing, issued regulations for tests to be passed by kite-balloon pilots. Those issued by the Army may be adopted. These are printed in the "Textbook of Military Aeronautics."

CHAPTER XXXII

EVOLUTION OF THE AERO MOTOR

The motor is the principal factor in aviation. The aeroplane is like a kite rising when moved against the air, but being heavier than a kite must come down as soon as motion ceases.

The ideal aero motor must be light, so it can be easily secured to the light frame of the aeroplane and will use little of the aeroplane's carrying capacity; powerful, so it can develop a certain speed and afford a large carrying capacity; reliable, so it will give an even speed and not stop during flights and force the aeroplane down; and economical in regard to fuel, thus avoiding the necessity of carrying large loads of gasoline or oil for long-distance flights. Having these qualities, it does not matter if it be rotary, i.e., having the cylinders disposed radially; star shape, rotating with the crank case around a stationary crankshaft; or reciprocating, i.e., having the cylinders disposed vertically, V shape, horizontally, and fan shape, remaining stationary while the crankshaft rotates. In the first the propeller is fastened to the crank case which revolves with the cylinders, in the second case it is fastened to the crankshaft.

The aero motor is practically but twelve years old. In the period from 1900 to 1905 there were no engines adaptable for use in flying machines on the market; those available at the time being bulky and weighing over twenty pounds per horse-power. It was this lack of a suitable motor that kept the early experimenters of flying machines to the ground, while the use of a good motor enabled the Wrights to make the first flight in 1903. But the Wrights did not make public their invention and it was not until 1906, when the Antoinette motor was developed to 50 horse-power in France that flights could be made in Europe, the Antoinette being used in 1906 by Santos-Dumont to make the first aeroplane flight ever made in public. Since then the evolution of motors has been constant. Nearly all the automobile and motor-boat motors have been put through a general rehauling to make them lighter, strong, more compact, and reliable, to suit the peculiar requirements of aircraft.

By July, 1909, nine different aero motors were used, rating up to 60 horse-power, their weight per horse-power ranging from five to eight pounds, as follows:

Motor	Lbs. per horse-power	Users
Antoinette .	5½	Used by Latham in all his flights in 1909
Anzani	6	Used by Bleriot in Cross-Channel flight in 1909
Curtiss	7	Used by Curtiss in Gordon-Bennett race 1909
E. N. V. ...	6½	Used by Bleriot in many flights
Gnome	6½	Used by Farman in record of 4:6:25 and others
R. E. P. ...	6	Used by Robert Esnault Pelterie
Renault	8	Just put on the market in 1909
Vivinus ...	8	Used by Sommer in record of 2:27:15 in 1909
Wright ...	7	Used by Wright Brothers in all their flights.

A year later the number of motors had increased to over twenty, ranging in horse-power as high as one hundred; a 100 horse-power Antoinette being used in Latham's monoplane, and 100 horse-power Gnome motor being used in the Bleriot monoplanes in the Gordon-Bennett race of 1910. The weight of these, as well as with the rest of those already mentioned, had been reduced from one to two pounds per horse-power, and they still lead par excellence.

By 1911, the number of motors had increased to over fifty, about thirty being European and twenty American. The Gnome was still in the lead, but it had close competitors in the R. E. P., Nieuport, Green, and Hall Scott engines.

During the period from 1912–14 aero motors developed rapidly, largely as the result of the impetus given in the nature of cash prizes offered by various organizations, especially in Germany, where the Kaiser offered a prize of 50,000 marks, and in Great Britain, Italy, France, and Russia.

The advent of the Mercedes in 1914 made possible a number of remarkable records. Reinhold Boehm, a German aviator for instance, on July 11 made a flight of 24 hours and 12 minutes without stopping. He flew an Albatross biplane equipped with a 6 cylinder, 75 horse-power Mercedes motor. On July 14, Heinrich Oelrich rose to a height of 26,246 feet, flying in an Albatross biplane equipped with a 100 horse-power Mercedes motor.

The Development of Aero Motors in the Great War

Germany supplied all the aero motors for the Central Powers in the Great War. The German aero motors are for the most part Mercedes, and Benz, 100–150 horse-power, Argus 150 horse-power.

The status and history of the development of the European aero motors used by the Allies may be gathered from the following quotations from the official report of the investigation of the British Royal Flying Corps rendered at the close of 1916. Italy has been using largely the Fiat and the Isotta Fraschini.

"The position at the outbreak of war was that the engines available did not exceed 80 horse-power. Soon afterwards there was the Canton-Unne (Salmson) of 140 horse-power. All these have since been discarded. The Royal Flying Corps also got within the first few months the 90 horse-power R. A. F. and the 120 horse-power Beardmore—the 90 horse-power R. A. F. in by far the largest numbers.

The Wright Brothers' 1906 type of aero motor which was a marvel at the time.

Later the Royal Flying Corps obtained the 100 horse-power Monosoupape and 110 horse-power Le Rhone, the latter in small quantities. Later still, the 110 horse-power Clerget, but substantially the highest horse-power engine which the Royal Flying Corps had in quantity for many months was the 90 horse-power R. A. F. Quite recently it has had the 140 horse-power R. A. F. the 160 horse-power Beardmore (originally known as Austro-Daimler, and now sometimes as the Austro-Daimler-Beardmore), and the 250 horse-power Rolls-Royce. So far only a few of these higher-powered engines have been delivered; but the output is increasing, and there are now other high-powered engines in sight. The Royal Flying Corps has, in effect, been carrying on with engines the bulk of which did not exceed 90 horse-power, together with a few very efficient 100 to 120 horse-power engines.

"The state of affairs thus disclosed was obviously unsatisfactory until recently, especially in view of the fact that the Germans have had from the first engines of considerably higher power, notably the Mercedes.

"In order to understand the position, we must go back to the spring of 1914. There was then held a naval and military engine competition for engines of from 90 horse-power to 200 horse-power, with a prize of £5000 for the best engine. There were 67 entries. Of these 23 were of engines from 125 horse-power to 200 horse-power. There was only one of the latter. Only nine engines came through the test. The prize was won by the 100 horse-power water-cooled Green engine. The highest-powered engine to come through was one of 120 horse-power.

"The R. A. F. before the war was designing a 200 horse-power, water-cooled engine, and had proceeded some way with the drawings by August, 1914. General Henderson, who was one of the judges at the engine competition, was of the opinion that high-powered engines would be required. He hoped, having regard to the entries for the competition and the high standing of some of the competitors, that private firms would proceed to develop and perfect high-powered engines, and he stopped the R. A. F.

The famous six-cylinder Mercedes Engine which has meant so much to German aviation.

designs and handed the drawings over to the Rolls-Royce Company and Messrs. Napier. The former declined to proceed on the R. A. F. lines, but designed independently the 250 horse-power Rolls-Royce engine, which is just beginning to be delivered. The Napier Company proceeded with the R. A. F. designs in collaboration with the R. A. F., and their joint efforts have produced a 200 horse-power engine, which is now being tested.

"The alleged delay in ordering better engines occurred in respect of the 110 horse-power Clerget, the 110 horse-power Le Rhone, and the 200 horse-power Hispano-Suiza engines.

"All these engines are admittedly good, and all have now been ordered in considerable quantities.

"There was, we think, no undue delay in ordering the 110 horse-power Clerget engine. There was very considerable delay in the case of the 110 horse-power Le Rhone engine, and slight delay in the case of the 200 horse-power Hispano-Suiza.

"We do not think that the delay in ordering the Le Rhone and the Hispano-Suiza engines was attributable to the fact that the directorate was trusting to the R. A. F. to produce equivalent or better engines. The only engine of its own which the R. A. F. was engaged upon was the 140 horse-power R. A. F., which could not have taken the place of either the 110 horse-power Le Rhone or the 200 horse-power Hispano-Suiza.

"We are, however, not so much concerned to see whether the rhetorical charge of 'too blind faith' is made out as to see whether any blame attaches to the directorate for the delay and whether the Royal Flying Corps suffered from it.

"In the case of the Hispano-Suiza engine, negotiations for its production were entered upon promptly and a draft agreement was sent to the company on November 3, 1915. The answer miscarried, and the matter dropped until February 22, 1916, when the directorate reopened the negotiations and carried them without delay to a successful issue. We think the four months' delay ought not to have occurred, although in this case no real harm accrued. The Hispano-Suiza engine was originally one of 150 horse-power. They were experimenting with one of 200 horse-power. The latter was bought

General Vehicle Company's Gnome motor.

for the Royal Flying Corps, and this engine was not in a sufficiently advanced stage to have been procured earlier.

"The history of the 110 Le Rhone engine is different. This engine was at first, and in 1915, procured through the French Government, but the French wanted it for themselves and were unable to continue to supply the Royal Flying Corps. It is a difficult engine to build, and a builder had to be found. This took time. Another fact was that it was doubtful whether both the 110 Clerget and the 110 Le Rhone would be wanted and some time was consumed in comparing the two engines. In the end the Le Rhone was ordered, but after a lapse of some twelve months or more from the time when there was difficulty in procuring them from the French Government. We think this delay was too long and that it would have been an advantage to the Royal Flying Corps to have been in possession of this engine in quantities earlier.

50–60 horse-power Clerget revolving engine showing the gearing by which the spider in Fig. 13 is rotated.

"It is the fact that the R. A. F.-Napier 200 horse-power engine has been ordered in large quantities and that these orders were placed before the engine had been proved. General Henderson explained to us that in war time one must sometimes gamble on an engine and trust to luck. This engine was being designed simultaneously with the 250 horse-power Rolls-Royce, upon which the directorate did not gamble. We think the reason why the R. A. F.-Napier engine was selected for the gamble was because it was—at any rate, partly of R. A. F. design, and that this is an instance in which great reliance has been placed on the R. A. F."

160 horse-power eighteen-cylinder Le Rhone engine. Both valves are operated by one rocker, and a "push-and-pull" rod. Cast iron liners are shrunk into the cylinders.

American Aero Motors

The development of American aero motors has been held back by the smallness of the demand at home, and the fact that the Allies would only place orders with concerns having facilities for large productions, and these extensive facilities could only be developed through the placing of large orders.

Therefore up to the close of 1916 only four types of motors were in use on United States Army and Navy aeroplanes, and only these four types could be considered as manufactured products, the Curtiss, the Hall Scott, the Sturtevant, and the Thomas.

A number of other very promising American motors were developed during 1915–16. American motors can be said to be equal to-day

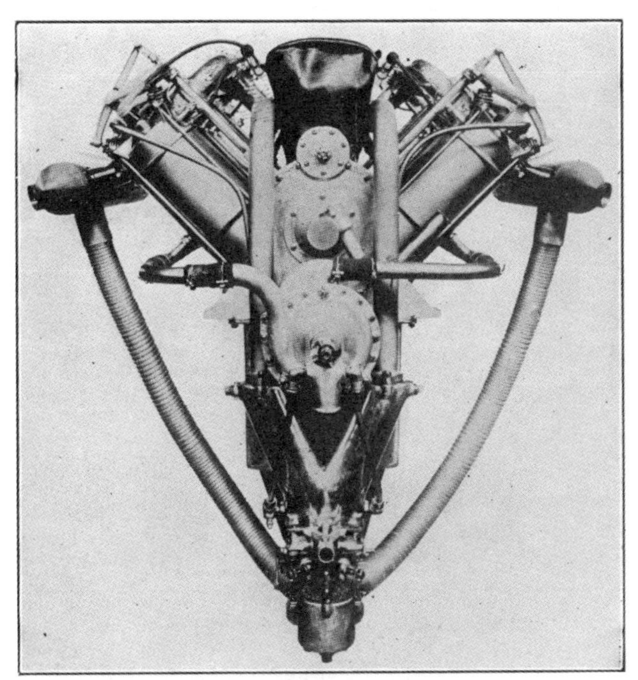

Curtiss V-2, 200 horse-power motor.

When the war was well under way the American motor makers suddenly found themselves in a very embarrassing position. The German Government refused to allow the Krupp works to export any of their product. For years American manufacturers had depended on the Krupp works for certain brands of steel, and suddenly the supply was cut off. The Krupp works had contracted to supply this country with large quantities of this high-power steel, and now it is known that this was really done for the purpose of preventing the American manufacturers from attempting to make it.

The motor makers were facing the ruin of their business when they found that the small stock in hand was the only steel that they had to go on with, and they issued calls for help to the various big steel mills. The result was that practically every one of the greater plants set to work to duplicate, if not better, the German product. Some of them worked for ten months, others for the last six, and after numberless experiments, at last six of the great mills succeeded in turning out an alloy that is, if anything, superior to the best that ever was exported to this country from the Krupp works.

Front end of the new A-5 Hall-Scott engine, showing the small head resistance offered by this type of engine.

to the best European products. The crucial period for American motors—and with European motors as well—was 1915, where large orders for motors were placed, but no Krupp steel was obtainable in manufacturing them.

The Krupp supremacy in the manufacture of certain types of steel, particularly the alloys used in the manufacture of crankshafts and other vital parts of motors, where the greatest strength, combined with the least weight, is desired, was maintained in past years, and, until the beginning of the war, by economic competition made possible by the fact that the Krupp people were subsidized by the German Government and could, therefore, underbid other steel concerns who had to depend on straight business for their existence.

The Curtiss twelve.

Until the fall of 1915 the crankshaft and

other high-grade steel parts turned out by the American steel mills were of very uncertain quality. Only about one piece in ten really lived up to the requirements, and the motor makers were having a serious time with it. Now, however, they can be sure of what they are using. Aviation motors are like the high-priced watches. One one-thousandth of an inch makes all the difference between a motor and a piece of junk, and steel that will stand the terrific strain is an absolute necessity.

The New Sturtevant eight-cylinder "Aluminum" Engine.

The accompanying table gives the characteristics of American motors being manufactured or developed at date of writing.

Thomas aeromotor, showing self-starter, double magnetos, Tachometer drive, stabilizer drive, gasoline and oil cooling circulating pumps.

Specifications for Aeroplane Motors Issued by the Navy Department

The office of Naval Aeronautics, Department of the Navy, issued the following general specifications for motors for aëroplanes on which bids were opened at 10 A.M., September 14, 1915.

Bids will be considered on motors differing in detail from those specified if complete specifications and drawings are submitted and if, in the opinion of the Government, the motors will give equal service to those specified.

All parts to be of the best material and workmanship. Engine to deliver not less than 100 brake horse-power for Item 1; 120 brake horse-power for Item 2; 140 brake horse-power for Item 3; 160 brake horse-power for Item 4; with muffler attached. Items

Sturtevant 140 horse-power motor.

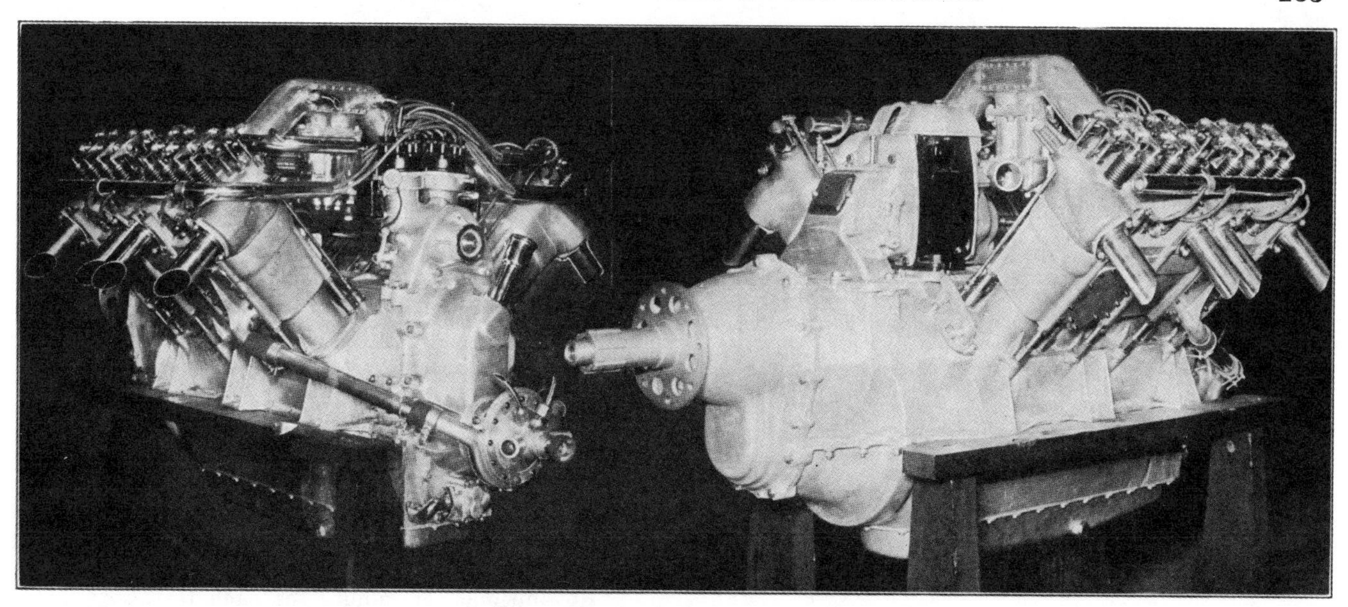

Aeromarine 8-cylinder aviation motor.

1, 2, and 3 will drive propellers not greater than 8 feet in diameter. They shall be well balanced and produce no excessive vibration at any power. To be capable of being throttled down to 20 per cent. of the revolutions per minute for full power. The weight of the engine complete, with ignition system, magnetos, carburetors, pumps, radiator, cooling water and propeller, not to exceed 5 pounds per brake horse-power. Engine to be fitted with some type of compression release as a means of stopping it. To be fitted with a practical means of starting from pilot's seat when installed in an aeroplane. All moving parts not lubricated by a splash or forced lubrication system to be readily accessible for inspection, adjustment, and oiling. Ready means shall be provided for checking and making adjustment to the timing of the engine. All parts to be machined all over where possible. Fillets to be of ample radius to insure strength. To have an accurate and positive lubricating system which will insure a uniform consumption of lubricating oil proportional to the speed of the engine. All parts subject to corrosion to be protected from the effects of salt water. To be fitted with an approved attachment for obtaining the revolutions per minute. To be provided with means for preventing fire in case the engine is turned upside down. A hand throttle lever and connections to carburetor to be provided that can be applied for convenient operation by the pilot. This lever to be designed with a positive means of retaining it at the throttle adjustment desired by the pilot. All

A recent Fiat motor.

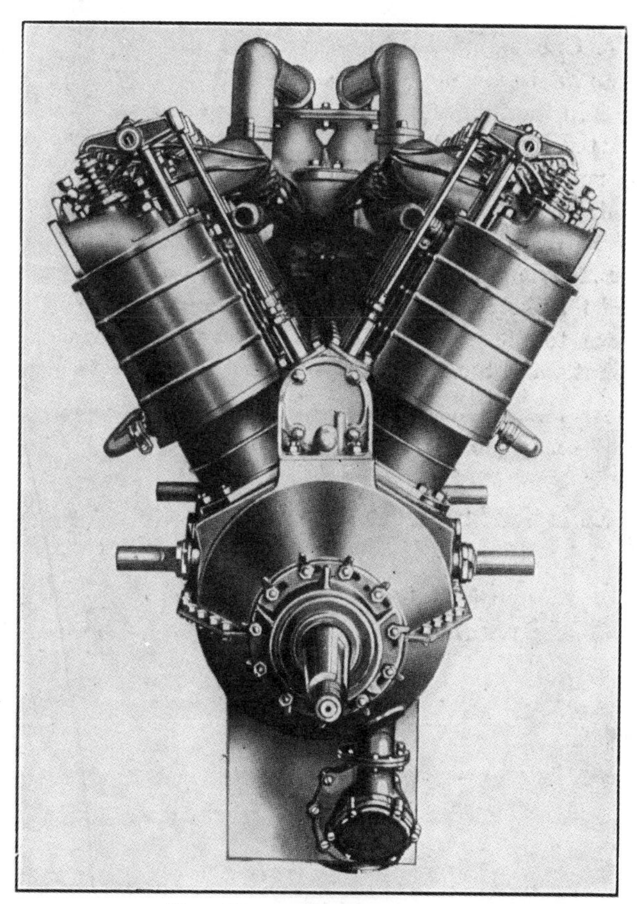

Van Blerck aviation motor.

Atwood 12-cylinder motor.

bolts and screws without any exception to be provided with an approved positive means for preventing backing out due to vibration. No soft solder to be used in any part of the power plant.

ACCEPTANCE TESTS

1. To determine power revolution curve, and fuel and lubricating oil economy.

2. To be run for five hours at full power, using one supply of lubricating oil.

3. Immediately upon completion of second test, to run a similar full-power run for five hours.

4. One motor to be selected at random from lot to make eight separate and distinct runs in the following manner: Engine to be started by the use of the starting gear: to be run for the first five minutes at 95 per cent. of full power while inclined at an angle of 15 degrees to represent climbing; then to be run for 22 minutes at loads varying between 75 per cent. and 95 per cent. of full power; then to be throttled down to 20 per cent. of full power revolutions per minute, and the engine to be tipped 0 degrees to simulate gliding, and to run in this position for two minutes; then to be leveled and run at full power for one minute. Engine to be then stopped, and after five minutes a similar test to be repeated, and this shall be continued until eight separate runs have been made.

Wisconsin aviation motor.

5. Motor to be run at full power for one-half hour under conditions approximating operations in the aeroplane in a heavy rainstorm.

NOTE.—No adjustments or alterations shall be made during any of the tests or intermissions. The failure to complete any test in a satisfactory manner shall require that that set of tests be made again.

The engine shall be capsized while running to demonstrate the means provided to prevent a fire.

During all tests an accurate record shall be kept of the following:

Total number of revolutions for each test.
Revolutions per minute for each test.
Lubricating-oil consumption for each test.

Trebert Revolving Engine.

Gasoline consumption for each test.
A record of thrusts for tests 2 and 3.

At the successful completion of these tests the engines shall be broken down and inspected. No parts shall show undue wear or deterioration, and the weights and balance of the separate parts shall be in conformity with the specifications.

SPECIFICATIONS FOR PARTS

Lower Crank Case.—To have an oil capacity for five hours running at full power, or fitted with an efficient automatic device to keep constant supply of oil at pump suction. To be fitted with a sight gage to show the height of the lubricating oil. Oil system to have a pressure gage capable of being mounted on an instrument board, and means to turn this system on and off at will. To be fitted with drain plugs. If possible, to be capable of removing from engine with engine installed on engine beds. To be fitted to prevent oil leaks at joints and connections. All fillets to be of ample radius to insure strength.

Upper Crank Case.—To be fitted with an efficient system of relieving pressure in the crank case. Wherever possible where webs are fitted which cannot be machined they should be of uniform thickness. Wherever attachments are made to the upper crank case they should be sufficiently reinforced to insure strength. All fillets to be of ample radius to insure strength.

Cylinders.—To be fitted to take two spark plugs. To be cooled to prevent excessive heating of valves. Cylinders to be counterbored to prevent piston wearing a shoulder at either end of the stroke. They shall be secured to the crank case so as to prevent oil leaking. To be fitted with an efficient detachable muffling system. To be attached to crank case with sufficient safety factors to insure against cylinders blowing off. Cylinder holding down bolts to be subjected to a uniform load on all members. Cylinders to be machined on the outside as well as inside.

Detroit gas turbine.

Spark plugs to be accessible and removable without removing any engine parts. Spark-plug points to extend into the combustion chamber. Any pockets in the head of the cylinder where burnt gas might collect to be avoided. Water jacket, if used, to be of a noncorrosive metal.

Pistons.—To be balanced and finished so that homologous parts shall be of uniform thickness. Sufficient thickness shall remain after the piston ring grooves have been cut to insure ample strength. Means shall be taken to prevent an excess of oil entering the combustion space. Wrist-pin bearings to be brushed with bronze. Means to be provided to prevent wrist pin or bushings from touching the cylinder walls. All pistons to be of the same weight and to balance at the same point.

Piston Rings.—To be of the leak-proof type, made in two parts, and not less than ¼ inch wide.

Beecher horizontal motor.

Connecting Rods and Bearings.—They shall be machined all over and balanced with liners (I-beam cross section preferred), crank-pin bearing cap and crank-pin bolts and nuts in place. All connecting rods to be of the same weight and to balance at the same point. Crank-pin bolt heads to be countersunk into connecting rod to prevent turning. Ends of wrist pin to be beveled.

Crank Shaft and Main Bearings.—Crank shaft to be carefully balanced and to be machined all over. If the crank shaft is made with a flange, to which it is the intention to secure the propeller, this flange shall be at least ½ inch thick. Special attention should be given to the strength of the crank webs.

Cam Shaft.—To be machined all over and balanced. Cams to be case-hardened to prevent pounding out. To be fitted with some means to decrease the amount of wear and friction on the cams and on the end of the push rods.

Push Rods, Valve Gear, Rocker Arms, Etc.—To be machined all over. To have an accurate and positive means of opening and closing valves. Clearance between valve stems and actuating mechanism of both intake and exhaust valves to be capable of adjustment. All parts to have sufficient strength factor to insure against distortion, such as bending of rocker arms and push rods, if employed. Rocker arms to be machined all over.

The 200 horse-power Maybach.

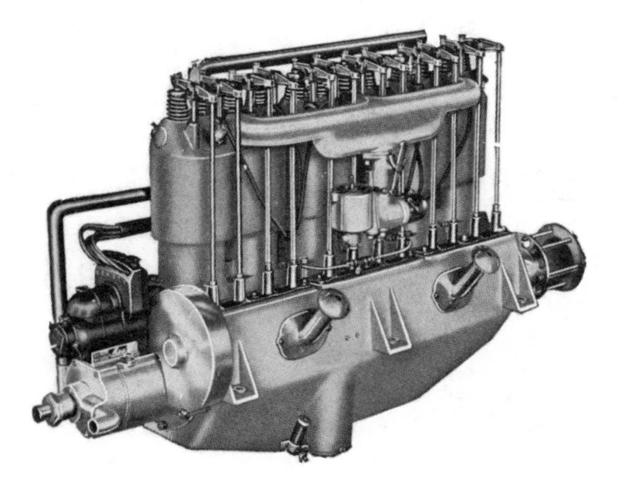

Aviation motor of the Maximotor Company.

Intake Manifold and Carburetor.—To be smoothly finished on the inside. No bolts or lugs to pass through or extend into the inside of the intake manifold. Sharp turns or bends to be avoided, and the distance from carburetor to any intake port to be approximately the same. Means to be provided for heating the mixture after entering the intake manifold. The use of two carburetors is preferred. The carburetor to be so lateced as to provide for the minimum amount of vibration (i.e., as near the line of the crank shaft as possible). The face of the cylinder to which the intake manifold is secured to correspond exactly to the cross section of the air-manifold flange. Means shall be provided for making accurate and permanent adjustments to the carburetor, and means to prevent these adjustments changing due to vibration. Means for adjustments from pilot's seat shall be provided to care for changes in altitude.

Magnetos.—Two magnetos to be used, either one capable of developing 90 per cent. of full power. To be waterproof. Positive means shall be provided for securing magneto leads to the distributor.

250 horse-power Packard motor.

Leads to spark plugs to be waterproof.

Angle of advance to be controlled from pilot's seat.

Circulating Pump.—To have a capacity sufficient to cool the engine when running at full power in air of 72° F.

A quotation (separate) on a list of selected spares to be included in the proposal.

All bids must contain the bidder's guaranty in regard to replacement of defective parts.

The following data and plans, in duplicate, must be submitted with all bids:

(a) Motor particulars: Make, model, cycle, bore and stroke, piston displacement, actual horsepower, speed, number and arangement of cylinders, cooling, lubrication, types of bearings, types of valves, method of starting, make of spark plug.

Carburetor: Make, number.

Gyro rotary aviation motor.

Magneto: Make, number.

Radiator: Make, number disposition.

Propeller: Make, diameter, efficiency.

(b) Material of the following parts: Cylinders, pistons, jackets, crank shaft, cam shaft, connecting rods, valves, all bearings, crank case.

(c) A weight schedule of the following: Power plant complete: Motor, propeller, transmission, radiator, cooling water, piping, etc.

(d) General arrangement plans: Plan, profile, and elevation. Curves of brake horse-power and revolutions per minute, also oil and gas consumption in pounds per brake horse-power per hour at full power (certified).

(e) Photographs of motor (five views—top, both sides, both ends).

Christofferson 135 horse-power motor.

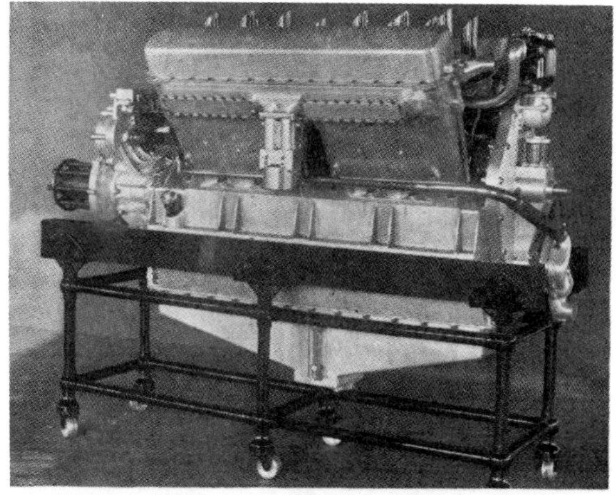

Knox 12-cylinder motor.

(f) A price list covering replacements for one year after delivery of the motor.

Alternate Bids.—The contractor will meet all detailed requirements of the original specifications, except in so far as they are modified by his bid.

Proposals must state distinctly whether the motors proposed are (1) as called for by the specifications, (2) as by modified specifications, (3) as by substitute specifications.

Each of the duplicate proposals must also be accompanied by full and detailed descriptions and drawings or illustrations showing the dètails of the motors it is proposed to furnish.

The name of the maker of the motors, the net weight, and the space occupied must be stated. Failure to comply with this requirement will render the bid liable to rejection.

If the motors proposed vary in any part from the specifications, special mention must be made of such points apart from the general description. When variations are not stated clearly and in detail, the contractor will be required to meet all details of the original specifications when the motors are inspected for final acceptance.

Inspection to be made at place of manufacture unless otherwise directed by the Bureau of Steam Engineering. Prompt inspection can be arranged if bidders will state on the blank lines below the name of the manufacturer as well as the place where the material will be manufactured, giving the exact address.

When the bidder and the manufacturer are the same, the exact address of the manufacturing establishment should be given, and not the office address.

If this information cannot be furnished in his bid, the contractor must, within five days after receipt of notice of award, furnish the Bureau of Steam Engineering with the foregoing information.

All handling of material necessary for purposes of inspection shall be performed and all test specimens necessary for the determination of the qualities of material used shall be prepared and tested at the expense of the contractor.

If contract is sublet, the contractor and subcontractor shall furnish the inspector representing the bureau concerned in their district quadruplicate copies

Martin 12-cylinder motor.

General Ordnance Company's 200 horse-power motor.

End view Duesenberg 12-cylinder aero engine.

End view Orlo model 0-8.

of all orders placed with manufacturers for materials, stating when possible the purpose of each item ordered and the specifications for the same. In all cases these orders shall contain the number of the original contract of which these constitute suborders.

In connection with the inspection of the material, if incorrect information is given, thereby causing one or more useless trips by the inspectors, the Government reserves the right to charge the expense of such useless trips to the contractor, and further inspection at the mills may be denied the contractor, at the option of the bureau.

THE SPECIAL ATTENTION OF BIDDERS IS INVITED TO THE FOLLOWING CONDITIONS:

In accordance with the requirements of the Naval Appropriation Act approved June 30, 1914, every bidder under this class, to receive consideration for his bid, must agree, if called upon to do so, to furnish the affidavit given below from a responsible member of the firm bidding or officer of the corporation. In case the bid is submitted by an agent or representative of the manufacturer, this affidavit must be furnished by the bidder and also a separate affidavit by the manufacturer whose product the bidder proposes to furnish; and in the case of a corporation a certified copy of the record of the action of the board of directors, showing the appointment of the officer making the affidavit, shall be appended.

Williams 125 horse-power motor.

AFFIDAVIT

................, being duly sworn, deposes and says: That he is the ——— of ——— company, as evidenced by the accompanying certified copy of the record of the action of the board of directors, and as such it is his duty to know and he does know and is thoroughly familiar with the business arrangements,

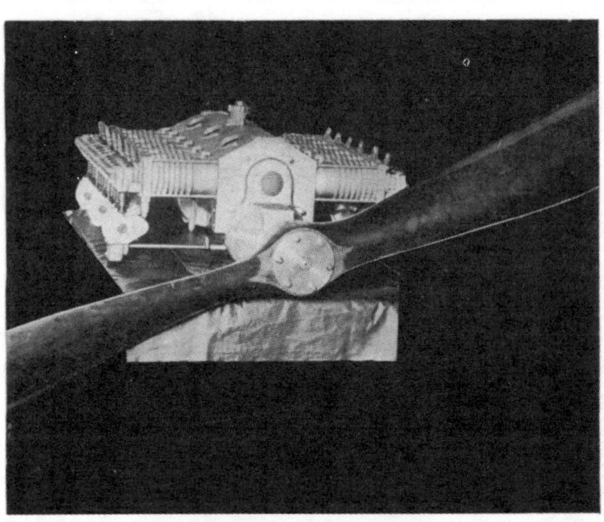

Ashmusen twelve-cylinder.

The Sterling-Sunbeam twelve.

and relations of said company; that he has carefully examined and is thoroughly familiar with the following provisions of the naval appropriation act approved June 30, 1914:

"That no part of any sum herein appropriated shall be expended for the purchase of structural steel, ship plates, armor, armament, or machinery from any persons, firms, or corporations who have combined or conspired to monopolize the interstate or foreign commerce or trade of the United States or the commerce or trade between the States and any Territory or the District of Columbia in any of the articles aforesaid"; that said company is not engaged in any such combination, agreement, conspiracy, or understanding as is prohibited by the above-quoted portion of said act; and that said company agrees to furnish whatever additional information the Navy Department may require to satisfy itself as to the status of the said company with respect to any combination, agreement, conspiracy, or understanding of the kind contemplated by the said act.

Sworn to and subscribed before me this day of

Notary Public.

hereby agree to furnish the above affidavit or affida-
I, we.
vits if called upon to do so by the Government.

....

....

NAME OF BIDDER.

The name of the bidder must be inserted in the above blank space. Failure to comply with this requirement will render the bid informal.

The contract awarded under this class will also contain the following provision:

Hispano-Suiza Model A.

The Isotta-Fraschini aeronautic motor, used extensively by the Italian navy. 200 horse-power.

This contract having been awarded, conformably to restrictive provisions in the Naval Appropriation Act of June 30, 1914, upon the express understanding that the party of the first part has not combined or conspired to monopolize the interstate or foreign commerce or trade of the United States or the commerce or trade between the States and any Territory or the District of Columbia in structural steel, ship plates, armor, armament, or machinery, and the.........of said company having furnished the Secretary of the Navy with an affidavit to this effect, it is hereby further covenanted and agreed, and this contract is upon the express condition, that in case it be ascertained at any time after the signing hereof that false representations were made in said affidavit with respect to the requirements referred to above of said statute, this contract may be annulled in whole or in part by the Secretary of the Navy at his discretion.

TABLE OF CHARACTERISTICS OF THE LEADING AMERICAN AEROPLANE MOTORS

Maker	*Model*	*H.P.*	*Bore & Stroke inches*	*Rev. P.M.*	*Cylinders*	*Carburetor*	*Magneto*	*Oiling*	*Wght. in lbs.*	*Self Starter*
Aeromarine	A-G 6	100	4 5/16x5 ⅛	1400	6-Vertical	(2) Zenith	(2) Dixie	Force-splash	425	...
Aeromarine	D-12	...	4 5/16x5 ⅛	1400	12-V	(2) Zenith	(2) Dixie	Force-splash	750	...
Aeromarine	8	100	3 ½x5 ⅛	1400	8-V	(2) Zenith	(2) Dixie	Force-splash	450	"Delco" electric
Ashmusen	...	70	3 ¾x4 ½	900	8-Horizontal	Ashmusen	Dixie	Force-feed	240	...
Ashmusen	...	105	3 ¾x4 ½	900	12-Horizontal	Ashmusen	Dixie	Force feed	345	...
Atwood	M-1	140	3 ½x4 ½	2500	12-60° V	Longnemare	(2) Dixie	Force-splash	625	"Bijur" electric
Backus	†	...	...	...	...	...	...	...	...	...
Bates	8-B	110	4 ½x5	1350	8-90° V	Schebler	Dixie	Force	410	"Snap"
Brennan	...	...	4 x5	...	8-V	...	...	...	...	...
Brennan	...	...	4 ½x5	...	8-V	...	...	...	...	...
Christofferson	...	130	4 ¾x6	1400	6-Vertical	Double	Double	Force	460	...
Curtiss	O X	90	4 x5	1400	8-90° V	(2) Zenith	Berling	Force	365	...
Curtiss	O X X	100	4 ¼x5	1400	8-90° V	(2) Zenith	(2) Berling	Force	412	...
Curtiss	V-2	200	5 x7	1400	8-90° V	(2) Zenith	(2) Berling	Force	690	"Perfect"
Curtiss	V-4	300	5 x7	1400	12-90° V	(2) Zenith	(2) Berling	Force	1086	"Perfect"
Detroit	Rotary	100	...	...	Rotary	...	...	...	...	...
Duesenberg	...	70	4 ⅜x6	1500	4-Vertical	...	...	...	365	...
Duesenberg	...	300	4 ⅞x7	1400	12-60° V	...	...	...	1040	Starter
Emerson	Speed	60	5 x5	1250	4-Vertical	Panhard	...	Oil in gas; & splash	250	...
Emerson	Speed	90	5 x5	1250	6-Vertical	Panhard	...	...	350	...
Frederickson	5-A	70	4 ½x3 ¾	1000	5-Revolving	Special	Single	...	180	...
Frederickson	10-A	140	4 ½x3 ¾	1000	10-Revolving	Special	Single	...		...
Gen. Ordnance	L. M.	200	4.75x6.5	1800	8-V	Zenith	(2) Dixie	...	876	...
Gen. Vehicle	Gnome	100	4.33x5.9	1200	9-Revolving	...	...	...	272	...
Gyro	K	90	4 ½x6	1250	7-Revolving	Spray	Bosch	...	215	...
Gyro	L	110	4 ½x6	1200	9-Revolving	Spray	Bosch	...	270	...
Hall-Scott	A-5	130	5 x7	1300	6-Vertical	(2) Zenith	(2) Dixie	Force	525	...
Hall-Scott	A-7 a	100	5 ¼x7	1400	4-Vertical	Zenith	(2) Dixie	Force	400	"Perfect"
Harriman	...	100	5 x5	...	6-Vertical	...	...	...	...	...
Johnson	...	90	5 x4	1400	6-V	...	...	...	298	...
Johnson	...	180	5 x4	1400	12-V	...	...	...	598	...
Kemp	J-8	80	4 ½x4 ¾	1150	8-V	(2) Zenith	Mea	Force	380	...
Kessler	...	200	4 ¾x5	...	12-60° V	(2) Zenith	...	...	...	...
Knox	...	300	4 ¾x7	1600	7-Horizontal	(2) Zenith	...	...	1400	"Bijur" electric
Macomber	Revolving	...	...	...	Radial	...	...	...	...	...
Manly	...	53	5 x5 ½	950	4-Vertical	...	...	...	151	...
Maximotor	A-4	50-	4 ½x5	1350	...	Master or Zenith	Berling or Bosch	Force	200	...
Maximotor	B-4	70	5 x5 ¼	1350	4-Vertical	Master or Zenith	Berling or Bosch	...	260	...
Maximotor	A-6	90	4 ½x5	1350	6-Vertical	Master or Zenith	Berling or Bosch	...	340	...
Maximotor	A 8 V	120	4 ½x5	1350	8-V	Master or Zenith	Berling or Bosch	...	500	...
Muffley	...	...	...	...	...	...	...	...	...	...
N. J. Aeroplane Co.	...	85	...	1200	9-Revolving	...	...	...	245	...
Nilson-Miller	†	...	...	...	...	...	...	...	...	...
Oldfield	15-A	150	2 ⅞x4 ⅝	1500	10-Revolving	Lee Oldfield	Atwater-Kent	Force & oil in gas	225	...
Orlo	O-8	125	4 ½x6	1400	8-90° V	Zenith	(2) Dixie	Force	475	...
Packard	...	...	4 x6	1400	12-40° V	...	...	...	800	"Delco" electric
Rausenberger	C-12	160	4 ⅛x6	1400	12-60° V	(2) Zenith	(2) Dixie or Bosch	Force-splash	750	"Cristensen"
Roberts	6 X	100	5 x5	1400	6-Vertical	Rayfield	(2) Optional	Oil in gas	325	"Cristensen"
Roberts	6 X X	165	6 x6	1400	6-Vertical	Rayfield	(2) Optional	Oil in gas	650	"Cristensen"
Roberts	E 12	350	6 x6 ½	1400	12-60° V	Rayfield	(2) Optional	Oil in gas	990	"Cristensen"
Robinson	Radial	...	...	...	...	...	...	...	...	...
Sterling	Sunbeam Coataler	320	4.3x6 ¼	1200	12-60° V	Special	(4) Special	Force	1000	Air starter
Sturtevant	D-6	80	4 ½x4 ½	1500	6-Vertical	Zenith	Bosch or Dixie	Force	325	"Cristensen"
Sturtevant	5 A-8	140	4 x5 ½	1200	8-V 90°	Zenith	Bosch or Dixie	Force	514	"Cristensen"
Sturtevant	5-A	...	...	...	...	...	...	...	...	...
Thomas-Morse	Thomas 8	135	4 x5 ½	1200	8-90° V	Zenith	(2) Splitdorf	Force	625	"Cristensen"
Thomas-Morse	Thomas 88	150	4 ⅛x5 ½	1200	8-90° V	Zenith	(2) Splitdorf	Force	525	"Cristensen"
Tone	2 V 9	180	4 ½x6 ¼	1800	9-V	Zenith	(2) Delco	Force-splash	576	"Delco" electric
Trebert	Revolving	...	...	...	16-Horizontal	...	...	...	350	...
*Van Blerck	F 8	124	4 ½x5 ½	1400	8-V	...	...	...	...	...
*Van Blerck	F-12	185	4 ½x5 ½	1400	12-V	...	...	...	...	...
Wells Adams	...	135	4 ½x6	1350	8-V	...	(2) Bosch	Force	400	...
Williams	...	125	...	...	8-90° V	...	...	...	...	...
Wisconsin	V 6	140	5 x6 ½	1400	6-Vertical	Zenith	(2) Bosch	Force	600	Air starter
Wisconsin	V 12	280	5 x6 ½	1400	12-60° V	Zenith	(4) Bosch	Force	1000	Air starter
Wright-Martin	Hispano Suize—B	75	4.72x5.1	1450	4-Vertical	Clandel	(2) Splitdorf	{ Castor oil Force Feed	315	{ Reduction Gear and Magneto
Wright-Martin	Hispano Suize—A	150	4.72x5.1	1450	8-V	Clandel	(2) Splitdorf		445	
Wright-Martin	Martin 8200	190	4 ⅝x7	1400	8-90° V	Double	Double	Pressure & splash	484	...
Wright-Martin	Wright	60	4 ⅜x4 ½	1400	6-Vertical	(2) Zenith	Bosch or Mea	Splash	305	...

NOTE: Blank spaces indicate information not given by the manufacturer. *—Temporarily discontinued. †—Motors built to special specifications only.

CHAPTER XXXIII

AERONAUTICS IN RELATION TO NAVAL ARCHITECTURE

By Naval Constructor H. C. Richardson, U. S. N.

Reproduced through the Courtesy of the Bulletin of Naval Architects and Marine Engineers.

The most intimate point of contact between aeronautics and naval architecture exists in the design of floats for aeroplanes, and I shall confine this paper to that subject, although the question of stream-line forms and the pressure and flow of air on such forms and on flat and cambered surfaces is also closely related to naval architecture.

Early in 1911 work was started at the model basin at the Washington Navy Yard, with a view to the development of floats for aeroplanes. This work has been constantly followed up by making tests of model floats of all descriptions, and particularly with models of floats actually furnished, or proposed, for use with naval aeroplanes. In addition to this, series of models embodying variations of particular features were tried from time to time, covering a wide range of possible useful features, and, where such models showed sufficient merit, full-size construction has been carried on, and the full-size floats have been given tests in actual service. A detailed account of the complete series of tests would be too voluminous for reproduction, and would cover many defective types, so that I shall enter only into a general discussion of types and the requirements for floats for naval aeroplanes.

It appears well to point out at once the differences existing between the conditions of use of aeroplane floats, and those met in the use of ordinary displacement vessels, or vessels of the hydroplane type. The principal differences arise from the fact that as the square of the speed increases the lifting power of the wings becomes more and more important, and the load carried by the aeroplane float constantly diminishes until at the get-away speed no load is any longer carried by the float. At the higher speeds near the get-away conditions, the form of the bottom of the float becomes unusually important, due to a prominent suction effect which is present if any downwardly bowed lines of flow are presented.

It might readily be imagined at first thought that what proves the best type for hydroplane hulls would naturally be the best type for aeroplane floats, but the reason just mentioned may operate seriously against certain types of hydroplanes for aeroplane use, for the hydroplane floats carry a diminishing load, are consequently less deeply immersed, and, therefore, present in general a different set of lines to the flow of the water than is the case with hydroplanes. Besides this, a hydroplane is never projected suddenly at high velocity into the water with any lateral velocity. This, however, is frequently the case with aeroplane floats, which must be able to make skidding landings, or get-aways, across the wind without danger.

The most prominent effect of the special conditions of operation of aeroplane floats is that of the greatly reduced resistance at high speeds. Plate A indicates decided differences existing under ordinary displacement conditions, and the conditions governing the use of aeroplane floats. On this plate curve *A* represents the resistance of a simple hull of constant displacement. Curve *B* represents the resistance of the same hull run at displacements corresponding to speeds. Curve *C* is the same hull as that for curve *B*, but is fitted with submerged blades; while curve *D* shows a typical resistance of a twin step float run at displacements corresponding to speed. The great advantage, so far as resistance is concerned, due to the lift of the wings, is clearly apparent from an inspection of these curves, and the addition of submerged blades clearly increases this advantage so far as resistance is concerned. But the performance of such blades is decidedly tricky, and although they have been used in a few instances they are considered tricky and unsatisfactory and unsuited to use in rough water.

A matter of considerable interest is that two different states of flow may be encountered at the same speed, depending on the manner in which the speed is approached. Thus, an aeroplane getting under way will require full power to attain the planing condition at about 25 miles per hour. However, once planing speed has been attained, it is possible to ease off the power very decidedly and still maintain planing at a lower speed.

When using water blades to assist in planing, a critical condition appears, and, if the blades are heavily loaded, this causes a sudden failure of lift as the blades approach the surface. When this happens the flow suddenly breaks from the back of the blade, causing a sheet of spray to rise at about 60 degrees to the surface of the water, and this appears to be independent of the sharpness of the edge of the blade itself. Following this failure in lift, the float then settles quickly until the original flow is reestablished, when the blade again lifts as before and proceeds to follow the cycle of performance just described.

The suction effect referred to was discovered in an attempt to reduce frictional resistance of the float to a minimum by the use of a parallel middle body having semi-circular midship sections with ogival ends. This form behaved admirably at moderate speeds, but when the get-away condition was approached, the model, instead of planing, indicated a strong suction effect and proceeded to lift sheets of spray well clear of the surface. Finally, at the get-away condition, with the model counterweighted to a zero displacement and just in contact with the surface, the suction influence of the curved bow and buttock lines was sufficient to drag the float down into the water until the deck was flush with the original surface, and a complete glassy sheet of spray was lifted several feet clear of the surface of the basin. The resistance was abnormally high. This effect in modified form has been found in several types of floats and in several instances has produced failure of designs, or seriously affected their performance. The effect may be readily understood by suspending a spoon, holding the end of the handle lightly between the fingers, and then letting the bottom of the bowl of the spoon touch a stream from a hydrant.

Having considered the special conditions which affect aeroplane floats, as distinguished from those affecting displacement forms, let us now consider what is required of aeroplane floats.

When adrift the floats must provide buoyancy and have sufficient reserve of buoyancy to provide against loss due to a damaged compartment.

Sufficient water-line inertia and freeboard are required to provide initial and reserve stability in a strong wind in the open sea. This problem is unusually high position of the center of gravity and of the requirement of keeping the wings and propellers well clear of the surface, and of keeping the center of gravity close to the center of lift. Those inclinations produce

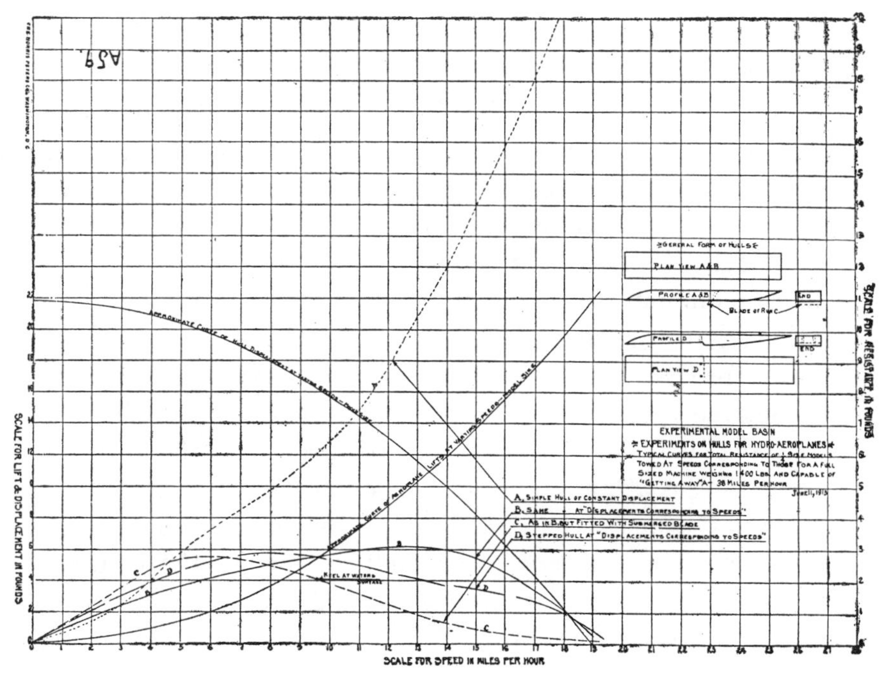

a rapid movement of the center of gravity, which must be exceeded sufficiently by the resultant change in the center of buoyancy. Sufficient initial stability must exist to provide against the force of the wind on the extended wings, which may be rolled to strong lifting angles.

When drifting it is desirable that the aeroplane should head into the wind. This is difficult to attain by control of form and may require the use of a sea anchor.

Under way at moderate speeds the aeroplane must steer readily in all directions and be able to manœuver in close quarters so as to get alongside the ship.

Water rudders are not always successful on such short floats as are usual, and are objectionable from handling the stowage points of view. If the water rudder is a part of the air rudder, action becomes confused in a cross wind, and sharp handling is necessary when the water rudder breaks clear of the surface.

Under way at high speed, particularly near the get-away speed or landing speed, the bow must have sufficient bearing power to prevent nosing under in a rough sea or badly judged landing. The form must also be such as to avoid undue pounding and the throwing of spray into the propellers.

In order to avoid the requirements of excessive power to drive the floats, planing should begin at moderate speeds—in the neighborhood of 25 miles an hour in a calm. Planing is readily attained by the use of large areas of planing bottom, but too rapid planing is not desirable, and a compromise is required because of the different water speeds involved when getting away in a calm or when heading up or down wind. If too much planing is present, the aeroplane may be tossed off rough water at less than flying speed, or before the air controls become effective. This always results in severe pounding and may cause the aeroplane to return to the surface in a dangerous attitude. If after planing is attained the attitude of the float relative to the surface can be modified, the wing can also be modified, and the actual get-away becomes controllable within limits. This latter feature is very desirable.

When getting away or landing, it is extremely desirable that the forces acting on the floats should be moderate and act with small moment arms about the center of gravity, so as to introduce moderate disturbing forces and require a minimum use of the air controls and maintain the proper attitude of the aeroplane.

In the air the floats should present a minimum resistance and interfere as little as possible with the flow of air to the supporting surfaces and the air controls. They also should not introduce disturbing influences on the center of pressure of the aeroplane, as this is also involved in the equilibrium of the aeroplane in flight.

All of the preceding qualities must be attained to as high a degree as practicable in floats, which at the same time must be rugged enough to stand severe punishment on the surface and still be of the highest possible construction. The solution is necessarily a compromise. Buoyancy and stability require forms and dimensions which conflict with the best aerodynamic forms and the requirements of moderate resistance and weight. Planing in flight also requires a form of bottom which does not readily meet stream-line requirements.

At the model basin at the Washington Navy Yard, a speed of over 15 knots is available; this admits of the use of one-ninth size models for a get-away speed of 45 miles per hour, and this size model is generally used.

The first condition to be provided is that of loading the model to the displacement corresponding to the speed. The assumption is made that the lift of the wings increases with the square of the speed. This is only approximately correct, but serves well for the purposes of comparison. It is not exact, for the attitude of the aeroplane is dependent on the amount of air control existing, and whether it is sufficient to overcome or modify the effect of the forces acting on the floats and thus obtain control of the angle at which the air surfaces act. As these effects cannot be anticipated, or the control be determined or applied, this assumption does not appear unreasonable, and the comparison of model and full-size performance appears to justify this and the other assumptions made.

The model is usually set at a trim of 3 degrees by the stern from the normal water line, as numerous experiments indicated this to be the average condition. A direct comparison was made of a one-quarter size float under the artificial conditions assumed and the same float fitted to a model aeroplane to the same scale. Eycept at low speeds, where the tip floats, which are not used in model tests, introduced increased resistance, and at high speeds where the complete model was not under flying control, the agreement was very satisfactory, and at the point of maximum resistance, which is of the greatest interest, was practically in exact agreement.

The models are first towed from a parallel motion system, which allows the model to rise or fall under the influence of the planing or suction effect while maintaining a constant trim relative to the surface. The rise or fall from the initial condition is recorded and indicates the planing power of the model. The gear is so arranged that the model is readily counterweighted to the desired degree for each condition.

Plate B is a complete record of test No. 1844. This test is that of a model of a twin-float design of a 2000 pound aeroplane. The model was one-ninth size and the corresponding speeds, therefore, were one third those for the full size. Get-away was assumed at 45 miles per hour.

Curve *B* shows the gross resistance of the model and towing gear, and curve *C* the net resistance of the model.

Curve *D* shows the change in trim at each condition due to the planing power.

Curve *F* shows the modification of the resistance curve when the model is towed free to trim, the model pivoting about a point corresponding to the relative position of the propeller thrust.

Curve *G* shows the water-planing power for this condition.

The performance indicated is typical. At first the water resistance rises rapidly, then as the influence of the wing life becomes appreciable and the model begins to plane, the resistance reaches a maximum at about 6 knots for the fixed condition, and 6 knots for the free to trim condition. The resistance then falls away steadily for both conditions, but in the case of the free to trim condition, holds up longer and usually has a secondary hump. As at high speeds the aeroplane is usually controllable, the fixed trim condition is attainable and, therefore, representative. At low speeds, however, the squatting effect may predominate, or may be even purposely produced to advantage, as the maximum resistance for the free to trim condition is practically always less than that for the fixed trim condition.

From a comparison of similar models of 2000 pound and 6000 pound floats, Froude's law of

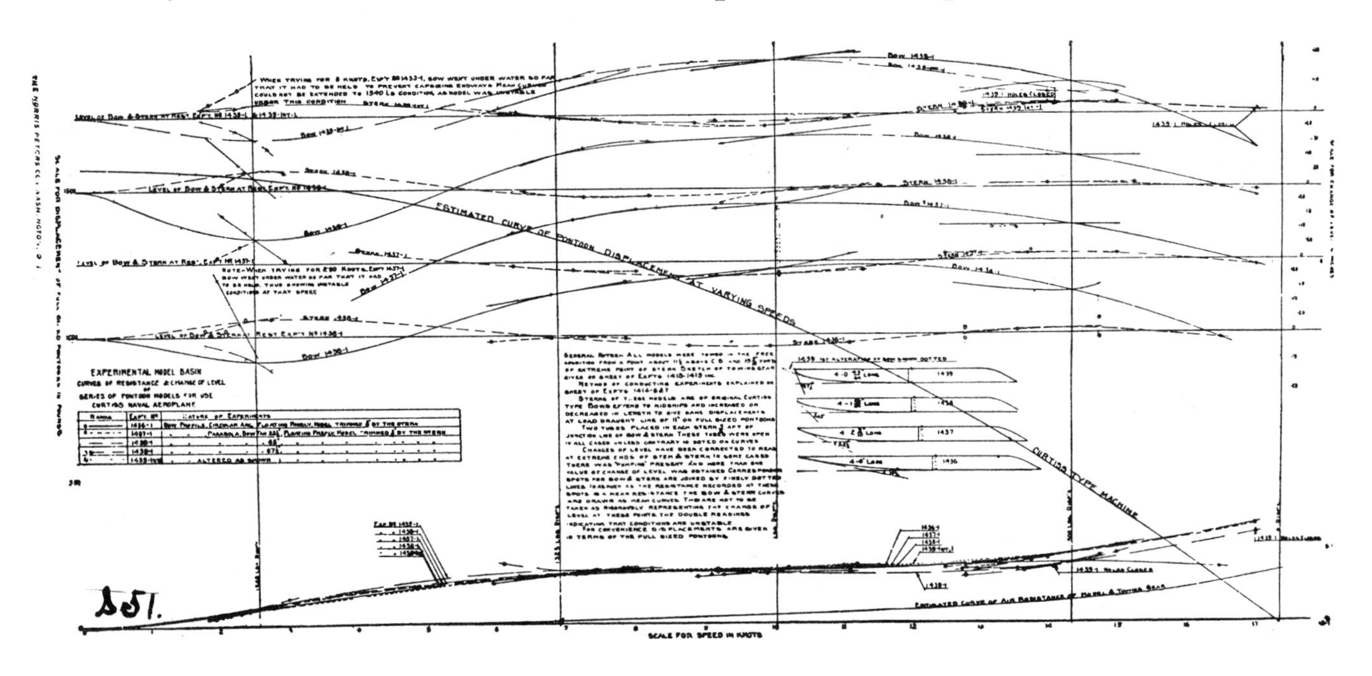

comparison appears to apply satisfactorily, and it also appears that dimensions proportioned to the displacement have planing power in proportion to the displacement, assuming the same get-away speeds in both cases and that the loads correspond to speeds.

Many forms and dispositions of floats have been proposed and tried, and for inland waters several successful forms now exist, but for work in the open sea no form is yet known to have all the required qualities.

The first success was attained by Curtiss with floats of an exaggerated blade form, but these were not as satisfactory as the normal rectangular section float, having a skidform profile. This is a simple and successful form for inland work, and it seems to matter little from a resistance point of view what form of bow curve is used. An examination of the curves of Plate C will show the close agreement in performance of bows having widely different forms of curvature. Various other forms of bows were tried —convex, concave, flat sloping bows, corrugated bows—and all show essentially the same characteristics.

The influence of steps has been tried—one, two, and up to six. From a resistance point of view, it appears unnecessary to provide more than two steps for aeroplane conditions. The first break is usually placed nearly under the center of gravity, and the second forms the stern of the float. If both steps are inclined in the same sense a biplane float is formed, but if the after step rises relative to the forward one, it more nearly approaches the monoplane or single-step form, which has desirable qualities. This is the form which has proved so successful in the flying boat and is the form used in model No. 1844, whose resistance curves have already been referred to. Ventilation of the steps facilitates quick planing and is useful, but is not essential if there is ample reserve of power.

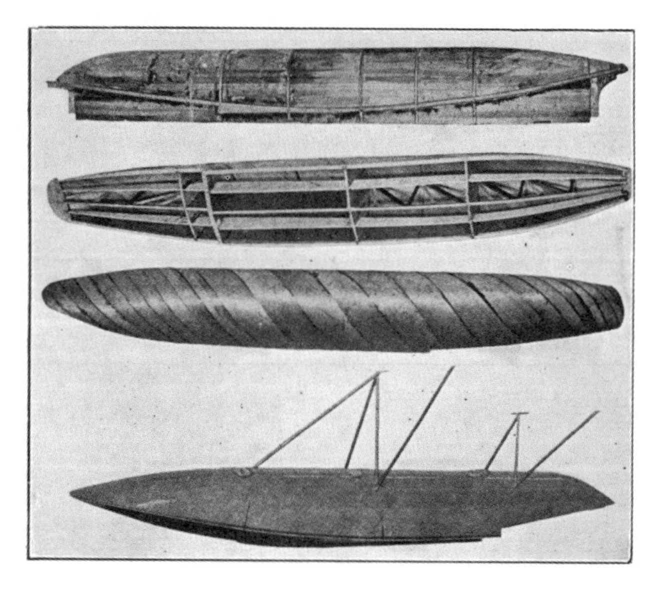

Interference between twin floats at the separation usual in practice appears inappreciable, except at very low speeds.

Flat-bottom floats pound heavily and the flat-bottom-step type porpoises strongly, unless it is handled very nicely even in smooth water.

The introduction of a moderate V-bottom greatly reduces the shock of landing on smooth or rough water. All straight V's throw a strong sheet of spray at the bow. This sheet of spray appears to be independent of the true bow wave. It can be suppressed by the use of hollow V-lines at the bow, but this form is difficult to build strongly and has had steering tendencies when trimmed by the bow or when making a skidding landing or get-away. A simpler method of suppressing this bow is the use of mud guards, so-called for obvious reasons.

The V-bottom also greatly reduces the danger of sticking a wing under on a skidding landing, or when running at high speed across the wind.

The use of fine lines at the bow or stern is not very general, because of the need of longitudinal stiffness. Such lines steer badly, and there is some evidence of the suction effect being present. It is not improbable, however, that a compromise between such lines and the V-bow may work out in solving the sea-going problem.

In construction details in this country the boat-builders' influence is just being asserted. The principal builders have followed a box-like construction not disposing of their material to advantage.

In recent efforts to develop a stream-line form of floats embodying as much as practicable the requirements enumerated, the following features have been developed:

Plate D shows views of these floats in different stages of construction. The general form

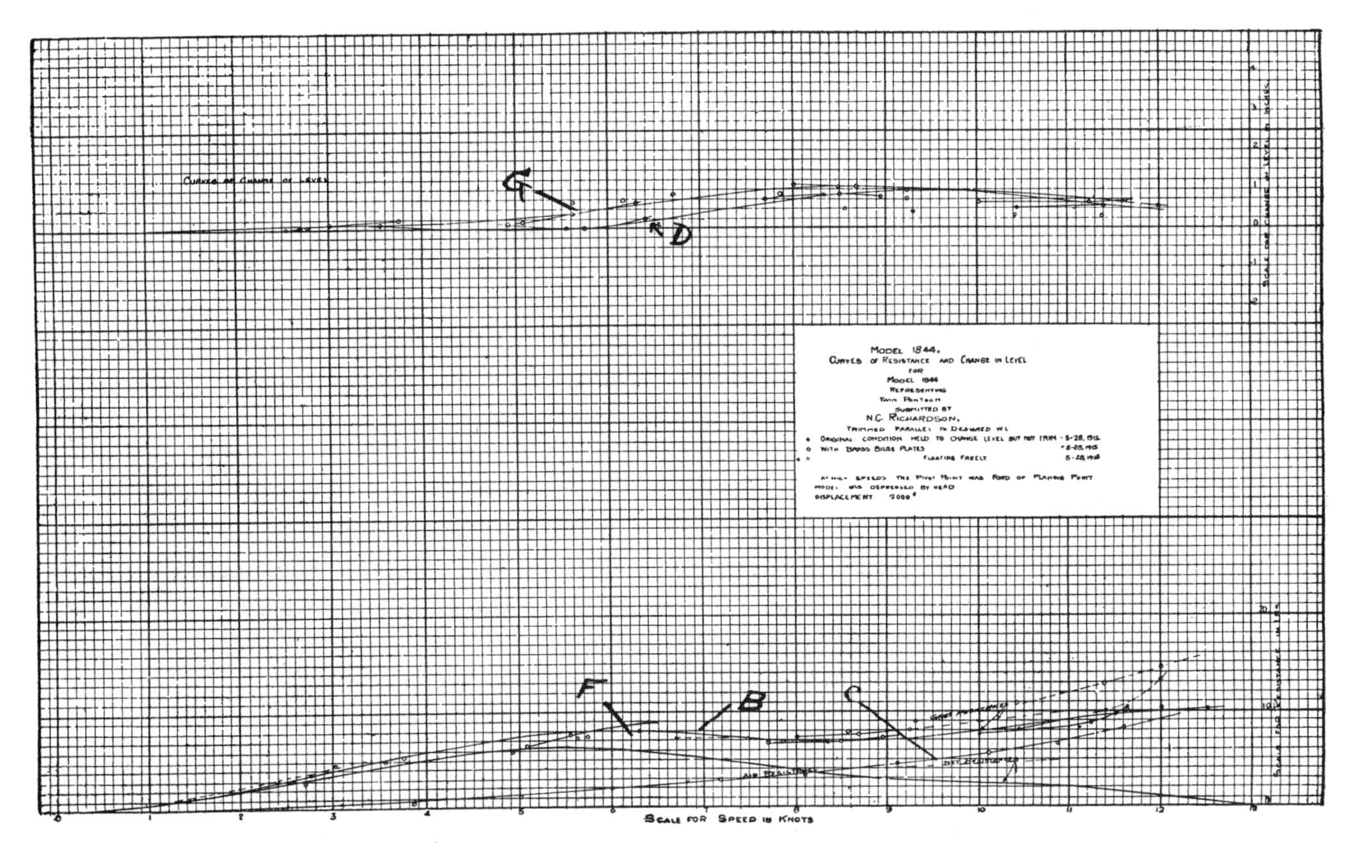

is as nearly stream line as practicable, having a curved deck comprised of two-ply spruce, applied diagonally, one across the other, with cotton sheeting and marine glue between the plies. This construction eliminates the usual deck joints and provides the arched form with diagonal bracing, continuous from chime to chime.

Through the center of the float a longitudinal truss provides the backbone of the construction. Two-ply bulkheads support the float attachments, and the principal bulkheads are tied to each other and to the step by a two-ply fore-and-aft bulkhead built into the center-line girder. These bulkheads are bound by oak ribs and intermediate ribs of oak support the top shell.

The bilge stringers and the keelson are continuous, the step being formed by tapered liners. The step is of oak, as when grounding, or when planing at high speed, the loads are concentrated at the step. The bottom is two-ply spruce, the inner ply running athwartships and the outer ply running diagonally outboard and aft from the keel to the bilge stringers.

Cotton sheeting and marine glue are used in all two-ply work and the plys are held together by special clinch nails of brass.

The bottom is supported by the keel and bilge stringers and an intermediate longitudinal which are in turn supported by the step and bulkheads. The stem and stern pieces are of spruce, shaped to take the longitudinal members and rabbeted to receive the shell planking. The system of bottom planking and longitudinals eliminates the use of bottom frames and really provides continuous framing in the planking itself, similar to Hands' original V. bottom construction. This type of bottom planking appears resilient and strong.

In assembly all butts are covered with glue. All woodwork is given two coats of shellac as soon as it is in place, and is finally given two coats of varnish to prevent the absorption of moisture.

All boundaries of the bottom are sheathed with brass strips set in marine glue. This affords protection and insures water tightness. The float attachments are locally reinforced and strengthened to transmit the loads to the important members.

The method of construction involves the use of a wooden form shaped to the molded lines and cut into sections to allow of removal.

The center-line girder is next assembled with the bow and stern blocks and the step. This is then assembled with the form in which the slots for the ribs and the bilge stringers have already been cut. The frames and stringers are then bent and set and faired flush with the surface of the form. Next the inner ply of planking, which has already been shellacked and varnished on the inner face, is applied and secured to the frames and longitudinals by means of screws. Steaming is unnecessary. In some cases it is necessary to humor the sharp curvatures at the ends by the use of hot water, applied with a swab.

This first ply of planking is then coated with marine glue and the sheeting is applied, followed by the second ply of planking, which is treated with marine glue as fast as each piece is shaped. The second ply is secured in a manner similar to the first ply. The seams of the first ply are traced on to the second ply in pencil, for the purpose of laying out the nails securing the two plys together. Once the second ply is laid, these nails are driven at intervals of about two inches along the margins of both plys of planking, the ends of the nails projecting into the wooden form. The form is now removed, the nails are clinched, and additional quilting nails are systematically driven and clinched. In the meantime the bulkheads and bottom planking have been assembled roughly to dimensions on a flat slab. The bulkheads are now neatly fitted and inserted and secured to the frames by screws and glue. Next, the intermediate longitudinals are placed and secured and all special bracing and fittings for the float attachments are placed and the bottom frames of the bulkheads secured. The bottom planking is now carefully fitted and the drain plugs installed in the proper locations. The hand holes and the holes for the ventilating tubes are laid off, cut, and reinforced, and then the bottom planking is applied and secured. The ventilating tubes are next secured. The entire float is now turned over to the painters, who carefully sandpaper the surfaces and apply two coats of shellac and two coats of varnish. After this is done the step casting, edge strips, false keels, mud guards, and deck fittings are applied.

Each of these floats, designed for 1000 pounds displacement and 60 per cent. reserve buoyancy, weighs, complete, 125 pounds. The principal dimensions are: Length, 15 feet; maximum beam, 24 inches; maximum draught, 14 inches.

The floats illustrated have been used in service with very satisfactory results. The aerodynamic properties of these floats are good. They have the least resistance when the deck is inclined at—3 degrees to the wind, and in this position have no lift; consequently, they present a minimum disturbance in the equilibrium of the aeroplane.

It may seem queer that metal floats have not come into general use. There are several reasons for this, however, the principal reason probably being that the form of floats is as yet nowhere near standardized, so that metal construction for special cases would prove excessively costly. But another reason is that when the attempt is made to build floats of the dimensions required on weights which compare favorably with wooden construction, it is found that the metal itself is so thin that it lacks the required stiffness to preserve its form under service conditions; further, that this thickness provides too little margin for the effects of corrosion, particularly in salt water; and, in particular, the bottom itself would require an elaborate system of support in order to prevent it bending badly under the heavy pressures encountered in service. However, as the number of aeroplanes in service becomes greater and the form of floats becomes to a certain extent standardized and larger, it is probable that metal construction can be satisfactorily solved.

CHAPTER XXXIV

AERODYNAMICS—EXPERIMENTAL RESEARCHES ON THE RESISTANCE OF AIR

By L. Marchis

Professor in the Faculty of Sciences, University of Paris, France.

(From the Second Annual Report of the National Advisory Committee on Aeronautics)

Classification of Experimental Methods

1. REACTIONS EXERTED BY THE AIR ON A BODY IN RELATIVE MOVEMENT WITH IT

When a body is in movement relative to the air with which it is surrounded, it is subject to a system of forces to which is given the name of "reactions exerted on the body by the air." These reactions are variable, especially as regards (1) the form of the body, (2) the position which it occupies in relation to the surrounding medium, (3) the various circumstances of its movement (time elapsed from origin of movement to present moment—velocity relative to the air), and, finally, (4) the mass of the fluid which surrounds the body in movement.

We shall not develop in detail the difficulties presented by each of these problems, of which certain have received only very imperfect solutions.

We shall, in what follows, consider only the case of a body surrounded completely by a great mass of air, relative to which it has a movement, established a long time previously, and of which the velocity and direction are constant and readily determined.

The reactions exerted by the air on the body in movement relative to it are reduced to a force and a couple. We shall assume that the body under experiment possesses, at the least, a plane of symmetry, thus eliminating the couple from the reactions of the air and reducing them to a single force, to which we shall give the name of "resistance of the air on the body in movement relative to it."

When we consider the movement of the body relative to the air which surrounds it, we have not only in view a movement of translation, but also a movement of simple rotation and likewise a movement of rotation combined with a movement of translation. In other words, we shall study here the problem of the propeller as well as that of the wings of an aeroplane.

2. MANNER OF PRODUCING THE MOVEMENT OF A BODY RELATIVE TO THE AIR WHICH SURROUNDS IT—BODY MOVABLE

Various experimental methods may be utilized in order to produce the movement of a body in reference to free air.

In an indefinite mass of air, at rest as a whole, the following types of movement may be given to the body:

(*a*) A movement of rectilinear translation;

(*b*) A movement of rotation about the axis of a mechanism;

(*c*) An oscillating movement, as in the case of a pendulum.

The methods by means of some form of mechanism or by means of a pendulum have been but little employed in France and we shall omit special reference to them.

The method employing the motion of translation may be applied in two forms:

(1) The body is allowed to fall freely in air, as calm as possible.

(2) The body is carried on some form of car which is moved in calm air.

In France the method of free fall has given rise to important investigations made by MM. Cailletet and Colardeau and especially by M. G. Eiffel.

The method by means of a car is now utilized by the Aerodynamic Institute of Saint-Cyr, at the laboratory of military aerostation of Chalais-

Photograph showing the arrangement of the model of an aeroplane when being tested at the United States Navy tunnels. The model is carried by a steel spindle which extends up through the top of the tunnel to the weighing balance which is placed overhead. For about two-thirds of the length in the tunnel the spindle is covered by a mask of stream-line form. This mask is secured to the ceiling of the tunnel and reduces the force acting on the spindle itself, and thus the spindle correction. The weighing balance consists of a weighing scale on the platform principle having three axes, two of them at the same horizontal line 61 inches apart, and the third vertically over one of the first, 48 inches above it. When a model is set at a given angle, the movements acting about each of these axes are measured by weighing them on the scale. With this data it is possible to compute horizontal and vertical components of the force acting on the model, that is, the drift and lift, and also to compute the line of application of the force. Tests are usually made at speeds of 40 miles an hour. At this speed and at the angle of least resistance an ordinary aeroplane wing model has a horizontal resistance of something less than one-tenth of a pound. It is therefore necessary that the balance should be capable of weighing a force with accuracy to about 2/1000ths of a pound.

The large size of the tunnel makes it possible to test full size radiators for aeroplane motors and comparative tests have recently been made on several types both as to air resistance and cooling capacity.

Meudon, and also by M. the Duke of Guiche. At Saint-Cyr and at Chalais-Meudon, the car is composed of a carriage moving on rails. M. de Guiche employs an automobile as a carrier.

A variant of the method of the car has been installed at the laboratory of military aviation at Vincennes. On a stretched cable a little hanging car rolls, carrying, attached below it, the objects under test with the necessary instruments.

The dimensions of the bodies on which the experiments are carried out may be of the order of those which are utilized in aviation itself. In other words, it is possible to operate upon equipment as used in actual aviation, or at least presenting dimensions differing but little from those used in practice.

From this point of view the method by displacement through the air opens up a field of investigation more extended than the method in which an artificial current of air is employed.

3. MANNER OF PRODUCING THE MOVEMENT OF A BODY RELATIVE TO THE AIR WHICH SURROUNDS IT—ARTIFICIAL CURRENT OF AIR

It is possible, in fact, to realize in an entirely different manner the relative movement of a body through the air.

Instead of moving the body under test, a fixed position is given to such body placed in an artificial current of air.

The body may then be disposed in the free air in front of the orifice through which the air enters under regulation by means of suitable devices. This method has been employed by M. Rateau.

The body under investigation may also be placed in an inclosure or integral part of the apparatus for the regulation of the current of air. It is placed, for example, in a part of a large cylindrical pipe which receives a current of air produced by a fan and of which the velocity, at a certain distance from the walls, has been rendered sensibly parallel to the pipe.

This method, furthermore, may be subject to certain variations:

(*a*) The body under investigation alone is placed in the inclosure in the interior of which the artificial current of air is produced. The apparatus for measuring the reactions of the air are on the exterior of this inclosure, their connection with the interior being made through the solid wall which limits the conduit.

This method is known under the name of the "tunnel method." It has not been largely employed in France. There exists at the present time at the Aerodynamic Institute of Saint-Cyr a tunnel of which the practical use has been interrupted by the present war.

(*b*) The apparatus employed for determining the circulation of the air is enlarged into a chamber of suitable size, traversed between two of its parallel walls by a cylinder of moving air. On the outside of the latter and within the chamber are located the experimenters with the necessary measuring apparatus.

We propose to call this the "Eiffel method."

In France this method has given very complete results. It is for us the characteristic method in connection with the use of an artificial current of air.

From the point of view of the convenience of carrying on the experiments, especially in large numbers, the last method is superior to the method by displacement in free air. The latter demands, in fact, that the external air shall be as calm as possible. This condition can only be realized on certain days and then only for certain hours of a given day. If along the right-line path of the body under investigation the wind should have everywhere the same intensity and the same direction, due allowance might be made for its existence.

But many investigations, notably those of M. Maurain at the Aerotechnic Institute of Saint-Cyr, show that at any given point in the air the wind is frequently subject to continued changes in direction and intensity.

But even if it allows the experimenter to regulate the conditions of any one investigation, the method by the use of the artificial current of air can only be applied to models reduced in size in comparison with actual practice in aviation. We shall see later the reason for this limitation.

One question immediately presents itself: How may the results obtained in the study of models be transformed in order to furnish information applicable to apparatus of full size? What is the law of similitude which makes possible the transformation of an investigation on a

Eiffel's Laboratory: The suction blower at the end of the wind tunnel, driven by 50 horse-power, which produces the artificial air current.

Experimental Chamber of Eiffel's Laboratory—In the right-hand corner is a Nieuport monoplane model mounted on arm which is connected to the balance shown above, which registers the air pressure.

small scale to corresponding phenomena on a large scale. This is the matter which we shall especially develop at a later point.

A further question presents itself: Do the methods mentioned above, namely, the displacement of the body under investigation and the method by the artificial current of air, lead to the same results? M. Eiffel maintains the affirmative, relying upon the fundamental principle of relative movement. M. de Guiche maintains the negative, arguing that the tunnel method does not realize fully the conditions which permit the application of such a principle.

We shall return to this question at a later point, in connection with the comparison of the results obtained by these two experimenters.

4. STUDIES OF AIRPLANES IN FREE FLIGHT

The methods which we have just considered require that the body under investigation be connected in a fixed manner with a support. The latter has, under good conditions, its dimensions reduced as much as possible. It is also removed as far as possible from the body under investigation, so that its presence will produce the minimum of disturbance. It is none the less true, however, that the airplane, thus studied, is not in the precise condition of free evolution in the open air.

For this reason investigations have been undertaken on airplanes during their free flight in the air. Unfortunately, the field of such investigation is limited. It cannot be carried through at the will of the experimenter; that is to say, of the pilot, who must first of all guard against danger of fall. Such experiments give complex results often difficult of analysis. Nevertheless it can not be denied that such results may have a very considerable practical value.

Experiments of this character were inaugurated in 1910 by MM. Gaudart and Legrand with a Voisin biplane. These experiments were, however, neither sufficiently systematic nor numerous to lead to significant results.

Quite otherwise are the researches made by

Commander Dorand, at Villacoublay, on a biplane of his own construction piloted by M. Labouchère. At the Institute of Saint-Cyr, MM. Toussaint and the Lieutenant of Aviation Gouin, have made important experiments on a Maurice Farman biplane and on a Blériot monoplane. Ingenious apparatus capable of registering the movement of the pilot was employed to furnish important indications regarding the operation of such actual aviation equipment.

5. THE TOTAL RESISTANCE OF THE AIR AND THE DETERMINATION OF THE PRESSURES AT EACH POINT OF THE SURFACE OF THE BODY UNDER INVESTIGATION

Let us return to the methods which, in a laboratory, may be employed in determining the resistance of the air upon a body in movement relative to it.

With regard to the method of measuring this resistance two types may be characterized:

(1) Determination, by means of a balance, of the total resistance on the entire body under investigation.

(2) Determination, at each point of the body, of the reaction exerted by the air at this point; a study, in some manner topographical in character, regarding the pressures resulting from the relative movement of the body and the air.

This investigation immediately leads, through a geometrical composition of the individual forces thus determined, to a knowledge of the complete resistance of the air.

The method by means of the balance has given wonderful results in the laboratory of M. Eiffel and at the Institute of Saint-Cyr. M. de Guiche has applied this method solely to the analysis of the distributed pressures.

Such is the general classification of the experimental methods at present in use in France for the study of the problems of aerodynamics. We proceed to give in detail the fundamental

Testing the model of a monoplane in Eiffel's wind tunnel.

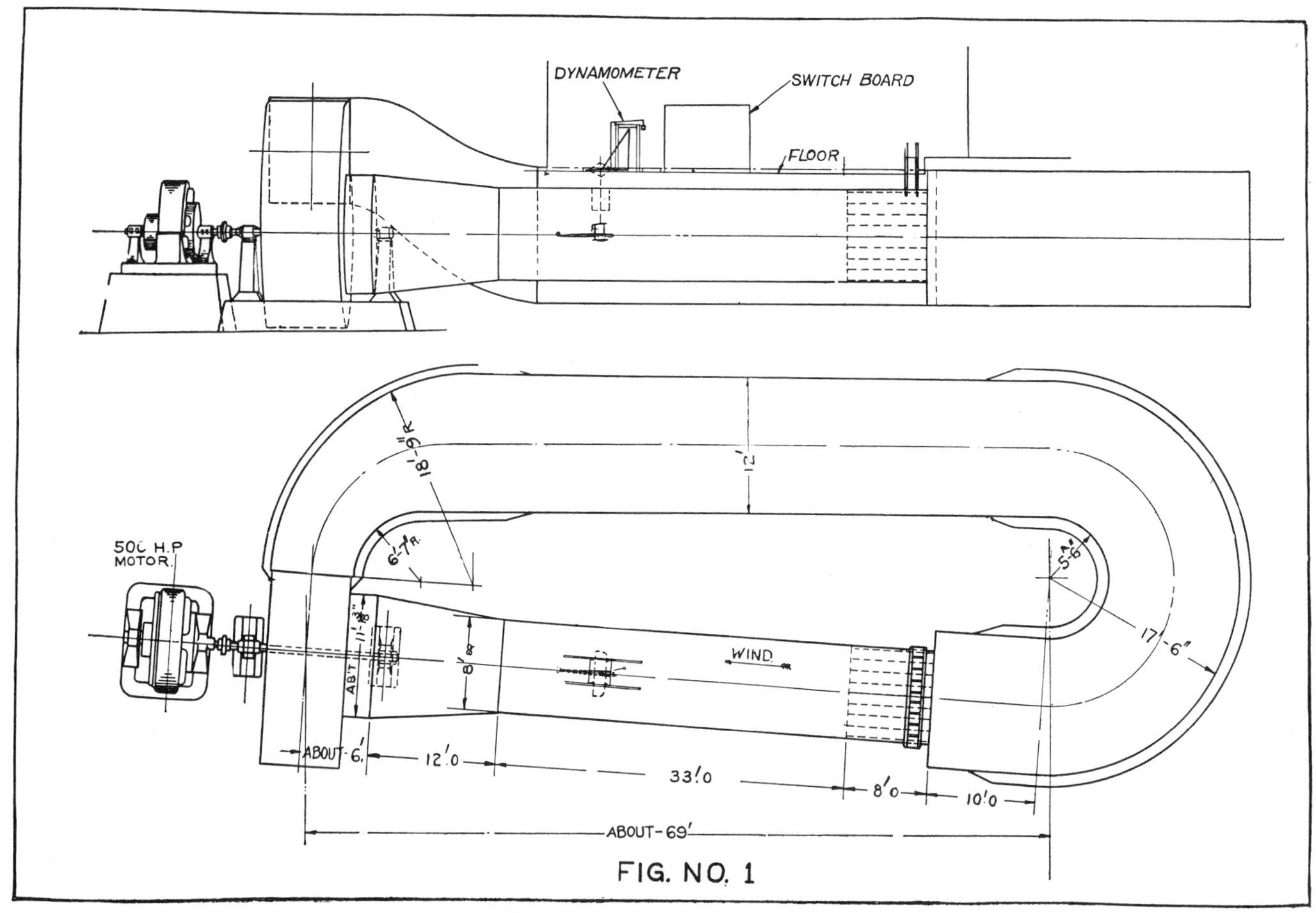

Plan and side view of the United States Navy wind tunnel for testing models.

principles of these investigations in a further study of the French aerodynamic laboratories.

Diagrams Representing the Results of Experiments

1. PROPOSED NOTATIONS

Let us consider a reduced size model of the body under investigation. Let λ be the ratio of the homologous linear dimensions taken in the body and in the model.

Let us suppose that the model is tried at a relative velocity V and that the results of the experiments are reduced to what they would be for a velocity V_1. If rv_1 and rv are the actions of the air on the model at speeds V_1 and V, we have the relation

$$\frac{rv_1}{rv} = \left(\frac{V_1}{V}\right)^2 \quad . \quad . \quad . \quad . \quad . \quad . \quad (1)$$

On the other hand, let us take the body under investigation. Let V be its velocity relative to the air, and suppose that the actions of the air are reduced to what they would be if the velocity had the value V_2. Denote by Rv and Rv_2 these values of the resistance of the air. We have the relation

$$\frac{Rv_2}{Rv} = \left(\frac{V_2}{V}\right)^2 \quad . \quad . \quad . \quad . \quad . \quad . \quad (2)$$

The relations (1) and (2) give:

$$\frac{Rv^2}{Rv} \times \frac{rv}{rv_1} = \left(\frac{V_2}{V_1}\right)^2 \quad . \quad . \quad . \quad . \quad (3)$$

But since rv and Rv are relative to the model and to the body under investigation at the same speed V, we have

$$\frac{rv}{Rv} = \frac{1}{\lambda^2} \quad . \quad . \quad . \quad . \quad . \quad . \quad . \quad (4)$$

Carrying this value into (3) we have

$$\frac{Rv_2}{rv_1} = \lambda^2 \left(\frac{V^2}{V^1}\right)^2. \quad . \quad . \quad . \quad . \quad . \quad (5)$$

In experiments in aerodynamics the values usually taken are $V_1 = V_2 = 10$ meters per second (32.8 feet per second). Measure is then taken of rv on the model or Rv on a body of normal size. Equations (1) and (2) then give

$$r_{10} = rv\left(\frac{10}{V}\right)^2 \quad . \quad . \quad . \quad . \quad . \quad . \quad (6)$$

$$R_{10} = Rv\left(\frac{10}{V}\right)^2 \quad . \quad . \quad . \quad . \quad . \quad . \quad (7)$$

The methods for the measurement of rv give at the same time:

(*a*) The component of rv along the direction of relative air movement.

(*b*) The component of rv normal to the direction of relative air movement.

With M. Eiffel, let us call r_x and r_y, F_x and F_y the components of r_{10} and R_{10} along and normal to the direction of relative air movement, respectively.

We have then the relations:

$$r_x = \text{component of } rv \text{ along direction of relative air movements} \times \left(\frac{10}{V}\right)^2 \quad (8)$$

$$r_y = \text{component of } rv \text{ along normal to direction of relative air movement} \times \left(\frac{10}{V}\right)^2 \quad . \quad . \quad . \quad . \quad . \quad (9)$$

$$F_x = \text{component of } Rv \text{ along direction of relative air movement} \times \left(\frac{10}{V}\right)^2 \quad . \quad . \quad . \quad . \quad . \quad . \quad . \quad . \quad . \quad (10)$$

$$F_y = \text{component of } Rv \text{ along normal to direction of relative air movement} \times \left(\frac{1}{V}\right)^2 \quad . \quad . \quad . \quad . \quad . \quad (11)$$

We may note that $r_x V^2$, $r_y V^2$, $F_x V^2$ and $F_y V^2$ are quantities of the order of force.

M. Eiffel calls R_x and R_y the components of airplanes, takes $V_2 = 1$ met./sec. $= 3.28$ ft./sec. and $V_1 = 10$ met./sec. $= 32.8$ ft./sec.

Equation (5) then gives

$$\frac{R_1}{r_{10}} = \left(\frac{\lambda}{10}\right)^2 \quad . \quad . \quad . \quad . \quad . \quad . \quad . \quad (12)$$

M. Eiffel calls R_x and R_y the components of R_1 along and normal to the direction of relative air movement.

We have then

$$\left.\begin{array}{l} R_x = r_x\left(\frac{\lambda}{10}\right)^2 \\ R_y = r_y\left(\frac{\lambda}{10}\right)^2 \end{array}\right\} \quad . \quad . \quad . \quad . \quad . \quad . \quad (13)$$

R_x and R_y are quantities of the same order as r_x and r_y.

If the model of the airplane is on a scale 1/10, $\lambda = 10$ and we have

$$\left.\begin{array}{l} R_x = r_x \\ R_y = r_y \end{array}\right\} \quad . \quad . \quad . \quad . \quad . \quad . \quad . \quad . \quad (14)$$

The numerical values calculated for the model apply directly to the airplane of normal size.

When the problem involves the wings of air planes, M. Eiffel places

$$\left.\begin{array}{l} K_x = \dfrac{\text{component of } r_y \text{ along direction of relative air movement.}}{SV^2} = \dfrac{r_x}{S \times 10^2} \\ K_y = \dfrac{\text{component of } r_v \text{ along normal to direction of relative air movement.}}{SV^2} = \dfrac{r_y}{S \times 10^2} \end{array}\right\} (15)$$

$$K_i = \sqrt{K_x^2 + K_y^2} \quad . \quad . \quad . \quad (16)$$

In this equation i is the angle between the direction of relative air movement and a reference line attached to the wing, generally the chord of the profile of the wing in its plane of symmetry.

K_x, K_y, K_i are quantities of the order of density.

S should be a mean between the surface of the wing exposed directly to the action of the air and surface on the back. Builders of airplanes usually consider S equal to the greatest projection of the wing on a horizontal plane.

2. STUDY OF THE WINGS OF AN AIRPLANE—POLAR DIAGRAMS OF M. EIFFEL

M. Eiffel represents the properties of the wings of an airplane by means of what he calls simple polar diagrams.

On two rectangular axes, he plots as abscissæ the values of K_x and as ordinates the values of K_y, the same scale being used for both. The curve thus traced in the $K_x K_y$ plane is called the "first simple polar." A point of the curve corresponds to a determinate value of the angle i. The radius vector from the origin to this point represents the quantity K_i. The angle of this radius vector with the axis of K_y is the angle between the resistance of the air and the normal to the direction of relative air movement. If this angle is denoted by θ we have

$$\text{tang. } \theta = \frac{K_x}{K_y} \quad . \quad . \quad . \quad . \quad . \quad (17)$$

The tangent drawn from the origin to the polar gives the value of θ, θ_m, for which the ratio $\frac{K_x}{K_y}$ is a minimum.

To each point of the curve corresponds a value of the angle i and a value of the angle θ. If $\theta = i$ the resistance of the air is normal to the chord of the profile; if $\theta < i$, the resistance of the air is forward of the normal to the chord. For $\theta > i$, it falls behind the chord.

This mode of representation (K_x and K_y represented to the same scale) is not suitable for the values of the angle i corresponding to the small values employed in aviation. In fact, for these values of the angle i the polar diagram approaches very close to a straight line slightly inclined to the axis of K_y. The comparison of one wing with another by simple superposition of diagrams is a delicate operation. In particular it is almost impossible to compare the wings regarding the minimum value of $\frac{K_x}{K_y}$.

Accordingly, M. Eiffel constructs what he calls the "second simple polar." He takes for K_x a scale five times larger than for K_y. In this mode of representation, a vector joining the origin with a point on the curve is no longer equal to K_i, and the angle of this vector with the axis of K_y is no longer the angle θ. However, the same as for the small values of i, less than 10°, K_i is very little different from K_y, and for the values K_i the ordinates of the new curve may be taken. It is convenient to add to this curve a scale representing values of $\frac{K_x}{K_y}$. On a parallel to the axis of K_x, passing through a point of K_y, values are plotted of $\frac{K_x}{K_y}$ corresponding to one of the intersections with the new curve of the radius vector starting from the origin and ending at this point. Let us call this line the axis of $\frac{K_x}{K_y}$. In order that $\frac{K_x}{K_y}$ may correspond to an angle i, it is necessary that the vector just named should cut the second polar curve. The minimum value of $\frac{K_x}{K_y}$ is then given by the point where the tangent from the origin to the polar curve meets the axis of $\frac{K_x}{K_y}$.

3. STUDY OF THE HORIZONTAL MOVEMENT OF AN AIRPLANE—THE LOGARITHMIC POLAR CURVE

In order to study the horizontal movement of an airplane, M. Eiffel has pointed out a very ingenious representation, to which he has given the name of logarithmic polar.

Let us consider a model of an airplane and let i be the angle made between the direction of relative air movement and a straight reference line intimately connected with the apparatus, for example, a straight line doubly tangent to the lower part of the principal planes, near the fuselage. To this value of the angle i, the experiment on the model will give corresponding values of the resistance of the air, of which the projections parallel and perpendicular to the air movement are r_x and r_y. To these, equations (13) serve to give the corresponding values of R_x and R_y relative to an airplane of full size.

Furthermore, let

Q = the weight of the actual airplane.

P = the power required to maintain horizontal flight with a relative velocity V.

The equations

$$\left.\begin{aligned} P &= R_x V^3 \\ Q &= R_y V^2 \end{aligned}\right\} \quad . \quad . \quad . \quad . \quad . \quad . \quad . \quad (18)$$

define the correlative values of P, Q, V, R_x and R_y, and hence of the angle i which corresponds to the horizontal flight of an airplane of determinate form (especially of an airplane in which the depth rudder occupies a determinate position when the axis of the propeller is parallel to the path of flight).

Let us consider such an airplane.

Equations (18) give immediately

$$\left.\begin{aligned} \log. R_x &= \log. P - 3 \log. V \\ \log. R_y &= \log. Q - 2 \log. V \end{aligned}\right\} \quad . \quad (19)$$

or

$$\left.\begin{aligned}\log. R_x &= \log. P - \frac{3}{\sqrt{13}} \times \sqrt{13} \log. V \\ \log. R_y &= \log. Q - \frac{2}{\sqrt{13}} \times \sqrt{13} \log. V\end{aligned}\right\} \quad (20)$$

The experiments on a model permit, for various values of i, the determination of corresponding values of R_x and R_y.

On two rectangular axes let us plot to the same scale, on the axis of abscissæ, distances proportional to the various values of log. R_x; on the axis of ordinates, distances proportional to the various values of log. R_y. We shall thus obtain in the plane of the axes a curve to which M. Eiffel has given the name of logarithmic polar. Each point on this curve corresponds to a determinate value of the angle i which is inscribed on the curve.

Let us consider a vector OM_i running from the origin O to a point M_i on the curve. This vector has for projections on the axes of coordinates the values log. R_x and log. R_y. But equations (20) show that this vector is the resultant of a broken line of which the vectors are

log. P directed along the axis of log. R_x.
log. Q directed along the axis of log. R_y.

$\sqrt{13} \times$ log. V directed in the third angle of the coordinate planes (— log. R_x, — log. R_y), and making with the axis of log. R_x an angle of which the cosine is equal to

$$-\frac{3}{\sqrt{13}} \quad \text{(See fig. 1.)}$$

If the two extremities, O and M_i, of the broken line are preserved, the segments may be run through in any order whatever. Thus, for example, we may have any one of the following orders:

log. P, $\sqrt{13} \times$ log. V, log. Q;
log. Q, log. P, $\sqrt{13} \times$ log. V;
$\sqrt{13} \times$ log. V, log. Q, log. P.

It is well known that starting from the point O one should, following the broken line, end at a point M_i of the logarithmic polar. The directions of the vectors are, furthermore, well known. If we take two of the vectors of the broken line, the trace of this line permits immediately the determination of the third.

We may thus solve graphically by means of the logarithmic polar a series of problems relating to the horizontal flight of an airplane when the axis of the propeller is parallel to the path. We might, for example, desire to know what weight should be given to the apparatus in order to obtain a given velocity with a given power.

In this problem the vectors log. P and $\sqrt{13} \times$ log. V are known in magnitude and direction; it is easy to trace them. From the extremity of the vector $\sqrt{13} \times$ log. V there is drawn a straight line parallel to the axis of log. R_y, which is continued to its point of intersection with the logarithmic polar. The vector log. Q is thus constructed; it gives the weight Q which is sought. At the same time, the point of intersection of this vector with the polar curve determines the angle i of the flight.

Let us now consider a velocity V_o which is, for example, the normal actual velocity of the airplanes (100 kilometers (62.1 miles) per hour). Then equations (18) may be written

$$\left.\begin{aligned}\frac{P}{V_o^3} &= R_x\left(\frac{V}{V_o}\right)^3 \\ \frac{Q}{V_o^2} &= R_y\left(\frac{V}{V_o}\right)^2\end{aligned}\right\} \quad . \ . \ . \ . \ . \ (21)$$

From these we derive

$$\left.\begin{aligned}\log. R_x &= \log.\left(\frac{P}{V_o^3}\right) - \frac{3}{\sqrt{13}}\sqrt{13} \times \log.\frac{V}{V_o} \\ \log. R_y &= \log.\left(\frac{Q}{V_o^2}\right) - \frac{2}{\sqrt{13}}\sqrt{13} \times \log.\frac{V}{V_o}\end{aligned}\right\} \quad (22)$$

On the axis log. V let us take a point V_o such that

$$OV_o = \sqrt{13} \times \log. V_o \quad \text{(fig. 1.)}$$

The vector V_oV then represents

$$\sqrt{13} \times \log.\frac{V}{V_o}.$$

Let us then carry this vector over to C_oB on the vector AB, and then project C_o to A_o on the axis log. R_x. Finally, lead the vector A_oB_o to the ordinate, parallel to the axis log. R_y. To the contour $OAB\ M_i$, in which

$$OA = \log. P,\ AB = \sqrt{13} \times \log. V,\ BM_i = \log. Q,$$

we thus substitute the contour $OA_oB_oM_i$, which is its equivalent since it has the same resultant, and which is such that

$$OA_o = \log.\left(\frac{P}{V_o^3}\right),\ A_oB_o = \sqrt{13} \times \log.\frac{V}{V_o}$$

$$B_oM_i = \log.\left(\frac{Q}{V_o^2}\right)$$

We have as a result:

Vector OA_o = vector OA + vector AA_o
Vector $AA_o = -3 \log. V_o$

$$\text{Vector } OA_o = \log. P - 3 \log. V_o = \log.\left(\frac{Q}{V_o^3}\right)$$

As we shall have constantly to consider a vector V_oV or A_oB or $\sqrt{13} \times \log.\frac{V}{V_o}$, it is natural to carry the point V_o to the origin.

When the velocity of the airplane is equal to V_o, the vector A_oB_o disappears; the points A and B_o become coincident with the point M_x (fig. 2). It is, in fact, easy to see that we have

$$M_xA_o = 3 \log.\frac{V}{V_o},\ B_oM_x = 2 \log.\frac{V}{V_o}$$

The coordinates log. R_x and log. R_y of the point have them for values

$$\left.\begin{aligned} \log. R_x &= \log.\left(\frac{P_o}{V_o^3}\right) \\ \log. R_y &= \log.\left(\frac{Q_o}{V_o^2}\right) \end{aligned}\right\} \quad . \quad . \quad . \quad (23)$$

From these equations we derive

$$\left.\begin{aligned} R_x &= \frac{P_o}{V_o^3} \\ R_y &= \frac{Q_o}{V_o^2} \end{aligned}\right\} \quad . \quad . \quad . \quad . \quad . \quad . \quad (24)$$

We are therefore able to develop a correspondence between a point M_x on the axis of abscissæ (fig. 1) and a value R_x, such that

$$OM_x = \log. R_x,$$

and a value P_o of the useful power such that

$$O\,M_x = \log.\left(\frac{P_o}{V_o^3}\right)$$

In other words, the axis of abscissæ may be graduated in terms of useful power.

In the same way we may graduate the axis of ordinates in terms of weight.

Let us now suppose that, in a problem, we have given the useful power P and the speed V. The axis of abscissæ, which is the scale for P, gives immediately the point A_o, such that

$$V_oA_o = \log.\left(\frac{P}{V_o^3}\right) \qquad \text{(See fig. 2.)}$$

The vector V_oV is such that

$$V_oV = \sqrt{13} \times \log.\frac{V}{V_o}.$$

The contour V_oA_o B may be traced. By carrying B_oM_i parallel to log. R_y and extending to the point of intersection with the polar curve, there is found the vector

$$B_o\,M_i = \log.\left[\frac{Q}{V_o}\right]$$

If this vector is led down from V_o on the scale of R_y, which is at the same time the scale of weight, the extremity of the segment gives immediately the weight Q which is sought.

In the system of units (meter, kilogram, second) generally used, P is expressed in kilometer-seconds, V in meter-seconds, Q in kilograms (weight). It is more convenient, for practical application, to graduate the scales for P and V in horse-power and in kilometers per hour. To this end it is sufficient to divide the indications of the first scale by 75 for horse-power and to multiply by 3.6 the numbers relating to velocity.

If the velocity W is less than V_o, the segment is directed opposite to the segment V_oV. The contour to consider is $V_o\,A_o'B_o'\,M_i$. (See fig. 2.) We have, in fact, in this case

$$\left.\begin{aligned} \log. R_x &= \log.\left[\frac{P}{V_o^3}\right] + 3 \log.\frac{V_o}{W} \\ \log. R_y &= \log.\left[\frac{Q}{V_o^2}\right] + 2 \log.\frac{V_o}{W} \end{aligned}\right\}$$

It is easily seen that these equations represent the projections on the two axes of the contour $V_o\,A_o'\,B_o'\,M_i$.

We are thus led to the following rule:

If we follow a broken line starting from the origin and ending on the polar curve, the direction in which each vector is traversed is the direction in which such vector should be placed, starting from the origin, on the corresponding scale in order to give its value.

It results immediately that if have the contour V_o *ABCD* (fig. 2), the vector *BC* corresponds to a speed greater than the vector *BD*, these two speeds being, furthermore, inferior to V_o.

Let us suppose, now, that the results found with the model do not correspond to the conditions which had been fixed *a priori* for the airplane. The question may then arise of changing proportionately the dimensions of the apparatus.

Let *N* be the ratio of the lineal dimensions of the second apparatus to those of the first; *N*, for example, might be 1.10 for an increase of 10 per cent in the dimensions.

The fundamental equations of horizontal flight for this new apparatus will be

$$\left.\begin{aligned} \frac{P}{V_o^3} &= R_x N^2 \left[\frac{V}{V_o}\right]^3 \\ \frac{Q}{V_o^2} &= R_y N^2 \left[\frac{V}{V_o}\right]^2 \end{aligned}\right\} \quad . \quad . \quad . \quad . \quad (25)$$

It is not necessary to construct special polar curves corresponding to various values of *N* in order to determine the value suited to this number. We find, in fact, from equations (25)

$$\left.\begin{aligned} \log. R_x &= \log.\left[\frac{P}{V_o^3}\right] - 3\log.\frac{V}{V_o} - 2\log. N \\ \log. R_y &= \log.\left[\frac{Q}{V_o^2}\right] - 2\log.\frac{V}{V_o} - 2\log. N \end{aligned}\right\} (26)$$

To the vectors,

$$\log.\left[\frac{P}{V_o^3}\right], \log.\left[\frac{Q}{V_o^2}\right] \quad \sqrt{13} \times \log.\frac{V}{V_o}$$

it is convenient to add a fourth vector,

$$\sqrt{8} \times \log. N.$$

This is directed along the line making the angle $5\frac{\pi}{4}$ with the axis abscissæ (Axis V_o *N*, fig. 2).

If *N* is greater than unity, the values of log. *N* are laid off along this axis; if *N* is less than unity, they are laid off along the line $\frac{\pi}{4}$ with the axis of the abscissæ.

If then the values fixed in advance are *P*, *Q*., *V*, it is sufficient, in order to have the value of *N* which will permit of realizing these values, to draw the fourth segment in a suitable direction until it meets the polar curve. The fourth segment indicates, furthermore, by its intersection with the polar curve, a suitable angle of flight.

Instead of terminating the polygonal contour running from the origin to a point of the curve by the vector $\sqrt{8} \times \log. N$, we may trace this segment first. In other words, instead of starting from the origin of coördinates as the origin of contour, we may start from a point situated on the axis of *N*. We then see immediately by the figure what becomes of the properties of an airplane, for which the dimensions have been multiplied by the number determined by the point on the axis of *N* which was taken for the point of departure. Every broken line drawn between this point and the polar curve gives the system of values *P*. *Q*, *V*, which corresponds to the modified apparatus.

We can not here indicate the solution of all the problems for which the consideration of the logarithmic polar provides. To this end, reference should be made to the work of M. Eiffel noted in the bibliography attached to this paper. However, we may note, in résumé, the results to which this study of the logarithmic polar leads.

For all the forms of apparatus studied by M. Eiffel, the logarithmic polar curves always present the same general characteristic, that of figure 3. Beginning with small values of the angles of incidence, we find the angles of horizontal flight for which the properties are the following:

(1) Angle i_1 for which R_x is minimum (fig. 3).

This angle is given by the point of contact of the tangent to the polar, parallel to the axis of ordinates.

Horizontal flight under this angle corresponds to the maximum speed for a given power, or to the minimum power for a given speed.

(2) Angle i_2 for which $\frac{R_x}{R_y}$ is a minimum (fig. 3).

This angle is given by the point of contact of tangent to the polar curve, drawn parallel to the axis of N.

Horizontal flight under this angle corresponds to the minimum tractive force required for a given weight, or to the maximum weight for a given tractive force.

(3) Angle i_3 for which $\frac{R_x^2}{R_y^3}$ is a minimum (fig. 3).

This angle is given by the point of contact of the tangent to the polar curve, drawn parallel to the axis V_oV.

Horizontal flight under this angle corresponds to the minimum power for a given weight, or to the maximum weight for a given power.

(4) Angle i_4 for which R_y is maximum (fig. 3).

This angle is given by the point of contact of the tangent to the polar curve, drawn parallel to the axis of abscissæ.

Horizontal flight under this angle corresponds to the maximum weight for a given speed, or to the minimum speed for a given weight.

It is seen that to each one of the angles of incidence, i_1, i_2, i_3, i_4, we may relate two magnitudes, of which one is maximum or minimum when the other is given. Each one of these angles is the most favorable angle regarding the two corresponding magnitudes.

As these polar curves, in their useful part, do not have a point of inflection, it follows that the nearer the angle of horizontal flight lies to one of the angles i_1, i_2, i_3, i_4, the better are the conditions with regard to the group of magnitudes which corresponds to these angles.

Let us take an example. The weight carried by an airplane is judged to be too small. It is desired to gain weight at the expense of speed, but at the same time preserving the same expenditure of power. It is sufficient to approach the point for which the weight will be maximum for a given power. It is well to give to the apparatus a construction such that horizontal flight (with the axis of the propeller in the direction of the path) is made under an angle as near as possible to the values indicated for i_3.

We have constructed the logarithmic polar curve for a given position of the depth rudder. We have, by means of this curve, studied the properties of horizontal flight for an apparatus under different angles of flight. We have then supposed that for all these angles the air resistance passed sensibly through the point of intersection of the axis of the propeller and of the vertical through the center of gravity. For each apparatus this is sensibly true for an average position for the depth rudder.

But we may approach still more closely to reality. Experiments made on a model with various positions of the depth rudder give the resultants of the air resistance which pass exactly through the point γ of intersection of the axis of the propeller and of the vertical through the center of gravity. We then have sufficient data for the following tables:

POSITION OF RUDDER	CHARACTERISTICS OF THE RESISTANCE OF THE AIR, PASSING THROUGH T
A	R_{xA}, R_{yA}, i_A
B	R_{xB}, R_{yB}, i_B
C	R_{xC}, R_{yC}, i_C

With these data we may construct the curves

$R_y = f_1\ (R_x)$ Ordinary polar curve.
$R_y = f_2\ (i)$
$\log. R_y = f_3\ (\log. R_x)$ Logarithmic polar curve.

A point of one of these curves gives, not only the values of R_x, R_y, i, but indicates at the same time the corresponding position of the depth rudder.

The Apparatus of Aviation

M. Eiffel has made, by the fan method, a great number of tests on models of certain forms of apparatus. From these tests we may deduce a certain number of rules, which we shall state at a later point; rules which may serve to establish the preliminary design of an airplane.

The interesting experiments at the Institute of Saint-Cyr on an airplane entire (by means of the car) or on an airplane in free flight are not yet sufficiently numerous to give ground for rules of construction for airplanes. However, these results merit statement.

(1) M. Eiffel has shown fully the use which may be made of a study of the logarithmic diagram for the conditions of operation of an airplane in horizontal movement. It is thus that he has studied the régime of maximum speed for a given power and also the economical régime.

The maximum speed for horizontal flight depends more especially on the engine installed on board the avion (see fig. 3, point i_1).

The economical régime, or régime of minimum power for given weight (see fig. 3, point i_3), is of great interest. In fact, when an avion rises with the maximum vertical speed, it is placed in conditions such that the useful power developed shall be minimum, the excess of power being utilized for raising the airplane to the greatest possible height.

The limiting speeds of an airplane for planing are:

(a) The maximum speed of normal horizontal flight.

(*b*) The speed corresponding to the minimum slope.

This minimum is defined by the minimum vaule of $\frac{R_x}{R_y}$. The angle which corresponds to this minimum is the best angle of planing of Col. Charles Renard.

The motive quality or sustaining quality of an airplane introduced by the Constructor Louis Bréguet has for value

$$g = \frac{\rho\sqrt{\frac{Q}{S}}}{\rho\frac{P_M}{Q} - V_m}$$

in which

ρ = efficiency of propeller.

$\frac{Q}{S}$ = weight in kilograms carried per square meter of surface.

$\rho\frac{P_M}{Q}$ = useful work (kilogram-meter-second) of the motor propeller combination per kilogram of weight carried. This power corresponds to the efficiency ρ of the propeller and to the full power P^M of the motor.

V = maximum vertical speed in meters per second.

(2) Ordinary monoplanes.

The following coefficients result from the experiments of M. Eiffel.

(*a*) The loads sustained in relation to the sustaining surface vary between 25 and 35 kilograms per square meter (5.12 to 7.17 pounds per square foot).

(*b*) The maximum speeds of horizontal flight are comprised between 26.4 and 33.3 meters per second (86.6 and 109.3 feet per second) or 95 and 120 kilometers per hour (59 and 74.6 miles per hour).

The speeds for the economical régime vary between 19.44 and 25 meters per second (63.8 and 82 feet per second) or 70 and 90 kilometers per hour (43.5 and 55.9 miles per hour).

Let us give the name "portance" to the ratio:

$$\frac{Q}{S} \times \frac{1}{V^2}$$

The portance for maximum speed of horizontal flight varies between 0.025 and 0.040. The portance for economical speeds varies between 0.040 and 0.070. The values utilized vary, therefore, between 0.025 and 0.070.

(*d*) The maximum useful power (maximum horizontal flight) per 100 kilograms (220 pounds) of weight carried varies between 8 and 11 horse-power.

The minimum useful power (economical régime) per 100 kilograms (220 pounds) of weight carried varies between 5 and 6 horse-power.

The useful power expended in raising 100 kilograms (220 pounds) with the maximum vertical speed varies between 1.5 and 6 horse-power.

(*e*) The maximum vertical speeds vary between 2.3 and 4.25 meters per second (7.55 and 13.94 feet per second).

(*f*) Let us assume 6 horse-power per 100 kilograms (220 pounds) for the economical régime.

Let there be an expenditure of 2 horse-power per 100 kilograms (220 pounds) for climbing.

This permits of raising 100 kilograms (220 pounds) a distance of 450 meters (147.6 feet) in five minutes.

In a preliminary design we may assume a useful power of 8 horse-power per 100 kilograms of weight carried.

If the propeller has a mean efficiency of 0.70, the power developed on the shaft is $8/0.7 = 11.5$ horse-power per 100 kilograms of weight carried.

In a preliminary design for a monoplane, it is necessary to count on 11 to 12 horse-power per 100 kilograms (220 pounds) of total weight carried, say 120 horse-power for an airplane of which the total weight in flying condition is equal to 1000 kilograms (2204 pounds). The consumption per horse-power will be 0.32 to 0.52 kilograms (0.71 to 1.15 pounds) of gasoline and oil, and the weight per horse-power of the engine-propeller equipment, 2 to 3.2 kilograms (4.41 to 7.06 pounds).

(*g*) The minimum values of $\frac{R_x}{R_y}$ are comprised between 0.16 and 0.20.

The best planing angles are comprised between 9° and 11.3° (mean angle = 10°).

The ratios of the limiting speeds for planing are comprised between 1.27 and 1.48.

(*h*) The values of the motive quality are comprised between 0.83 and 1.05.

(3) Biplanes.

(*a*) The loads carried in relation to the carrying surface vary between 15 and 30 kilograms per square meter (3.07 and 6.15 pounds per square foot).

(*b*) The maximum speeds for normal horizontal flight are comprised between 19.44 and 27.8 meters per second (63.8 and 91.2 feet per second), or 70 and 100 kilometers per hour (43.5 and 62.1 miles per hour). The economical speeds very between 13.9 and 22.2 meters per second (45.6 and 72.8 feet per second), or 50 to 80 kilometers per hour (31 and 49.7 miles per hour).

(*c*) The values of the portance for maximum speeds are comprised between 0.035 and 0.045 and the values for economical speeds between 0.060 and 0.065.

The values utilized lie between 0.035 and 0.065—that is to say, within narrower limits than for monoplanes.

(*d*) The maximum useful power per 100 kilograms (220 pounds) of weight carried varies between 5 and 7 horse-power.

The minimum useful power per 100 kilograms (220 pounds) of weight carried varies between 4 and 5 horse-power.

The useful power expended in lifting 100 kilograms (220 pounds) with the maximum vertical speed varies from 0.5 to 2.5 horse-power.

(*e*) The maximum vertical speeds vary between 0.5 and 1.6 meters per second (1.64 and 5.25 feet per second).

(*f*) Let us assume 5 horse-power per 100 kilograms (220 pounds) minimum useful power and 2 horse-power per 100 kilograms (220 pounds) for useful power required for climbing; it is seen that, for 100 kilograms (220 pounds) of weight carried, there will be required a useful power of 7 horse-power, or a power of 10 horse-power absorbed by the shaft, assuming 0.70 for the mean efficiency of the propeller. This indicates a power of 100 horse-power for an airplane of 1000 kilograms (2204 pounds).

For a biplane as compared with a monoplane, there is therefore required less power for the same weight carried.

(*g*) The minimum values of $\frac{R_x}{R_y}$ are contained between 0.142 and 0.228. The best planing angles range between 8° and 11°. The ratios of the limiting speeds of planing are comprised between 1.08 and 1.22.

(*h*) The values of the motive quality are comprised between 0.75 and 1.17.

(4) Hydravions (seaplanes).

(*a*) The loads in relation to the carrying surface vary between 30 and 40 kilograms per square meter (6.15 and 8.19 pounds per square foot).

(*b*) The minimum useful power per 100 kilograms of weight carried is 5 to 6 horse-power for hydravions with floats and 4 to 5 horse-power for hydravions with a boat fuselage.

For the first it is necessary to provide 12 to

13 horse-power (on account of the surface tension which must be overcome as the floats leave the surface of the water) for the power developed by the engine on the shaft per 100 kilograms of weight carried, or 104 horse-power (say an engine of 120 horse-power) for an equipment weighing 800 kilograms. The weight of the engines in flying condition represents about 45 per cent. of the total weight of the entire equipment.

For hydravions with a boat fuselage it is necessary to count on 13 or 14 horse-power per 100 kilograms of weight carried for the power developed by the engine on the shaft, or 560 horse-power (two engines of 300 horse-power) for an equipment weighing 4000 kilograms (weight of engines = 45 per cent. of the total weight of the equipment).

Memoranda:

CHAPTER XXXV

GOVERNMENT AND CIVILIAN ORGANIZATIONS DEVELOPING NAVAL AERONAUTICS IN THE UNITED STATES

United States Navy Department

Secretary of the Navy, Honorable Josephus Daniels; *Assistant Secretary of the Navy,* Honorable Franklin Delano Roosevelt; *Chief of Bureau of Navigation,* Rear Admiral Leigh Carlyle Palmer; *Chief Bureau of Operations,* Rear Admiral William S. Benson; *Chief of Bureau of Steam Engineering,* Engineer in Chief R. S. Griffin; *Chief Bureau of Construction and Repair,* Chief Naval Constructor David W. Taylor; *Comdt. of Marine Corps,* Maj. Gen. G. Barnett; address Navy Department, Washington, D. C. (See chapter on United States Navy Aeronautics for history of United States Navy Aeronautics. See also chapter on "Aerial Defenses Needed for the Fifteen Naval Districts."

Division of Naval Militia Affairs

Captain Thomas Pickett Magruder, United States Navy, *Chief of Division;* Lieutenant Commander Allen Buchanan, United States Navy, *Assistant to Chief of Division (Personnel);* Lieutenant (Junior Grade) Francis Thornton Chew, United States Navy, *Assistant to Chief of Division (Personnel);* Ensign Francis Gaines Blasdel, United States Navy (retired), *Assistant to Chief of Division (Matériel);* Allen Edmund Mechem, *Chief Clerk.* Address: Navy Department, Washington, D. C.

National Naval Militia Board

(Appointed May 29, 1914, by Secretary of the Navy, in accordance with Section 17 of the Naval Militia Act of February 16, 1914.)

Atlantic Coast, Northern District: Commodore Robert P. Forshew, New York Naval Militia; Commander Joseph M. Mitcheson (retired), Pennsylvania Naval Militia.

Atlantic Coast and Gulf of Mexico, Southern District: Captain Caleb D. Bradham, North Carolina Naval Militia.

Great Lakes District: Captain Edward A. Evers, Illinois Naval Militia.

Pacific Coast District: Lieutenant John A. McGee, California Naval Militia.

See chapter on "Naval Militia Aeronautics" for history of Naval Militia Aeronautics.

The Aircraft Production Board and Council of National Defense

ADDRESS: MUNSEY BUILDING, WASHINGTON, D. C.

The Aircraft Production Board was created on May 21, 1917, to act in the closest cooperation with the War and Navy Departments in carrying out the Government's aircraft policy. The members of the Board are: Howard E. Coffin, chairman; Brigadier General George O. Squier, Chief Signal Officer, U. S. A.; Rear Admiral David W. Taylor, Chief of the Bureau of Construction, U. S. N.; Major Sidney D. Waldon, S. O. R. C.; E. A. Deeds and R. L. Montgomery. In the announcement of the creation of the Aircraft Production Board there was stated that the Aircraft Production Board was to carry out America's war policy in the air, which involves a program of turning out in American factories of thousands of aeroplanes, including both training and battle types, and the establishment of schools and training fields with sufficient capacity not only to man these machines but to supply a constant stream of aviators and mechanics to the American forces in Europe.

Historic.—A movement to bring about closer cooperation between the various branches of the Government, and to coordinate our national resources was first suggested some three years ago. In the spring of 1915, several patriotic organizations, including the Aero Club of America, the National Security League, the Navy League, the Army League, the National Aerial Coast Patrol Commission, cooperating in the Conference Committee on National Preparedness, advocated the establishment of the Council of National Defense to comprise cabinet officers with the chairmen of the most important committees of Congress and various distinguished civilian experts. The Government was quick in realizing the value of such a council to coordinate the national resources for national defense, and the council was authorized by the Sixty-fourth Congress.

Authorization.—The purpose of the Council is set forth in the congressional authorization which was part of the Army Appropriation Bill (H.R. 17498), 1916, as follows:

"That a Council of National Defense is hereby established, for the coordination of industries and resources for the national security and welfare, to consist of the Secretary of War, the Secretary of the Navy, the Secretary of the Interior, the Secretary of Commerce, and the Secretary of Labor.

"That the Council of National Defense shall nominate to the President, and the President shall appoint, an advisory commission, consisting of not more than seven persons, each of whom shall have special knowledge of some industry, public utility, or the development of some natural resource, or be otherwise specially qualified, in the opinion of the Council, for the performance of the duties hereinafter provided. The members of the advisory committee shall serve without compensation, but shall be allowed actual expenses of travel and subsistence when attending meetings of the commission or engaged in investigations pertaining to its activities. The advisory commission shall hold such meetings as shall be called by the council or be provided by the rules and regulations adopted by the council for the conduct of its work.

"That it shall be the duty of the Council of National Defense to supervise and direct investigations and make recommendations to the President and the heads of the executive departments as to the location of railroads with reference to the frontier of the United States so as to render possible expeditions, concentration of troops and supplies to points of defense; the coordination of military, industrial, and commercial purposes in the location of extensive highways; the mobilization of military and naval resources for defense; the increase of domestic production of articles and materials essential to the support of armies and of the people during the interruption of foreign commerce; the development of seagoing transportation; data as to amounts, location, method and means of production, and availability of military supplies; the giving of information to producers and manufacturers as to the class of supplies needed by the military and other services of the Government, the requirements relating thereto, and the creation of relations which will render possible in time of need the immediate concentration and utilization of the resources of the nation.

"That the Council of National Defense shall adopt rules and

regulations for the conduct of its work, which rules and regulations shall be subject to the approval of the President and shall provide for the work of the advisory commission to the end that the special knowledge of such commission may be developed by suitable investigation, research, and inquiry and made available in conference and report for the use of the council; and the council may organize subordinate bodies for its assistance in special investigations, either by the employment of experts or by the creation of committees of specially qualified persons to serve without compensation, but to direct the investigations of experts so employed."

The first step to organize the Council of National Defense was taken by President Wilson on October 11th, 1916, by the appointment of seven civilian members of the advisory board of the Council as follows:

Daniel Willard, Chairman; Bernard M. Baruch, Howard E. Coffin, Hollis Godfrey, Samuel Gompers, Dr. Franklin H. Martin, Julius Rosenwald.

Director of the Council and of the Advisory Commission—Walter S. Gifford.

Secretary of the Council and of the Advisory Commission—Grosvenor B. Clarkson.

Chief Clerk and Disbursing Officer, E. K. Ellsworth.

In announcing the appointments, President Wilson issued the following statement:

The Council of National Defense has been created because the Congress has realized that the country is best prepared for war when thoroughly prepared for peace. From an economic point of view there is now very little difference between the machinery required for commercial efficiency and that required for military purposes. In both cases the whole industrial mechanism must be organized in the most effective way. Upon this conception of the national welfare the Council is organized, in the words of the act, for "the creation of relations which will render possible in time of need the immediate concentration and utilization of the resources of the nation."

The organization of the Council likewise opens up a new and direct channel of communication and cooperation between business and scientific men and all departments of the Government, and it is hoped that it will, in addition, become a rallying point for civic bodies working for the national defense. The Council's chief functions are:

1. The coördination of all forms of transportation and the development of means of transportations to meet the military, industrial and commercial needs of the nation.

2. The extension of the industrial mobilization work of the Committee on Industrial Preparedness of the Naval Consulting Board. Complete information as to our present manufacturing and producing facilities adaptable to many-sided uses of modern warfare will be procured, analyzed and made use of.

One of the objects of the Council will be to inform American manufacturers as to the part they can and must play in national emergency. It is empowered to establish at once and maintain through subordinate bodies of specially qualified persons an auxiliary organization composed of men of the best creative and administrative capacity, capable of mobilizating to the utmost the resources of the country.

The National Advisory Committee for Aeronautics

OFFICE: MUNSEY BUILDING, WASHINGTON, D. C.

The National Advisory Committee on Aeronautics was established by Act of Congress, March 3, 1915, to meet the growing demand for a greater activity in aeronautics, by the Military Departments of the Government. The purpose of the committee is set forth in the congressional authorization as follows:

Authorization.—An Advisory Committee for Aeronautics is hereby established, and the President is authorized to appoint not to exceed twelve members, to consist of two members from the War Department, from the office in charge of military aeronautics; two members from the Navy Department, from the office in charge of naval aeronautics; a representative each of the Smithsonian Institution, of the United States Weather Bureau, and of the United States Bureau of Standards; together with not more than five additional persons who shall be acquainted with the needs of aeronautical science, either civil or military, or skilled in aeronautical engineering or its allied sciences: Provided, That the members of the Advisory Committee for Aeronautics, as such, shall serve without compensation: Provided further, That it shall be the duty of the Advisory Committee for Aeronautics to supervise and direct the scientific study of the problems of flight, with a view to their practical solution, and to determine the problems which should be experimentally attacked, and to discuss their solution and their application to practical questions. In the event of a laboratory or laboratories, either in whole or in part, being placed under the direction of the committee, the committee may direct and conduct research and experiment in aeronautics in such laboratory or laboratories: And provided further, That rules and regulations for the conduct of the work of the committee shall be formulated by the committee and approved by the President.

That the sum of $5000 a year, or so much thereof as may be necessary, for five years is hereby appropriated, out of any money in the Treasury not otherwise appropriated, to be immediately available, for experimental work and investigations undertaken by the committee, clerical expenses and supplies, and necessary expenses of members of the committee in going to, returning from, and while attending meetings of the committee: Provided, That an annual report to the Congress shall be submitted through the President, including an itemized statement of expenditures.

The creation of the National Advisory Committee for Aeronautics was the first important movement inaugurated for effective cooperation between the government departments interested in aeronautics, and with the industrial interests of the nation, especially those engaged directly in the manufacture of aircraft. The committee represents all interested departments of the United States Government and, in addition, has the authority to officially coordinate their efforts and to cooperate with the manufacturers and designers of aircraft.

The committee was organized on April 23, 1915, with temporary headquarters in the State, War, and Navy Building. The rules and regulations, which were adopted by the committee at this meeting, and later approved by the President, provide for the annual election of a chairman and a secretary, and also authorize an executive committee consisting of seven members of the advisory committee. To the executive committee full authority has been delegated to control the administration of the affairs of the committee and to supervise all arrangements for research and experiment undertaken or promoted by the committee. The committee, however, will not expend public money for the development of inventions or experimenting with inventions for the benefit of individuals or manufacturers.

The full committee meets twice a year—in April and October. The executive committee holds regular meetings monthly, and special meetings when necessary.

The organization of the committee, as of May 15, 1917, is as follows: William F. Durand, chairman; S. W. Stratton, secretary; Joseph S. Ames, Capt. V. E. Clark, John F. Hayford, Charles F. Marvin, Hon. Byron R. Newton, Michael I. Pupin, Brig.-Gen. George O. Squier, Rear-Admiral D. W. Taylor, Lieut. J. H. Towers, Charles D. Walcott.

To facilitate the work of the committee, the following subcommittees have been established:

Aerial Mail Service—Brig.-Gen. George O. Squier, U. S. A., chairman.

Aero Torpedoes—Lieut.-Commander John H. Towers, U. S. N., chairman.

Aircraft Communications—Dr. Michael L. Pupin, chairman.

Aeroplane Mapping Committee—Brig.-Gen. George O. Squier, U. S. A., chairman.

Bibliography of Aeronautics—Prof. Charles F. Marvin, chairman.

The Naval Consulting Board of the United States at the first meeting in Secretary Daniels' office, October 6, 1916. The names are as follows:

1 Thomas A. Edison
2 Secretary Daniels
3 Elmer A. Sperry
4 Hudson Maxim
5 Miller R. Hutchinson
6 Alfred Craven
7 Peter Cooper Hewitt
8 Benjamin B. Thayer
9 Joseph W. Richards
10 Arthur G. Webster
11 Andrew M. Hunt
12 Andrew L. Riker
13 Henry A. Wise Wood
14 Rear-Admiral Robert S. Griffin
15 Rear-Admiral Victor Blue
16 R. G. Lamme
17 Rear-Admiral David W. Taylor
18 Spencer Miller
19 Frank J. Sprague
20 Lawrence Addicks
21 Howard E. Coffin
22 Rear-Admiral Joseph Strauss
23 Thomas Robbins
24 Wm. L. Emmet
25 L. H. Baekeland
26 Wm. L. Saunders
27 W. R. Whitney
28 Captain Mark J Bristol
29 Robert S. Woodward
30 Rear-Admiral Wm. S. Benson
31 M. B. Sellers

Buildings, Laboratories, and Equipment—Dr. Joseph S. Ames, chairman.

Design, Construction, and Navigation of Aircraft—Brig.-Gen. George O. Squier, U. S. A., chairman.

Governmental Relations—Dr. Charles D. Walcott, chairman.

Nomenclature for Aeronautics—Dr. Joseph S. Ames, chairman.

Patents—Dr. Charles D. Walcott, chairman.

Physics of the Air—Prof. Charles F. Marvin, chairman.

Power Plants—Dr. S. W. Stratton, chairman.

Relation of the Atmosphere to Aeronautics—Prof. Charles F. Marvin, chairman.

Standardization and Investigation of Materials—Dr. S. W. Stratton, chairman.

Foreign Representatives—Dr. Charles D. Walcott, chairman.

Naval Consulting Board of the United States

OFFICE: MUNSEY BUILDING, WASHINGTON, D. C.

The Naval Consulting Board was authorized by Secretary of the Navy Daniels in August, 1915, who also appointed Mr. Thomas A. Edison, chairman of the Board for life. The purpose of creating the Board was, in the words of Secretary Daniels, "to make available the latent inventive genius of our country to improve our navy."

The first meeting of the Board was held in the offices of Secretary Daniels on October 6, 1916. The members and navy officials who attended the meeting are shown in the accompanying illustration. Aeronautic matters are attended to by the Committee on Aeronautics, of which Mr. Elmer A. Sperry is chairman, succeeding Mr. Henry A. Wise Wood, resigned, whose place on the Naval Consulting Board was filled by the appointment of Mr. Bion J. Arnold.

The Naval Consulting Board has done much work of national importance, including the mobilization of industries.

The United States Coast Guard

The Coast Guard is part of the Department of the Treasury, of which Hon. William Gibbs McAdoo is Secretary and Hon. Byron R. Newton is Assistant Secretary. The offices of the Coast Guard are at the Munsey Building, Washington, D. C. The administrative officers are: Captain Commandant, Ellsworth P. Bertholf; Chief of Division of Operations, Oliver M. Maxam; Chief of Division of Material, G. H. Slaybaugh; Superintendent of Construction and Repair, Senior Captain Howard Emery; Engineer in Chief, Charles A. McAllister; Inspector, Senior Capt. D. P. Foley.

At a meeting of the Board of Governors of the Aero Club of America in 1911, Mr. Alan R. Hawley, the President of the Club, pointed out the possibility of increasing the efficiency of the Life Saving Service and Revenue Cutter Service by employing aeroplanes, and emphasized the desirability of concentrating efforts to carry out this idea, which would bring about the use of aeroplanes for utilitarian purposes. This proposal was entered in the minutes of the meeting. In December, 1913, the special number of "Flying," to celebrate the tenth anniversary of the first flight, printed an extensive article by Mr. Henry Woodhouse, pointing out the possible developments which would bring about the use of aeroplanes for utilitarian purposes, and pointed out that one of the prospective fields for the employment of aeroplanes for utilitarian purposes would be the Life Saving Service and the Revenue Cutter Service.

As soon as Mr. Byron R. Newton, the present Assistant Secretary of the Treasury, was appointed to office, he very enthusiastically began to consider the possibility of increasing the efficiency of the Revenue Cutter Service and the Life Saving Service, which are under the jurisdiction of the Treasury Department, and finding that the aeroplane would, more than anything else, increase the efficiency of the Revenue Cutter Service and the Life Saving Service. which were combined under the name of the Coast Guard, he began to consider the plan to give an aviation section to the Coast Guard.

At about this time, Captain Thomas of the United States Navy discussed the possibilities of employing aeroplanes in the Coast Guard with Captain Benjamin H. Chiswell, of the Cutter *Onondaga,* who became very enthusiastic and began to figure out plans to bring about this development. Captain E. P. Bertholf, who commands the Coast Guard became interested in the plan, and arrangements were made to send two coast guard officers, Second Lieutenants Elmer F. Stone, and Charles E. Sugden, to the Pensacola Naval Aeronautical Station for instruction in aviation: The cooperation of the Curtiss institution was secured, and the Coast Guard was offered every facility for experimentation at the Curtiss Aviation Camp at Newport News, Virginia. Lieutenant Stone was first designated as observer, and was replaced by First-Lieut. Norman B. Hall. The Sperry Gyroscope Company offered the use of the gyroscope compass and a radio apparatus for experimentation, and Lieutenant Hall has since been developing the scientific and mechanical part of the proposition.

Then a tentative bill was drawn by Captain C. A. McAllister, Chief Engineer of the Coast Guard Service, which was introduced in Congress by Congressman Andrew J. Montague, of Virginia.

Upon the earnest recommendation of the Secretary of the Treasury, and of the other administrative officials of the Coast Guard, Congress very wisely authorized, in the Naval Appropriation Act, approved August 29, 1916, aviation facilities for this humanitarian branch of the public service. The Act, in general, provides that for the purpose of saving life and property along the coasts of the United States and at sea contiguous thereto, and to assist in the national defense, there shall be established not exceeding ten aviation stations at such points on the Atlantic and Pacific coasts, the Gulf of Mexico and the Great Lakes, as the Secretary of the Treasury may deem advisable. One of these stations may be fitted up as a school of instruction, and provision is made for the necessary increased personnel of the Coast Guard by authorizing not more than fifteen additional commissioned officers and forty warrant officers and enlisted men for aviation duty. The Secretaries of War and Navy are also authorized at the request of the Secretary of the Treasury, to receive officers and enlisted men of the Coast Guard for instruction in aviation at any aviation school maintained by the Army and Navy. While engaged in aviation duty all officers and men will receive the additional pay and allowances now or hereafter authorized for officers and men of the Navy detailed for such duty.

Unfortunately in the closing hours of Congress after the Naval Appropriation Bill had become a law, the necessary appropriations for this extension of the facilities of the Coast Guard were not made available. However, this fact will not necessarily impede progress in the institution of this important auxiliary, as already a commissioned officer, Capt. B. M. Chiswell, U. S. C. G., has been detailed in charge of aviation matters, and several junior officers have been assigned to naval and commercial schools for the necessary instruction. In the immediate future a board of officers is to be appointed by the Coast Guard to select a site for an experimental station, to determine the type, size, and equipment of the necessary hangars, and in a general way to outline a general plan for the institution of the authorized aviation facilities. By the time Congress reconvenes in December this board will have reported and made its recommendations. Deficiency appropriations will then be submitted to Congress and as soon as made available active steps will be taken towards as rapid development of this function of the Coast Guard as is possible under the circumstances.

The Aero Club of America and Constituent Aero Clubs

The Aero Club of America has been the most important factor in the developing of naval aeronautics in America. In the early days of the hydroaeroplane, when this type of aircraft was considered as a freak, it was the Aero Club of America and the constituent aero clubs that stood sponsor for it, a number of their members purchasing hydroaeroplanes and flying boats for pleasure. It was also the Aero Club of America that spent thousands of dollars to conduct a campaign of education to make known the need of aerial defenses for the fifteen naval districts. The chapters on the "Evolution of the Seaplane," "Naval Militia Aeronautics," and the "Aerial Coast Patrol" give details of some of the most substantial work of the Aero Club of America and its constituent aero clubs to develop naval aeronautics.

The Aero Club of America was founded in 1905 and is the sole representative of the International Aeronautic Federation and the Pan-American Aeronautic Federation in the United States, thereby being affiliated with the aero clubs of thirty-eight countries.

The officers and governors of the Aero Club of America are:

Alan R. Hawley, President; Henry A. Wise Wood, Vice-President; Cortlandt F. Bishop, Vice-President; Charles Jerome Edwards, Vice-President; Godfrey L. Cabot, Vice-President; Charles Elliot Warren, Treasurer; William Hawley, Secretary. Governors: Cortlandt F. Bishop, James A. Blair, Jr., Robert J. Collier, Howard E. Coffin, W. Redmond Cross, Charles Jerome Edwards, Brig.-Gen. Robert K. Evans, U. S. A.; Max C. Fleischmann, John Hays Hammond, Jr., Alan R. Hawley, Major F. L. V. Hoppin, Henry B. Joy, Albert Bond Lambert, W. W. Miller, Capt. James E. Miller, S. O. R. C.; George M. Myers, Harold F. McCormick, Rear Admiral Robert E. Peary, Raymond B. Price, Allan A. Ryan, Alberto Santos-Dumont, Evert Jansen Wendell, Henry A. Wise Wood, Henry Woodhouse.

The chairmen of the Aero Club of America's committees are:

Admission............Major F. L. V. Hoppin
Aerodynamics............Dr. A. F. Zahm
Aeronautical Map............Rear Adm. Robt. E. Peary
Affiliated Clubs............Russel A. Alger
Auditing............Harrington Emerson
Collier Trophy............W. Redmond Cross
Contest............Alan R. Hawley
Dirigible & Kite Balloon............Henry Woodhouse
Entertainment............Chas. Jerome Edwards
Finance............Allan A. Ryan
Foreign Relations............Cortlandt F. Bishop
Grievance............Evert Jansen Wendell
House............Alan R. Hawley
Law............W. W. Miller
Library............Chas. Jerome Edwards
Marine Flying............Henry A. Wise Wood
Governmental Affairs and Relations............W. Redmond Cross
Military & Naval............Col. Cornelius Vanderbilt
Pan-American............Albertos Santos-Dumont
Public Safety............Gen. Theo. A. Bingham, U. S. A.
Publicity............Henry Woodhouse
Spherical Balloon............George M. Myers
Technical............Raymond B. Price
Trans-Atlantic Flight............Cortlandt F. Bishop

The membership of the seven committees having most to do with military aeronautics are:

Committee on Governmental Affairs and Relations: W. Redmond Cross, Chairman; Maj. Raynal C. Bolling, S. O. R. C.; Howard E. Coffin, Alan R. Hawley, Capt. James E. Miller, S. O. R. C.; Raymond B. Price, Henry A. Wise Wood, Henry Woodhouse.

Contest Committee: Alan R. Hawley, Chairman; W. Redmond Cross, Lieut. Godfrey L. Cabot, N. M. M.; Capt. Philip A. Carroll, S. O. R. C.; Capt. D. de F. Chandler, U. S. A.; A. B. Lambert, Capt. J. C. McCoy, S. O. R. C.; Lieut.-Comdr. H. C. Mustin, U. S. N.; Henry A. Wise Wood, Henry Woodhouse.

Military and Naval Aviation Committee: Cornelius Vanderbilt, Chairman; Lieut. Vincent Astor, N. M. N. Y.; Maj. Raynal C. Bolling, S. O. R. C.; Captain Mark L. Bristol, U. S. N.; Maj. Charles de F. Chandler, U. S. A.; Comdr. R. K. Crank, U. S. N.; Brig.-Gen. Robert K. Evans, U. S. A.; Rear Admiral Bradley A. Fiske, U. S. N.; Lieut. Lee H. Harris, N. M. N. Y.; Lt. V. D. Herbster, U. S. N.; Major F. L. V. Hoppin, N. G. N. Y.; Capt. James E. Miller, S. O. R. C.; Lieut.-Com. H. C. Mustin, U. S. N.; Rear Adm. Robert E. Peary, U. S. N.; Brig.-Gen. George O. Squier, Chief Signal Officer, U. S. A.; Lt. Comr. John H. Towers, U. S. N.; Major Charles Elliot Warren, Officers' Reserve Corps.

Marine Flying Committee: Henry A. Wise Wood, Chairman; Frederick M. Bourne, Lieut. Godfrey L. Cabot, N. M. M.; James Deering, W. Earl Dodge, Lieut. F. Trubee Davison, N. R. F. C.; James Elverson, Jr.; John Hays Hammond, Jr.; Harry S. Harkness, Henry B. Joy, A. L. Judson, Harold F. McCormick, Ogden Mills Reid, William E. Scripps, Rodman Wanamaker, Harry Payne Whitney, Eugene S. Willard.

Dirigible and Kite-Balloon Committee: Henry Woodhouse, Chairman; Capt. Thomas S. Baldwin, S. O. R. C.; Capt. Mark L. Bristol, U. S. N.; Major Charles de F. Chandler, U. S. A.; Brig.-Gen. Robert K. Evans, U. S. A.; Rear Admiral Bradley A. Fiske, U. S. N.; Lieut. V. D. Herbster. U. S. N.; Otto H. Kahn, Major Frank P. Lahm, U. S. A.; Lieut. L. H. Maxfield, U. S. N.; Herman A. Metz, Samuel C. Morehouse, E. R. Preston, Raymond B. Price, A. Leo Stevens, Professor David Todd, Ralph H. Upson.

Aeronautical Map and Landing Places Committee: Rear Admiral Robert E. Peary, Chairman; Bion J. Arnold, A. H. Ackermans, August Belmont, James Gordon Bennett, Cortlandt F. Bishop, Captain Mark L. Bristol, U. S. N.; Capt. W. Starling Burgess, S. O. R. C.; Lieut. Godfrey L. Cabot, N. M. M., President Aero Club of New England; President Manuel Estrada Cabrera of Guatemala; Capt. Joseph E. Carberry, U. S. A.; Capt. W. I. Chambers, U. S. N.; J. Parke Channing, Roy D. Chapin, Alexander Smith Cochran, Robert J. Collier, Glenn H. Curtiss, Comdr. Cleveland Davis, U. S. N.; F. E. deMurias, Charles deSan Marzano, Charles Dickinson. President of Aero Club of Illinois; F. G. Diffin, W. Earl Dodge, Harrington Emerson, John Hays Hammond, Jr.; W. Averill Harriman, Wm. Hawley, Henry B. Joy, Otto H. Kahn, Frank S. Lahm, A. B. Lambert, Henry Lockhart, Jr.; Ensign Robert A. Lovett, N. R. F. C.; Harold F. McCormick, Capt. J. C. McCoy, S. O. R. C.; Herman A. Metz, Eugene Meyer, Jr.; Capt. James E. Miller, S. O. R. C.; Lieut.-Comdr. Henry C. Mustin, U. S. N.; George M. Myers, President Aero Club of Kansas City; George W. Perkins, Augustus Post, Col. Samuel Reber, U. S. A.; Thomas F. Ryan, Alberto Santos-Dumont, Mortimer L. Schiff, Frank A. Seiberling, William G. Sharp, U. S. Ambassador to France; Lawrence B. Sperry, Brig.-Gen. George O. Squier, U. S. A.; Chief Signal Officer Joseph A. Steinmetz, President Aero Club of Pennsylvania; James S. Stephens, Lieut.-Com. J. H. Towers, U. S. N.; K. M. Turner, George W. Turney, Inglis M. Uppercu, Col. Cornelius Vanderbilt, W. K. Vanderbilt, L. A. Vilas, George Von Utassy, Rodman Wanamaker, Evert Jansen Wendell, Hugh L. Willoughby, Henry A. Wise Wood, Henry Woodhouse, Maj. Orville Wright, S. O. R. C.; William Wallace Young, A. Francis Zahm.

The Aero Club of America maintains a club House and headquarters at 297 Madison Avenue, New York City, and Washington offices in the Union Trust Building, Washington, D. C.

The aero clubs affiliated with the Aero Club of America are: Aero Club of Baltimore, Aero Club of Buffalo, Colorado Aero Club, Aero Club of Connecticut, Aero Club de Cuba, Aero Club of Dayton, Aero Club of Illinois, Aero Club of Michigan, Aero Club of New England, Aero Club of New York, Aero Club of Ohio, Aero Club of Pennsylvania, Aero Club of Pittsfield, Queen City Aero Club, Aero Club of St. Louis, Aero Club of Washington, Aircraft Club of Peoria, Harvard Aeronautical Society, Kansas City Aero Club, Milwaukee Aero Club, Pacific Aero Club, Rochester Aero Club, Aero Club of Northwest, Western Aero Association, Wichita Aero Club, New York Flying Yacht Club, Aeronautical Society of California, Aero Club of Iowa.

The National Aerial Coast Patrol Commission

ADDRESS: UNION TRUST BUILDING, WASHINGTON. D. C.

The National Aerial Coast Patrol Commission was founded in 1915. (See Chapter on "Aerial Coast Patrol" for report of the Commission's work.)

Central Committee: Hon. Thomas R. Marshall, Vice-president of the United States, Honorary Chairman; Rear Admiral Robert E. Peary, U. S. N. retired, Chairman; Senator Morris Sheppard, of Texas; Senator James E. Watson, of Indiana; Representative Julius Kahn, of California; Representative Murray Hulbert, of New York; Hon. Byron R. Newton, Assistant Secretary of the Treasury; Hon. William M. Ingranam, Assistant Secretary of War; Mr. Alan R. Hawley, President, Aero Club of America; Mr. Henry Woodhouse, member Board of Governors, Aero Club of America, Delegate on Industry and Education Pan-American Aeronautical Federation; Lieut. F. Trubee Davison, Flying Reserve, U. S. Navy, Organizer, Aerial Coast Patrol Unit No. 1; Dr. E. Lester Jones, Superintendent U. S. Coast and Geodetic Survey; Dr. H. C. Frankenfield, Chief Forecaster U. S. Weather Bureau; Hon. Emerson McMillin, Mr. John Hays Hammond, Jr.

Secretary: Mr. Earl Hamilton Smith, Washington Representative Aero Club of America.

State members: The Presidents of the Aero Clubs affiliated with the Aero Club of America.

The Adjutants General and the Commanding officers of the Naval Militia of the States.

Board Cooperating with the Commandant of the Third Naval District in the Organizational of the Naval Reserve Forces

ADDRESS: 297 MADISON AVENUE, NEW YORK.

The Board, which was appointed on March 2d at the request of the naval authorities, and is cooperating with the other Committees working in organizing the naval reserves of the Third Naval District, consists of:

Chairman, nominated by Rear Admiral Usher, Mr. Alan R. Hawley, President of the Aero Club of America, etc.

Operations and Coordination.—Rear Admiral Robert E. Peary, Chairman of the National Aerial Coast Patrol Commission, member of the Board of Governors, Aero Club of America; Eugene S. Willard, member Board of Governors and Chairman of New York State Enrollment Committee, Naval Training Association of the U. S.; member Executive Committee, Power Craft Association, etc.

Enrollments.—Lieutenant F. Trubee Davison, organizer Volunteer Aerial Coast Patrol Unit No. 1; Lewis S. Thompson, Augustus Post, Evert Jansen Wendell, Clinton David Backus.

Instruction.—Robert A. Lovett, member of Volunteer Aerial Coast Patrol Unit No. 1; co-organizer of Aerial Coast Patrol Unit No. 2; F. C. G. Eden, who is in charge of the Dodge Aviation Training Camp for college men; Congressman F. H. La Guardia.

Aviation Training Camps.—Lewis S. Thompson, in charge of the aviation training camp of Volunteer Aerial Coast Patrol Unit No. 1; W. Earl Dodge, founder of the Dodge Aviation Training Camp for College men; Harold Irving Pratt, Harry Frank Guggenheim, Congressman Murray Hulbert, J. F. Knapp, Rodman Wanamaker.

Flying Equipment.—Henry Woodhouse, member Board of Governors, Aero Club of America, Chairman Committee on Aeronautics, National Institute of Efficiency, etc.; Lieut. F. Trubee Davison, W. Earl Dodge, Harold Irving Pratt, Howard S. Borden, August Belmont, Rodman Wanamaker.

Torpedoplanes, Aerial Guns and Explosives.—Rear Admiral Bradley, A. Fiske, retired, President Naval Institute of the United States; member of Committee on Naval Aeronautics, Aero Club of America, etc.; Frank M. Leavitt.

Naval Anti-Aircraft Defenses.—Eugene S. Willard, Schuyler Skaats Wheeler, Captain Robert A. Bartlett, Henry Woodhouse, Charles Elliott Warren.

Communications (Radio, Signaling, etc.).—John Hays Hammond, Jr., member Board of Governors, Aero Club of America; Conference Committee on National Preparedness, etc.; Lawrence B. Sperry, Schuyler Skaats Wheeler, William Dubellier.

Observation Balloons.—Rear Admiral William N. Little, retired, member of numerous organizations.

Secretary of the Committee.—Henry Woodhouse.

(The first name in each sub-committee is the Chairman of that committee.)

This committee has rendered some valuable reports and has done much to establish seaplane stations and encouraging civilians to join the Naval Reserves and arranging for the training of aviators for coast patrol work.

The Aircraft Manufacturers' Association

The Aircraft Manufacturers' Association was organized at the close of 1916. At the elections of officers held May 9, 1917, the following officers were elected: Honorary President, Glenn H. Curtiss; President, Frank H. Russell, Burgess Company, Marblehead, Mass.; Vice-President, Albert H. Flint, L-W-F Engineering Co., College Point, L. I.; Treasurer, Inglis M. Uppercu, Aeromarine Plane & Motor Co., New York City; Secretary, Benjamin L. Williams, New York City; Assistant Treasurer, A. H. Flint. Address 1501 Fifth Avenue, New York City.

Advisory Committee: Albert H. Flint, H. B. Morse, Benjamin Foss.

Committee on Patents: Curtiss Aeroplane & Motor Corp. Chairman, Thomas-Morse Aircraft Co., Sturtevant Aeroplane Co.

Committee on Materials: L-W-F Engineering Co. (Chairman), Curtiss Aeroplane & Motor Corp., Burgess Co., Standard Aero Corp., Sturtevant Aeroplane Co., B. F. Sturtevant Co., Aeromarine Plane & Motor Co., Thomas-Morse Aircraft Co.

Membership Committee: Benjamin Foss, H. B. Mingle, Mr. A. H. Flint.

Standardization Committee: Curtiss Aeroplane & Motor Corp. (Chairman), L-W-F Engineering Co., Burgess Co., Standard Aero Corp., Sturtevant Aeroplane Co., B. F. Sturtevant Co., Aeromarine Plane & Motor Co., Thomas-Morse Aircraft Co.

Publicity, Advertising and Censorship Committee: Fay L. Faurote.

The Society of Automotive Engineers

The Society of Automotive Engineers was the result of the amalgamation, in 1916–17, of the Society of Automobile Engineers and the American Society of Aeronautic Engineers. The officers are:

George W. Dunham, president; Jesse Vincent, vice-president; Charles M. Manley, vice-president; Herbert Chase, treasurer; Benjamin B. Bachman, Harry L. Horning, Charles W. McKinley, Fred E. Moskovics, David Beecroft, John G. Utz, Wm. H. Van Dervort, Russel Huff.

COUNCIL

Finance Committee: H. M. Swetland, Christian Girl, F. C. Glover, Elmer A. Sperry, H. R. Sutphen.

Address: 29 West 39th Street, New York City.

The National Special Aid Society

NATIONAL HEADQUARTERS, 259 FIFTH AVENUE, NEW YORK.

The National Special Aid Society, which was founded in January, 1915, and incorporated in July, 1916, has cooperated extensively with the Aero Club of America and the National Aerial Coast Patrol Commission in advancing the cause of aerial preparedness and has established "The Aviation Treasure and Trinket Fund." The purpose of the Fund is "to meet the Red Cross needs of the air services; the care of those dependent in case of disaster and the long list of the flyers wants in so far as we are permitted and are able."

The officers of the National Special Aid Society are: Mrs. William Alexander, president; Mrs. Charles Frederick Hoffman, vice-president; Mr. Leroy W. Baldwin, treasurer; Mr. Warren A. Mayou, assistant treasurer; Mrs. Henry A. Wise Wood, secretary; Mr. George F. Sweeney, executive secretary.

The members of the Board of Directors are: Mrs. William Alexander, Mrs. John E. Alexander, Mrs. William Allen Bartlett, Mrs. Charles H. Ditson, Mrs. Eliot Butler Whiting, Miss Isabelle H. Hardie, Mrs. Charles F. Hoffman, Mrs. William W. Hoppin, Jr., Mrs. Franklin D. Pelton.

The members of the General Committee are: Mrs. Henry M. Alexander, Mrs. William Alexander, Mrs. J. Stewart Barney, Mrs. O. H. P. Belmont, Mrs. William H. Bliss, Mrs. Oliver B. Bridgman, Mrs. I. Townsend Burden, Jr., Mrs. Frederic Foster Carey, Mrs. Francis Carolan, Mrs. Henry Ives Cobb, Jr., Mrs. William H. Crocker, Mrs. Frederick Y. Dalziel, Mrs. Howard Davison, Duchesse de Chaulnes, Miss Elsie de Wolfe, Mrs. William K. Dick, Mrs. Charles H. Ditson, Mrs. Elisha Dyer, Mrs. Harold Godwin, Mrs. Edwin Gould, Mrs. George Jay Gould, Mrs. Jay Gould, Mrs. Charles E. Greenough, Mrs. P. Cooper Hewitt, Miss Louise Iselin, Mrs. Bradish G. Johnson, Miss Luisita A. Leland, Mrs. Goodhue Livingston, Mrs. John A. Logan, Jr., Mrs. Charles McNeill, Mrs. John G. Milburn, Mrs. John Purroy Mitchel, Mrs. Paul Morton, Mrs. Frederick Neilson, Mrs. Charles de L. Oelrichs, Mrs. Charles M. Oelrichs, Mrs. Frederic Pearson, Mrs. Franklin D. Pelton, Mrs. Henry Pierrepont Perry, Mrs. Allison Wright Post, Mrs. Charles A. Post, Mrs. James Brown Potter, Mrs. William Potter, Mrs. Alexander D. B. Pratt, Mrs. Pulitzer, Mrs. Roche, Mrs. William Rockefeller, Mrs. Theodore Roosevelt, Mrs. George Rose, Mrs. William F. Sheehan, Miss Evelyn Rives Smith, Mrs. Ormond G. Smith, Mrs. Vivian Spencer, Mrs. Leonard Thomas, Princess Pierre Troubetzkoy, Mrs. Allen Gouveneur Wellman, Mrs. Charles W. Whitman, Mrs. M. Orme Wilson, Mrs. Frank Spencer Witherbee.

The "Aviation Treasure and Trinket Fund" of the National Special Aid Society which is patterned after the "Silver Thimble" fund of Great Britain, is supervised by the Aviation Committee of the National Special Aid Society, of which Mrs. H. P. Davison is honorary chairman and Mrs. William A. Bartlett is chairman.

Conference Committee on National Preparedness

HEADQUARTERS: FORTY-SECOND STREET BUILDING, NEW YORK CITY.

The Conference Committee on national preparedness was founded in May, 1915 for the purpose of coordinating the efforts of the organizations working to build up our national defenses. The following organizations became affiliated with the Conference Committee:

The Aero Club of America, the Woman's Section of the Movement for National Preparedness, Amalgamated with the National Civic Federation, the American Legion, the American Red Cross, American Society of Aeronautic Engineers, Army League, Automobile Club of America, Institute of Radio Engineers, National Society for the Advancement of Patriotic Education, National Aerial Coast Patrol Commission, National Security League, the National Special Aid Society, the United States Power Squadrons.

The Conference Committee has carried on an extensive campaign of education on behalf of National Preparedness and Universal Training.

Officers: Henry A. Wise Wood, Chairman; Alexander M. White, Vice-Chairman; Raymond B. Price, Treasurer; James E. Clark, Recording Secretary.

The Pan American Aeronautic Federation

UNITED STATES HEADQUARTERS: 297 MADISON AVENUE, NEW YORK.

The Pan-American Aeronautic Federation was founded as a result of the conference of National Aero Clubs of North, Central and South American countries held at Santiago, Chili, March, 1916. In accordance with Article 2, of its statutes, the purposes of the P. A. A. F. are:

(a) Spreading the knowledge of aeronautics by publications, conferences and having aeronautic exhibitions of all kinds.

(b) Fostering the establishment of schools for training pilots for aeroplanes, balloons and dirigibles.

(c) Fostering the establishment of schools for mechanics for aeroplanes and dirigibles.

(d) Fostering the establishing on the American continent of aerotechnical laboratories for testing and improving aeronautic material and conducting of all kind of research.

(e) Fostering the study of the atmosphere of the American Continents in cooperation with the observatories of the different countries.

(f) Fostering the making and issuing of aeronautic and topographical maps to be used in the service of aeronautics in different countries.

(g) Establishing aerodromes and setting apart proper landing places for aircraft throughout the different countries.

(h) Fostering the study of the history, theory and application of aeronautics pertaining to aerial navigation; including the publication of literary works on same and introducing the study of aeronautics in American Universities.

(i) Studying and analyzing the progress of aerial navigation in the different countries.

The officers of the Federation are: Honorary president for life, Mr. Alberto Santos-Dumont (United States); president, Mr. Jorge Matto Gormaz (Chile); first vice-president, Mr. Cortlandt F. Bishop (United States); second vice-president, Mr. Marechal Borman (Brazil); third vice-president, Mr. Joaquin C. Sanchez (Uruguay); fourth vice-president, Mr. Amador F. Del Solar (Peru); general secretary, Mr. Alberto Mascias (Argentina); informing secretary, Armando Venegas (Chile); treasurer, Mr. Severo Vaccaro (Argentine); directors, Mr. Manuel Seminario (Ecuador); Roberto Araya (Paraguay); Colonel Mr. Carlos Nunez del Prado (Bolivia).

The United States delegates to the Pan-American Aeronautic Federation are: *Scientific:* Orville Wright; *For Sports:* Alan R. Hawley; *Juridical:* Emerson McMillin; *Military:* Rear Admiral Robert E. Peary; *Public Education and Industries:* Henry Woodhouse.

Memoranda:

CHAPTER XXXVI

RULES GOVERNING TESTS FOR THE FEDERATION AERONAUTIQUE INTERNATIONALE PILOT CERTIFICATES

Since the earliest stage of aviation in America certificates for balloon, aeroplane, and dirigible pilots have been granted under the rules and regulations of the International Aeronautic Federation (Federation Aeronautique Internationale), which controls all sporting aeronautic events. The Federation which was founded in 1905, and has its headquarters in Paris, France, is represented in other countries by their national aeronautic bodies. In the United States its representative is the Aero Club of America with headquarters at 297 Madison Avenue, New York City. Founded in 1905, the Aero Club has taken the leading part in the upbuilding of our aerial defenses. The club appoints representatives to witness all tests, and all pilot licenses are issued by its authority. The Aero Club of America may grant aeronautical and aviation pilots' certificates to persons who are over eighteen (18) years of age, citizens of the United States, or citizens of a country not represented in the Federation Aeronautique Internationale, or citizens of a country represented in the Federation Aeronautique Internationale, with the permission of the representative organization of the applicant's nationality.

The following are the rules under which certificates are granted by the Aero Club of America:

1. A person desiring a pilot certificate must apply in writing to the secretary of the Aero Club of America. He must state in his letter the date and place of his birth, and enclose therein two unmounted photographs of himself about $2\frac{1}{4}$x$2\frac{1}{2}$ inches, together with a fee of five dollars. In case the applicant is a naturalized citizen of the United States he must submit proof of naturalization.

2. On the receipt of an application the secretary will forward it promptly to the Contest Committee, which, in case of an application for an aviator's certificate, will designate a representative to supervise the test prescribed by the International Aeronautical Federation, and will advise the representative of the name and location of the applicant and, through the secretary, advise the applicant of the appointment of the representative to take the test.

3. In case the application is for a spherical balloon or for a dirigible balloon pilot certificate the applicant will be fully advised by the Contest Committee.

4. All applications for aviator's certificates must reach the secretary a reasonable time in advance of the date that the applicant may expect to take the required test.

5. No telegraphic applications for certificates will be considered.

Applicants for each class of certificate must be of the age of 18 years, and in the case of dirigible certificates 21 years, and must pass, to the satisfaction of the properly designated representatives of the Aero Club, the tests prescribed by the Federation Aeronautique Internationale as follows:

Spherical Balloon Pilot's Certificate

Candidates must pass the following tests:

(A) Five ascensions without any conditions.

(B) An ascension of one hour's minimum duration undertaken by the candidate alone.

(C) A night ascension of two hours' minimum duration comprised between the setting and the rising of the sun.

The issue of a certificate is always optional.

Dirigible Balloon Pilot's Certificate

Candidates must be 21 years of age.

They must hold a spherical balloon pilot's

FÉDÉRATION AÉRONAUTIQUE
INTERNATIONALE
AERO CLUB OF AMERICA

No. 716

The above-named Club, recognized by the Fédération Aéronautique Internationale, as the governing authority for the United States of America, certifies that

Thorne Donnelley

born 19th day of July 1895 has fulfilled all the conditions required by the Fédération Aéronautique Internationale, for an aviator pilot, and is brevetted as such.

Dated 9th May 1917

Alan R. Hawley
President.

William Hawley
Secretary.

Signature of pilot:

Facsimile of the Pilot certificate. It is bound in leatherette and two additional pages contain information in different languages for the authorities.

certificate and furnish proof of having made twenty (20) flights in a dirigible balloon at different dates.

They must also undergo a technical examination.

In case, however, the candidate does not already possess a spherical balloon certificate, he must have made twenty-five (25) ascensions in dirigibles before he can apply for a certificate.

The application for the certificate must be countersigned by two dirigible balloon pilots, who have been present at at least three of the departures and landings of the candidate.

The issue of the certificate is always optional.

Aviator's Certificate

1. Candidates must accomplish the three following tests, each being a separate flight:

A and B. Two distance flights, consisting of at least 5 kilometers (16,404 feet) each in a closed circuit, without touching the ground or water, the distance to be measured as described below.

C. One altitude flight, during which a height of at least 100 meters (328 feet) above the point of departure must be attained; the descent to be made from that height with the motor cut off. A barograph must be carried on the aeroplane in the altitude flight. The landing must be made in view of the observers, without restarting the motor.

2. The candidate must be alone in the aircraft during the three tests.

3. Starting from and landing on the water is only permitted in one of the tests A and B.

4. The course on which the aviator accomplishes tests A and B must be marked out by two posts or buoys situated not more than 500 meters (547 yards) apart.

5. The turns round the posts or buoys must be made alternately to the right and to the left so that the flight will consist of an uninterrupted series of figures of 8.

6. The distance flown shall be reckoned as if in a straight line between the two posts or buoys.

7. The landing after the two distance flights in tests A and B shall be made:

(*a*) By stopping the motor at or before the moment of touching the ground or water;

(*b*) By bringing the aircraft to rest not more than 50 meters (164 feet) from a point indicated previously by the candidate.

8. All landings must be made in a normal manner, and the observers must report any irregularities.

The issuance of the certificate is always optional.

Official observers must be chosen from a list drawn up by the governing organization of each country.

Hydroaeroplane Pilot's Certificate

The tests to be successfully accomplished by candidates for this certificate are the same as those for an aviator's certificate, except that starting from and landing on the water is permitted for all of the tests.

Licenses

Every person holding a pilot certificate of the Federation Aeronautique Internationale may obtain the license issued optionally by the Contest Committee of the Governing Board to its own citizens or those under its jurisdiction. This license constitutes the title "qualification" which alone allows the holder to act as pilot in events governed by the present regulations of the Federation Aeronautique Internationale. It is independent of those set forth in Article 28 of the Statutes of the Federation Aeronautique Internationale.

Application for License

All applications for licenses must contain the following particulars: name and surname, date of birth, nationality, origin and number of pilot certificate.

Every request must be accompanied by the certificate.

Validity and Withdrawal of a License

A license shall be valid until the thirty-first of December of each calendar year.

It may not be withdrawn either temporarily or definitely by the Contest Committee except after approval by the national governing body.

Expert Aviator

The Aero Club of America, having established the grade of expert aviator, has instructed its Contest Committee to prescribe the qualifications for that grade and to make the necessary rules.

The following is published for the information and guidance of all concerned:

The Aero Club of America may grant a certificate as expert aviator to all aviation pilots holding certificates under the Regulations of the Federation Aeronautique Internationale, who are over 21 years of age, and have been recommended for this by the Contest Committee. An aviator desiring this certificate must apply in writing to the secretary of the Aero Club of America giving the sum of five dollars. He must pass, to the satisfaction of properly designated representatives of the Aero Club of America, at a place and date fixed by the Contest Committee, such tests as may be prescribed for the calendar year in which he may take his tests.

Holders of the expert certificate are permitted to fly over cities in straight flight and the privilege of using Governor's Island, New York as a temporary landing station is extended to them by Chief of the Department of the East.

Tests for Calendar Year 1917

Each applicant must pass a thorough physical examination by a reputable, competent physician, designated by the Contest Committee of the Aero Club of America. The applicant must posses normal heart and lungs as well as normal sight and hearing and shall be free from all nervous affections. In case the physician is in doubt as to the physical stability of the applicant, an examination shall be made immediately following a trial flight to determine this point.

After passing the physical examination the applicant must pass the following tests:

1. A cross-country flight, from a designated

starting point to a point at least 25 miles distant and return to the starting point without alighting.

2. A glide, without power, from a height of 2500 feet, coming to rest within 164 feet of a previously designated point without the use of brakes.

3. A figure eight around two marks, 1640 feet apart. In making turns, the aviator must keep all parts of his apparatus within semi-circles of 164 feet radius from each turning mark as a center.

The issuance of the certificate is optional with the Aero Club of America.

Memoranda:

CHAPTER XXXVII

IDENTIFICATION MARKS FOR AIRCRAFT

Distinctive marks for identifying aircraft have been adopted by the nations at war. Such marks are painted on both upper and under side of the wings and on the real controls, that they may be clearly distinguished from every point of view. There have been few cases of a nation using a captured aircraft to deceive its enemy. The United States Army has used a star and the Navy an anchor as distinguishing marks. The American seaplane *Vera Cruz* flew the American flag.

An Austrian seaplane with wings marked with the Iron Cross is shown in chapter on Anti-Aircraft Guns.

On May 20, 1917, Secretary Daniels announced the adoption of an efficient insignia for all aeroplanes, seaplanes, captive balloons, and dirigibles. The insignia of the flying corps combines the red, white, and blue of the national emblem, consisting of a white star with red center on a blue circular background. The official order for the adoption of the new aero insignia for the Navy follows:

"A five-pointed white star inside of a blue circumscribed field, with the center of the star red. The diameter of the circumscribed circle will be equal to the chord of the wing on which the insignia is placed. The diameter of the inner circle will not extend to the inner points of the star by an amount equal to one twenty-fourth of the diameter of the circumscribed circle. The inner circle will be painted red; that portion of the star not covered by the inner circle will be painted white, and that portion of the circumscribed circle not covered by either inner circle or star will be painted blue. The shades of red, white and blue will be the same as those used in the American Flag."

The order in regard to naval craft continued: "One of each of these insignia will be placed on the upper surface of each upper wing, and one of each in a corresponding position on the lower surface of each lower wing. Both sides of that portion of the rudder which is in rear of the rudder post will be painted with three equally wide bands, parallel to the vertical axis of the aeroplane and colored red, white and blue of the same shades as mentioned hereinbefore, the blue band being nearest the rudder post, the white band in the center, and the red band at the tail of the rudder.

"One of these insignia will be placed on top and one on bottom of gas bag of dirigible balloon, the center of each insignia being in the vertical plane through the fore and aft axis of the gas bag. The center of insignia on top will be sixty feet from forward end and the center of insignia on bottom will be just forward of suspension band. The circumscribing circle of insignia for dirigible will be five feet

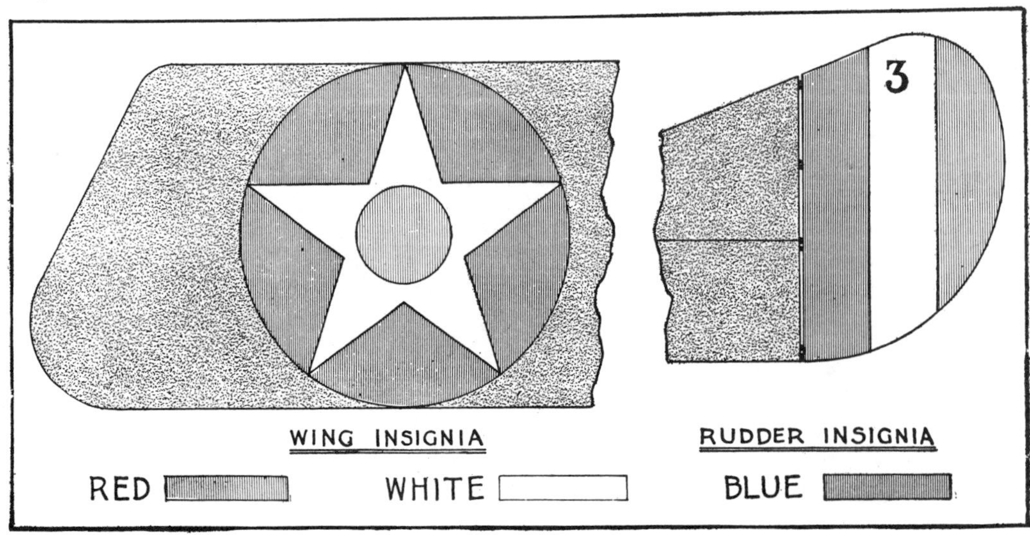

Regulation Insignia for United States Navy Seaplanes.

Identification marks painted on the aeroplane wings by the countries at war to distinguish them. The United States Aviation Section has adopted a star, to be painted on aeroplane wings.

in diameter. The rudder of each dirigible will be marked in a manner similar to that applied to aeroplanes, except that stripes will not exceed five feet in length or eighteen inches in width. If there is more than one rudder only the outboard side of each outboard rudder will be

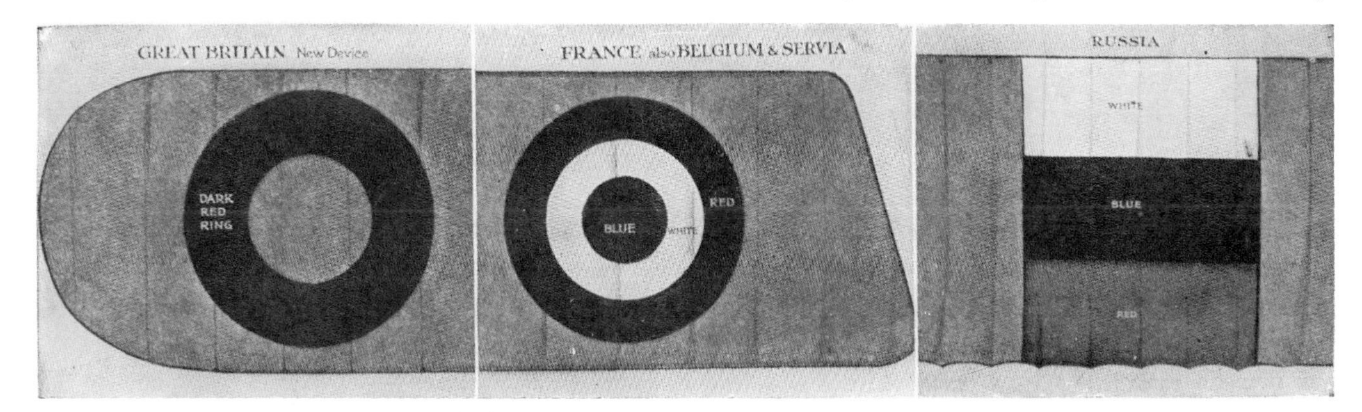

marked. One of these insignia will be placed on top and bottom of gas bag of captive balloon, the center of each insignia being in the vertical plane through the fore and aft axis of the bag, and two and one-half feet aft of the seam joining nose to main body. The circumscribing circle will be five feet in diameter. The building number of each aircraft will be placed in figures three inches high on each side of the rudder, at the top of the white band hereinbefore mentioned. No other markings shall be placed on any Navy aircraft, except such as may hereafter be prescribed. All Navy aircraft will be immediately marked in accordance with this order, and in future, specifications for Navy aircraft will require that the contractors place the building number and distinguishing insignia on all aircraft, and on such spare parts as bear these marks in completed aircraft."

Flag used on early American seaplanes.

Identification mark on British seaplanes.

INDEX